THERMAL
IMAGING
TECHNIQUES

THERMAL IMAGING TECHNIQUES

*Proceedings of a Conference
Held October 4-5, 1962
at Arthur D. Little, Inc.,
Cambridge, Massachusetts*

Edited by
Peter E. Glaser
Arthur D. Little, Inc.

and

Raymond F. Walker
National Bureau of Standards

Springer Science+Business Media, LLC

1964

Library of Congress Catalog Card Number 64-19979

ISBN 978-1-4899-5647-7   ISBN 978-1-4899-5645-3 (eBook)
DOI 10.1007/978-1-4899-5645-3

© 1964 Springer Science+Business Media New York
Originally published by Plenum Press in 1964.
Softcover reprint of the hardcover 1st edition 1964

INTRODUCTION

The principle of focusing the image of the sun or an incandescent source onto the surface of a body in order to heat it to a high temperature has been known since ancient times. However, it is only in recent years that serious consideration has been given to the utilization of this heating technique in industrial and domestic applications, and in research in high-temperature chemistry and physics. Early investigators used the sun as the source of energy almost exclusively, and several conferences to discuss the progress made with the solar heating devices have been held in various parts of the world. For research purposes, however, solar furnaces have certain disadvantages—e.g., they are "outdoor" devices, much dependent on the weather and climate. In research increasing attention has therefore been paid to possible alternative sources that would make operation under more conventional laboratory conditions possible.

The carbon arc is one such source which evolved from this activity, and is in fact now most commonly used for research purposes. The introduction of the arc has led to the study of a wide range of physical and chemical properties of substances using imaging techniques, and a conference was therefore organized to assess both the progress which has been made and the problems which remain. The proceedings of the conference form the substance of this book.

The conference is believed to have been the first to be devoted primarily to the application of imaging techniques to research, and also the first in which the use of electrical sources was of more dominant interest than that of solar furnaces. In many instances, however, the experimental techniques described in this volume would be equally applicable with either type of source.

Thermal imaging techniques have in principle a number of advantages for research purposes. One advantage is that they permit experiments to be carried out in the 1000–3500°C temperature range under extremely pure conditions or in strongly oxidizing or reducing atmospheres—chemical conditions which are not easy to reproduce using more conventional electrical or chemical heating techniques. This particular advantage arises because the sample can be completely isolated from the heat source, and except for the radiation falling on it, disturbing electric fields or reactive furnace gases and components can be excluded. At the same time any desired atmosphere can be chosen to surround the sample, or it can be exposed to a high vacuum.

While arc imaging techniques are attractive in principle, early research-ers found that difficult problems arise when one wishes to maintain a sample under a constant heat flux or a steady temperature for long periods of time. Here the problem of temperature measurement also introduced several diffi-culties and uncertainties. Attempts have been made, therefore, to find new sources which would not suffer from the shortcomings of the arc, and several of the papers presented at the conference are concerned with this problem. Many ingenious devices have been constructed to measure the heat flux incident upon a sample, and several of these are described in the following papers. However, the difficulty of measuring temperatures accurately is still not completely solved. It seem probable from the evidence presented here that instrumentation is no longer the major problem in accurate temperature measurement. It is a better understanding of the temperature gradients and thermal conditions prevailing at the heated sample surface which is now required.

One of the striking results of the conference was the evidence produced on the range of usefulness of imaging techniques. The sparsity of data on the physics and chemistry under the conditions which are the particular forte of image furnaces makes it profitable to obtain even qualitative information only roughly related to the International Practical Temperature Scale. For this reason more extensive use has been found for thermal imaging techniques than might be indicated by a consideration of the practical problems involved in applying the techniques. Examination of the following papers will show that not only have attempts been made to study a wide range of physical and chemical properties of substances, but imaging techniques have also been applied to such difficult-to-control processes as crystal growth and zone refining.

These applications demonstrate that imaging techniques have graduated from the stage of being laboratory curiosities and can take their place with such competing high-temperature techniques as electron beams and lasers. The aim of publishing the proceedings of the conference is to make both the progress and the outstanding problems in the field more widely known, with the hope that still more fruitful application of the techniques will result and that study of associated problems will be stimulated.

ACKNOWLEDGMENTS

On behalf of all participants at the Conference it remains to express our appreciation of the efforts of many people who made the Conference possible. Particular acknowledgment is due to the following, who not only presided over individual sessions of the Conference, but added their comments and criticisms of the papers as part of the editorial review: Professor J. Drowart, Universite Libre de Bruxelles, Belgium; Dr. A. J. Drummond, The Eppley Laboratory Inc., Newport, Rhode Island; Dr. Nevin K. Hiester, Stanford Research Institute, Menlo Park, California; and Dr. W. W. Lozier, National Carbon Co., Parma, Ohio. Thanks are also due to the authors of the papers who reduced to minimum proportions the task of editing their manuscripts. Finally a special word of appreciation is due John Criden whose efforts were invaluable at every stage in the organization of the conference.

Peter E. Glaser
Arthur D. Little, Inc.

Raymond F. Walker
National Bureau of Standards

CONTENTS

Imaging Furnace and Radiation Source Development

Instrumentation and Measurement Techniques

Crystal Growth

Experimental Procedures

Imaging Furnace

and Radiation Source Development

Chapter 1

An Arc Imaging Furnace for Solid Propellant Ignition Studies

Gerald E. Dolan*
Thiokol Chemical Corporation
Elkton Division
Elkton, Maryland

A. INTRODUCTION

The arc imaging furnace has proven to be a versatile instrument for studying solid propellant ignition. The features which make this instrument attractive are: (1) reproducible flux levels, (2) clean source of heat, and (3) what may be termed 'instant' heat, i.e., the high blackbody temperatures may be turned on and off instantly without any appreciable heat losses, which for other high-temperature devices constitute a significant source of error.

B. BACKGROUND

From an historical aspect, the ignition of solid propellant has been a *hit or miss proposition—an art instead of a science. In the early days if reliable ignition was not achieved, more igniter material was added. Although pyrotechnic devices are still quite widespread, improved reliability* has resulted with the use of PYROGEN® igniters (a registered trademark of Thiokol Chemical Corporation for a small rocket motor which exhausts into and ignites a larger unit). Hypergolic ignition, which has received considerable attention recently, utilizes an easily oxidized (fuel-rich) surface which is sprayed with a strong oxidizing agent. Although reliability makes hypergolic ignition desirable, this method suffers from excessive ignition delays and the need for additional hardware to assure contact between the two materials.

Experience has shown that ignition is achieved principally by conduction and radiation. For instance, a small strand of composite solid propellant may be ignited in about 2 to 5 sec by placing it in contact with a lighted cigarette. The heat is transferred to the surface of the propellant by the impingement of hot gases and/or hot particles. Hot-gas ignition is normally achieved by squib ignition of nitrate esters (nitroglycerine—nitrocellulose compositions) or other pyrophoric materials which in turn impart sufficient energy to bring about steady-state ignition of the propellant. In hot particle ignition, incandescent particles come in contact with the propellant surface, and a steady-state burning is achieved from the many "starts" induced by the particles.

*Development Chemist.

3

In dealing with the mechanisms involved in propellant ignition, the modern concept recognizes that it is a transient process whereby the chemical oxidation reaction is carried out in the vapor phase. In its simplest form ignition is achieved in three basic steps:

1. The energy is transferred to the surface from an external source, causing decomposition of the solid oxidizer and polymeric binder.
2. An exothermic reaction takes place in the gas phase (flame zone) if the threshold ignition energy has been exceeded for the particular propellant.
3. The heat from this exotherm represents the source of heat for further volatilization of the surface, which ultimately leads to a steady-state burning condition.

The type of heat source is of no particular consequence except as it affects the pressure within the system. A high pressure makes ignition easier, because it brings the flame zone closer to the surface. This is one factor which accounts for reliable ignition with the use of PYROGEN® units. The hot combustion products from the PYROGEN® pressurize the motor and supply the necessary heat requirements for sustained propellant ignition. The proximity of the flame zone (in the vapor phase) to the propellant surface and consequent ease of heat transfer between the two are enhanced by a pressure build-up. Actually it is possible to extinguish the burning propellant by a sudden reduction in pressure. The propellant surface is cooled by adiabatic expansion of the gases (caused by a reduced chamber pressure) which cuts off the source of vaporized material for continued sustained combustion.

In the relatively short life span of modern solid rocketry, two principal methods have been employed for evaluating propellant ignitability: (1) the autoignition test and (2) the black powder relative ignitability test.

In the autoignition test, a cube of propellant is dropped into a molten bath of Woods metals, and the time for ignition is determined. A series of such tests at strategically selected bath temperatures makes it possible to define the 5-sec or 5-min autoignition temperature. Although this test is useful for defining safety limits, the problems involved in the ignition of rocket motors are beyond its scope.

The black powder relative ignitability test developed at Picatinny Arsenal represents a decided improvement in studying propellant ignitability [1]. For this test, various quantities of black powder are fired in a pressurized closed bomb containing a uniform-size propellant sample. This is essentially a go–no-go type test, which is statistically oriented so that the quantity of black powder for 50% ignition and the standard deviation from the mean may be determined. This test is desirable, because it is sufficiently sensitive to monitor the effects of surface treatments and because it duplicates closely what is experienced in the motor chamber.

Recently, at Princeton University [2], a shock tube was used to study propellant ignition requirements. The shock tube operates on the principle of adiabatic compression of gases to obtain temperatures at the surface which are sufficient to ignite the propellant. A long tube with a rupture disk located strategically at one end is pressurized. The other end, which contains the sample, is either evacuated or filled with a test gas at a comparatively low pressure. A study of the shock-wave patterns with various test gases in the system has shed light upon the requirements for propellant ignition. McAlevy and Summerfield have demonstrated that the ignition of solid propellant is carried out in the vapor phase and the use of the test gas, oxygen, lowers the time-delay requirements for propellant ignition.

C. ARC IMAGE FURNACE AS A TOOL FOR IGNITION STUDIES

The arc imaging furnace has proven to be a versatile instrument in the study of propellant ignition. The early investigations with the arc image technique were performed by Byers and Fishman of Stanford Research Institute ([3]). Their work was characterized by investigations of ignition delays and determination of the threshold ignition energy as affected by: (1) aging of solid propellant, (2) pressure, (3) propellant surface condition, (4) various environmental gases, and (5) various flux levels. Their treatment of propellant ignition is fairly comprehensive, and a detailed review is beyond the scope of this paper. Lowry and Asphend are also active in the field of investigating propellant ignition with the arc imaging furnace ([4]).

Early in 1961, the Elkton Division of Thiokol Chemical Corporation set up an arc imaging furnace for solid propellant ignition studies. The features of the instrument were:

1. High and intermediate flux level capabilities.
2. Rotating shutter disks for accurate control of pulse duration.
3. Versatile sample holder for pressure tests and provisions for accurately focusing the radiant energy.

Ordinarily, the high flux level is used for propellant ignition studies, because the heat losses are minimized by operating at the high amperage levels. This high flux is obtained with the use of the "Ultrex 300" negative

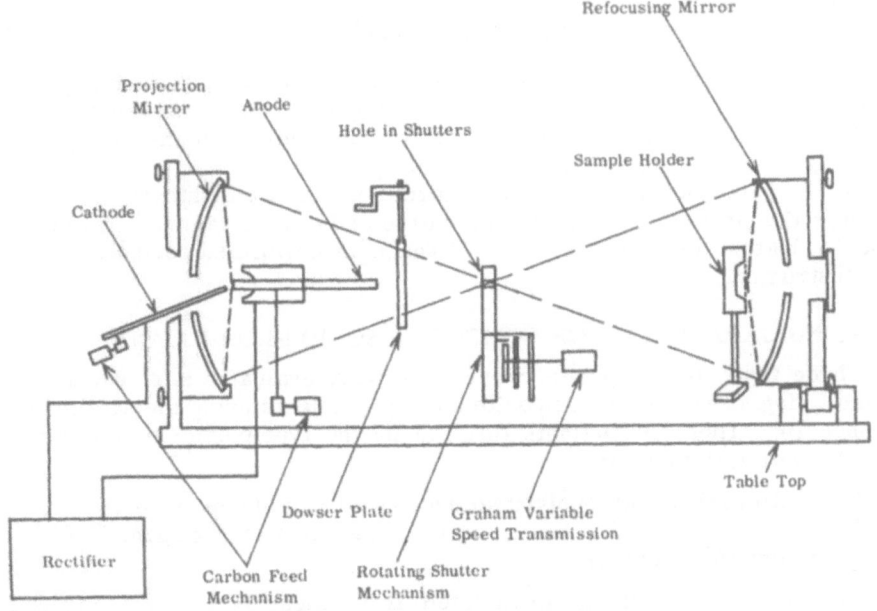

Fig. 1-1. Schematic of arc imaging furnace.

carbon electrodes with proper adjustments to the rectifier and automatic feed mechanism.

The test method developed for propellant ignitability was the go-no-go type series of tests, statistically designed for determination of the threshold ignition energy. This value, which represents the 50% fire point, is obtained by varying the energy pulse duration from one test to the next. The energy pulse duration is controlled by three rotating disks located at the crossover focus point as illustrated in Fig. 1-1. The middle disk makes one rotation for every 16 revolutions of the high-speed disk and 16 revolutions for each revolution of the slow disk. The speed of the disks is controlled with a Graham variable transmission. A 2-in.-diameter hole is located at the edge of the rim of each disk, and when the slow disk moves into the twelve o'clock position, all three holes become aligned to allow the passage of an energy pulse to the sample in the holder. The propellant sample, $\frac{3}{8}$ in. in diameter and $\frac{1}{4}$ in. thick, is positioned on a stage enclosed in a sample holder. The sample is further enclosed with $\frac{3}{4}$-in.-OD pyrex glass tubing which allows the transmission of radiant energy. A small pen light is precisely positioned in place of the negative carbon electrode in order to determine the sample location which will receive the greatest amount of energy.

The furnace energy output is determined by means of a copper button calorimeter. A small disk of copper, the same diameter as the propellant sample, is placed in the sample holder. A fine-wire thermocouple is soldered to the back of the button so that the temperature of the button can be determined immediately after the energy pulse has passed.

To reduce the emissivity, the button is coated with carbon by use of a reducing gas flame. The furnace output is determined from the following equation:

$$F = MC_p \Delta T / A_s t \text{ [cal/cm}^2 \text{ - sec]} \tag{1}$$

where F is the flux; M is the weight of copper button, in g; C_p is the specific heat of copper button in cal/g-°C; ΔT is the change in button temperature, in °C; A_s is the surface area exposed to radiant energy, in cm^2; and t is the time duration of the energy pulse, in sec.

A Tetronix oscilloscope with Polaroid camera attachment is used to monitor the profile of the emf–time curve and to determine accurately the rise in button temperature. The furnace is calibrated periodically with this instrument.

D. AGING CAPABILITIES OF COMPOSITE SOLID PROPELLANTS

Three types of composite propellants were evaluated in a recent study of the aging behavior of composite solid propellants using the arc imaging furnace [5]. These propellants differed by the type polymeric fuel binder used in their formulation:

1. Hydrocarbon and a polybutadiene–acrylic acid copolymer.
2. Polyurethane with a castor oil cure and plasticizing agent.
3. Polysulfide rubbers.

Samples of propellants formulated from these binder systems were subjected to accelerated aging by placing them in ovens at elevated temperatures. After various aging periods these propellants were tested with the arc imaging furnace and compared with the unaged samples for evidence of change in propellant ignitability. Surface response curves for the three propellants are shown in Figs. 1-2, 1-3, and 1-4.

Of the three, the hydrocarbon propellant ignited most easily after being subjected to accelerated aging under high-temperature conditions. The polyurethane base propellant exhibited an unusual aging behavior. During the period of from one to four weeks and at conditioning temperatures of 130 to 170°F, the ignition requirements were reduced. However, samples aged for four to twelve weeks required 50 to 100% more energy for ignition. Finally, after 12 to 16 weeks at 210°F, a drop-off in ignition requirements was experienced. Likewise, the polysulfide propellant exhibited a significant decrease in ignition requirements after 12 weeks storage at 210°F.

E. CORRELATION OF ARC IMAGING FURNACE TEST DATA WITH IGNITION REQUIREMENTS FOR DEVELOPMENT TEST MOTORS

In conjunction with the aging studies, an attempt was made to correlate arc image threshold ignition energy data with actual engine firings.

For this study, a 5-in.-diameter development test rocket motor with a cylindrical center perforate grain design was used. In this series of tests, the quantity of igniter materials was progressively decreased from one

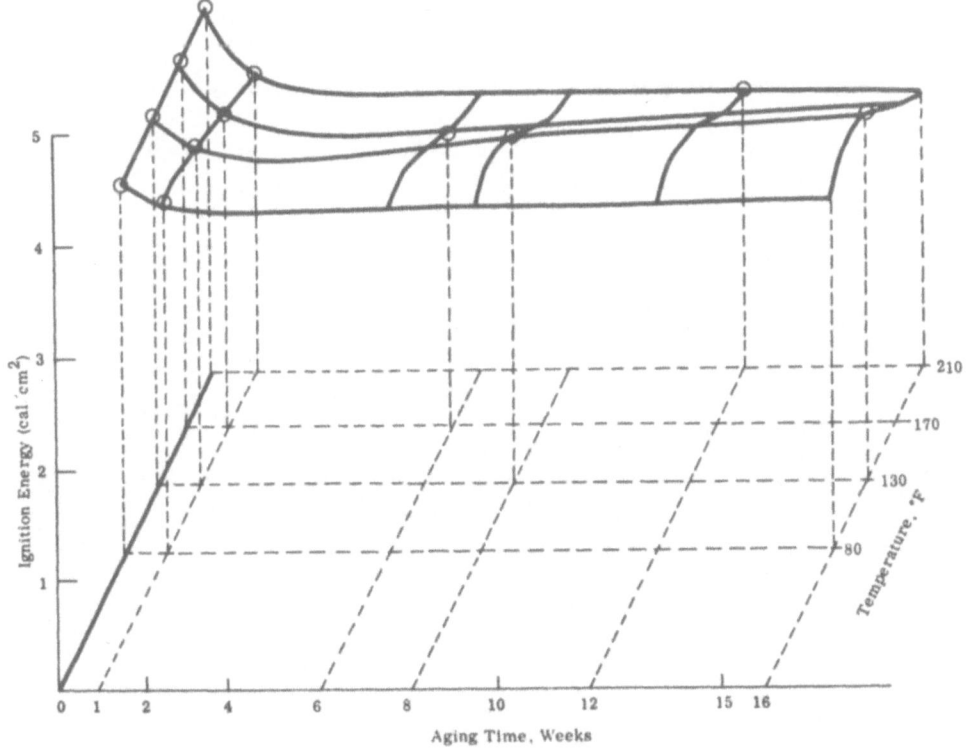

Fig. 1-2. Surface response curve AIF ignitability studies—hydrocarbon propellant.

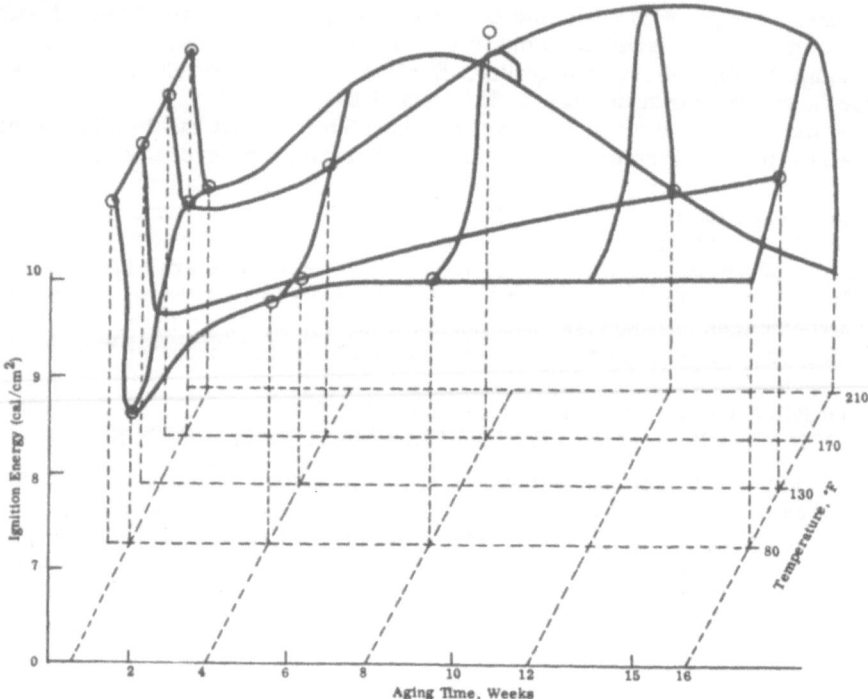

Fig. 1-3. Surface response curve AIF ignitability studies—polyurethane propellant.

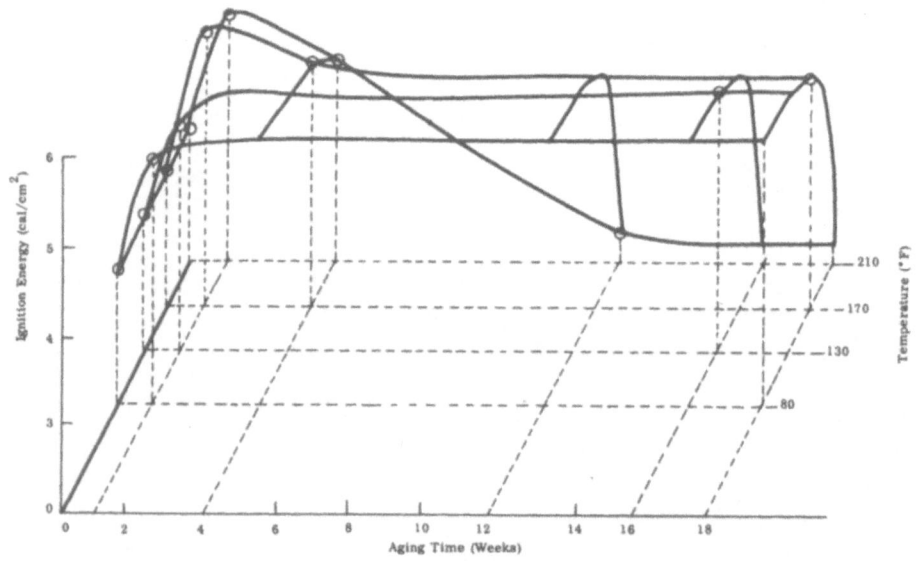

Fig. 1-4. Surface response curve AIF ignitability studies—polysulfide propellant.

TABLE 1-I

Correlation of AIF Data and Motor
Ignition Requirements

Threshold Ignition Energy (cal/cm^2)			
Propellant type	AIF data	5-in. motor data	Ratio of motor to AIF values
Hydrocarbon	3.29	10.4	3.2
Polyurethane	9.54	34.8	3.6
Polysulfide	3.50	—	—

firing to another until excessive igniter delay times were encountered. The excessive delays were assumed to represent the threshold ignition energy for propellant ignition in the motor. When the threshold ignition energies from the motor firings were compared with the arc imaging furnace data, a fairly constant factor was obtained with hydrocarbon and polyurethane propellants. This correlation of motor data with the arc imaging furnace results is illustrated in Table 1-I. The values for 5-in.-motor data were obtained from the heat of decomposition of the quantity of igniter material per unit area of grain surface. The motor data for the polysulfide propellant were incomplete so that no correlation was possible.

The fact that the ratios of motor data to arc imaging furnace data are fairly close for the two propellants shows that the arc imaging furnace technique is a valuable instrument for checking the ignitability of composite propellants. It also represents a valuable guide for a rocket engineer who designs the ignition requirements for a new propulsion system.

F. PROPELLANT IGNITABILITY AS AFFECTED BY SURFACE TREATMENTS

An investigation was undertaken to determine what effects were produced on propellant ignition when a mold release agent was applied to the mandrel core. The release agents facilitate the core removal after the propellant has cured. The mandrel is used to mold the initial grain configuration which controls the burning surface area. The arc imaging furnace tests showed that a silicone release agent inhibited the burning surface area considerably, whereas a conventional paste wax had no appreciable effect. As shown in Table 1-II, an energy increase of approximately 46% was required to obtain propellant ignition with a hydrocarbon base propellant.

TABLE 1-II

Radiant Energy for Threshold Ignition
(150 A Input)

Sample treatment	Threshold ignition energy (cal/cm^2)
None	5.10 ± 0.44
RTV (silicone)	7.56 ± 1.14*
Paste wax	5.14 ± 0.43

*2-sigma variance limits.

TABLE 1-III

Comparison of Ignitability of Fresh and
Water Quenched Propellant Surfaces

Surface condition	Threshold ignition energy (cal/cm^2)
Water quenched surface (burning extinguished)	6.6 ± 0.74*
Fresh surface	6.3 ± 0.39

*2-sigma variance limits.

Another study of the effects of surface conditions was made with a poly-urethane propellant. The ignitability of a fresh propellant surface was compared with what was experienced with a quenched propellant surface. The purpose of the program was to study means of terminating thrust in a rocket motor. The propellant was ignited and subsequently extinguished by adiabatic expansion of the gases and quenching with water. The sudden gas expansion was achieved by removal of the nozzle with explosive bolts.

As expected, the dried quenched surface, which showed effects of scorching, was found to be slightly more difficult to ignite than the fresh surface. The differences are not believed to be significant, since the two sigma variance limits overlap. A tabulation of the results is presented in Table 1-III.

G. EFFECTS OF PRESSURE AND VACUUM AGING ON PROPELLANT IGNITABILITY

An extensive study was made of the effects of pressure and vacuum aging on polyurethane base propellant ignitability. An increase in pressure contributed to the ease of ignition. Conversely, ignition requirements were much greater for exposures made under vacuum conditions. The results are illustrated in Fig. 1-5.

As the pressure decreases, the propellant becomes more difficult to ignite. The reliability of the value at 10 mm Hg is questionable because the ignition was not sustained. The value represents the energy required to volatilize or pyrolyze more than half of the sample. An analysis of the test showed that the products of original volatilization and pyrolysis clouded the glass by condensation and prevented or significantly reduced the transmission of energy to the surface to affect ignition. Also, the vacuum condition reduces the proximity of the flame zone in the vapor phase to the propellant surface and adversely affects the ignitability.

It should be noted that, normally, propellants are not expected to ignite under absolute vacuum conditions, although they are subjected to extremely low pressures in space flights. Reliable ignition performance has been obtained with various Thiokol® retrograde rocket motors for deorbiting such vehicles as the Discoverer series and the Mercury Man-In-Space programs. The motor case was pressurized by the PYROGEN® unit prior to propellant ignition.

In order to determine the effects of space aging on polyurethane composite propellants, samples were subjected to 10^{-8} mm Hg vacuum conditions for a two-week period. No significant change in the weight or ignitability was apparent when samples were tested under ambient conditions.

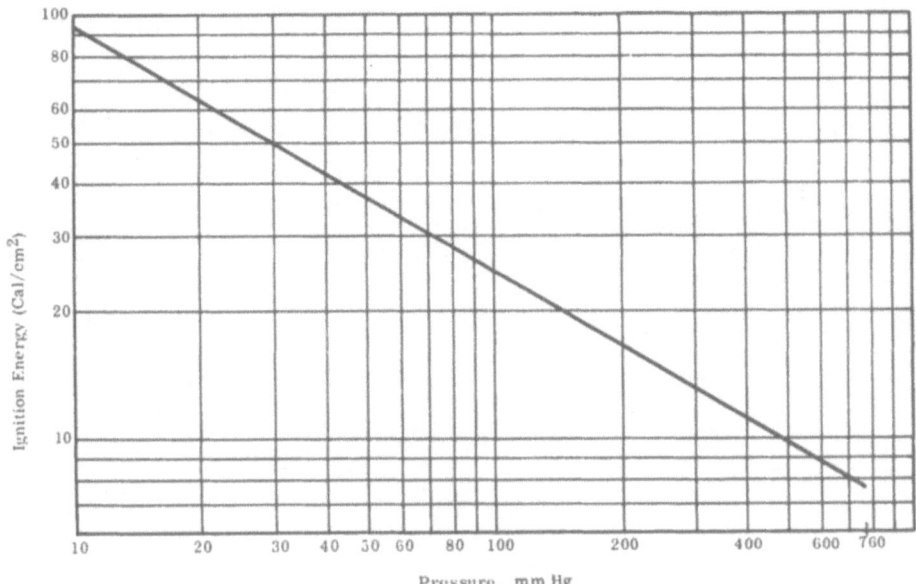

Fig. 1-5. Effect of vacuum pressure on polyurethane propellant ignitability.

H. CONCLUSIONS

In view of the work which has been accomplished with the arc imaging furnace, the following conclusions are presented:

1. The arc imaging furnace is a useful tool for studying the ignition characteristics of various propellant systems subjected to different accelerated aging and pressure conditions.

2. A correlation of propellant ignitability was obtained between the arc imaging technique and motor performance data.

3. The importance of the surface condition of the propellant for reliable ignition has been emphasized by experiments connected with this instrument.

4. Experiments with the apparatus on polyurethane propellant under vacuum conditions indicated that an increase in energy was required for ignition. However, after extended vacuum storage, the ignitability of the polyurethane propellant remained unaffected.

REFERENCES

1. Grande, Cecilio, "New Methods for the Measurement of Relative Ignitability and Ignition Deficiency," Picatinny Arsenal, Dover, New Jersey, Samuel Feltman Ammunition Laboratory. Ordnance Project No. TA1-502, Army Project No. 5804-01-040 (February, 1958).
2. McAlevy, Robert F., and Summerfield, Martin, "The Ignition Mechanism of Composite Solid Propellant, Princeton University, Princeton, New Jersey (June 1, 1961).
3. Byers, Rodney E., and Fishman, Norman, "Solid Propellant Ignition Studies with High-Flux Radiant Energy as a Thermal Source," Progress in Astronautics and Rocketry, Vol. 1 (Academic Press, New York, 1960), p. 673.
4. Lowry, E. M., and Asphend, R. K., "Solid Propellant Ignition Studies Using the Arc-Image Apparatus under Dynamic Conditions," Bermite Powder Company, Saugus, California (November 27, 1961), (Confidential).
5. Sienicki, E. A., et al. "A Study of Methods of Improving the Shelf Life of Solid Propellant Rocket Engines," Report No. E 115-61, Volumes I and II, Thiokol Chemical Corporation, Elkton Division, Elkton, Maryland (May 1961), (Confidential).

Chapter 2

An Arc Imaging Furnace System for Studying the Ignition of Solid Propellants

A. L. Camus

Arthur D. Little, Inc.
Cambridge, Massachusetts

A. INTRODUCTION

A solid propellant ignition study arc imaging system was recently developed. This system may be divided arbitrarily into four separate parts: (1) The radiant energy source and optical system; (2) the timing shutters, which control the exposure time; (3) the sample bomb; and (4) the event recording system, which records the exposure and ignition delay times, controls the automatic start of the recorder chart, and permits calculating the energy received by a sample.

The capabilities of such a system are as follows: Minimum exposure—2.5 msec. Maximum exposure—up to several minutes. Maximum average flux at the sample location over a $3/8$-in.-diameter spot: 295 cal/cm^2-sec, with a peak of 400 cal/cm^2-sec over a 1-mm-diameter spot. The flux is achieved with an ADL-Strong arc imaging furnace (Ignition Study Model AIF-3), and adjustable from zero to maximum. Environmental conditions—from vacuum to 350 psi, by connecting the sample bomb to the appropriate source of vacuum or pressure. In this case, the available flux is reduced by 10 to 20%, depending on the type of glass and the size of the tube used to enclose the sample. Chart record, with a time pulse to permit measurement of exposure and ignition delay times. A 10-msec event occupies a length of $1/2$ in. on the chart.

B. SHUTTER SYSTEM

The shutter system is mounted at the crossover point (see Fig. 2-1). This system is similar to the one used at Stanford Research Institute and described by Beyer and Fishman [1]. Briefly, it consists of a set of three coaxially mounted, 13-in. discs that rotate at three different speeds in the ratio of 1:16:256. A 2-in.-diameter hole is located near the rim in each of these discs, and radiant energy from the arc source is transmitted and refocused onto the sample at the moment when the three holes are aligned with the optical axis of the arc furnace. The three discs are gear-synchronized and are driven by a variable-speed transmission. Thus, the exposure time is a function of the speed setting and is easily adjustable. The shutter system is enclosed in a steel housing to provide for the operator safety.

C. SAMPLE BOMB

The sample bomb is attached to the sample positioner at the reimaging focal point. The sample is mounted inside a glass tube that is sealed at both ends by rubber-gasketed plugs. These plugs are machined such that the space enclosed by the tube can be evacuated or pressurized. The upper plug holds a $\frac{3}{8}$-in.-diameter stainless steel ring, which is the actual sample holder. Soft samples may be cut directly by using the ring like a biscuit cutter. Another plug supports a copper disc calorimeter, which occupies the same position as the sample. Knowing the physical constants of the calorimeter, one can then calculate the exposure energy from the results of a test exposure.

D. EVENT RECORDING SYSTEM

1. Exposure Time and Ignition Delay Time

A photomultiplier tube views the sample through a collimating tube and through the center hole of the reimaging mirror. During an exposure the radiation is reflected by the glass tube containing the sample and is sensed by the phototube. Its output signal is amplified and recorded on a direct-recording oscillograph (Honeywell Model 906C). Timed pulses from a multivibrator fire a flash tube every 1 sec, 0.1 sec, or 10 msec and create time reference grids on the recording chart (see Fig. 2-2).

If the sample ignites, the light it emits during burning will also be sensed. The amplified signal will appear on the chart; in Fig. 2-2 it is the irregular portion of the trace lasting for 50 msec after exposure ended. Any delay in ignition can be easily measured.

The ratio of the radiation intensity to the light emitted by the sample may be quite large. A clipping network in the detector circuit prevents overloading of the amplifier and damage to the galvanometer.

2. Automatic Chart Drive

Accurate measurement of events as short as 2.5 msec is made possible by use of a 50-in./sec chart speed. To conserve chart paper, detents are located on the slow- and medium-speed shutter discs. These detents energize the gate circuits of two 2N877 silicon controlled switches. Once the gate of the solid-state switch is energized, even by a pulse as short as 200 to 300 μsec in the case of the minimum exposure, the switch turns on and stays on until the anode voltage is turned off. A mercury-wetted contact relay is used as a load, and its contacts close the recorder chart drive remote control.

Different exposures require different shutter speeds. By positioning the chart drive detent according to the exposure needed, it is possible to start the chart motor approximately 0.5 sec before exposure. The chart gets to full speed in about 0.3 sec; thus, only about 15 in. of chart are lost in every test.

For a minimum exposure of 2.5 msec, the fast shutter rotates at 3000 rpm, and it takes 12 sec for this shutter to get to speed from standstill. During that time, the medium-speed disc, whose detent actuates the chart drive, makes 40 revolutions and closes its switch 40 times, while the slow

Fig. 2-1. Shutter system (housing removed).

shutter makes 2.5 revolutions. For these reasons, three detents are used. The first flashes a warning light from $2\frac{1}{2}$ to $3\frac{1}{2}$ sec before actual exposure. On seeing this light the operator opens the douser of the arc lamp and energizes the solid-state switch circuit. The second detent readies the circuit of the third detent through an auxiliary relay. The third detent energizes the remote chart drive control. To stop the chart the operator has only to release the push button he depressed when the light flashed. A chart is provided that shows both the detent position for any particular exposure and the dial setting of the variable speed drive.

3. Calorimeter Measurements

We have found that the values of flux calculated from the calorimeter constants and the temperature rise are not necessarily the values for which the sample ignites; a calorimeter gives an average value of flux, whereas the sample ignition may be due to the higher peak at the center of the image.

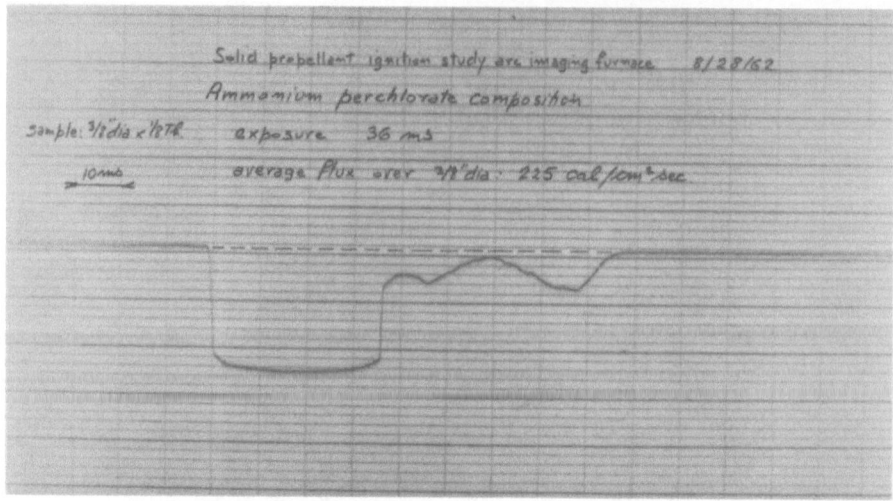

Fig. 2-2. Typical oscillogram for propellant ignition.

If we assume an ideal calorimeter, i.e., one with negligible losses and an absorptivity of 1, we have

$$\text{Flux} = \frac{(MC_p)\,T}{At} \tag{1}$$

where flux is average flux over the area A in cal/cm^2-sec; T is the temperature rise, in °C; t is the exposure time, in sec; A is the area of calorimeter, in cm^2.

$$MC_p = M_1 C_{p1} + M_2 C_{p2}$$

where M_1 represents the weight of the copper disc in grams, C_{p1} the specific heat of copper in cal/gr/°C, M_2 the weight of the material used to solder the thermocouple leads to the disc, and C_{p2} the specific heat of the solder.

The temperature rise in equation (1) is determined from the emf trace on the chart, but one must take into account the fact that this emf corresponds to the difference between the temperature attained by the calorimeter and ambient temperature ([2]). Tests of four different copper calorimeters yielded the results shown in Table 2-I. (See Fig. 2-3 for typical trace.)

The thermocouple leads were soft-soldered to the copper disc using a 60 tin–40 lead solder. The values for total MC_p were calculated using the

TABLE 2-I

Calorimeter No.	Diameter, in.	Nominal thickness, in.	Total MC_p, cal/°C	Copper–constantan thermocouple gauge number
I	$\frac{3}{8}$	0.006	80.6×10^{-4}	30
II	$\frac{3}{8}$	0.010	160×10^{-4}	30
III	$\frac{3}{8}$	0.020	293×10^{-4}	30
IV	$\frac{3}{8}$	0.006	85.4×10^{-4}	36

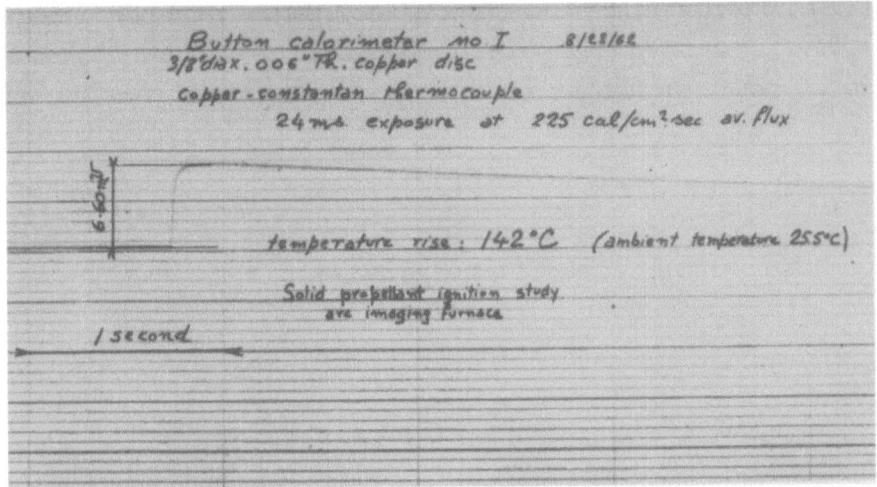

Button calorimeter no I 8/23/62
3/8"diax. .006" Th. copper disc
Copper-constantan thermocouple
24 m.s. exposure at 225 cal/cm²·sec av. flux

temperature rise: 142°C (ambient temperature 255°c)

Solid propellant ignition study
arc imaging furnace

1 second

Fig. 2-3. Response of button calorimeter to test exposure.

specific heats of copper (0.0939 cal/g-°C), tin (0.0577 cal/g-°C) and lead (0.0320 cal/g-°C) at 100°C [3]. The fact that the specific heat is slightly temperature-dependent was not taken into consideration.

The values of total MC_p listed above are not exactly proportional to the indicated nominal thickness, because thickness variations of ± 0.001 in. or more were found and because of nonlinear influence of the thermocouple leads.

With each calorimeter, the temperature rise was found to be proportional to the exposure time, within small experimental errors. It was observed that the temperature at the thermocouple measuring junction was still rising, in some cases, after the exposure was completed, but the final temperature rise was proportional to the exposure duration. This means that the temperature rise is proportional to the energy received by the calorimeter. Thus, if one wants to find the transmission factor of the glass tube in which the sample is enclosed, he has only to make the calorimeter test twice—the first time with the glass tube on, the second time without glass tube—and take the ratio of the corresponding temperature rises. The detailed measurement technique is given in reference (4), which correlates the results obtained from a disc calorimeter with those obtained from an absolute calorimeter.

ACKNOWLEDGMENT

The author wishes to thank Mr. Patrick Murphy for his contribution during the assembly and instrumentation of this work.

REFERENCES

1. Beyer, Rodney B., Fishman, Norman, "Solid Propellant Ignition Studies with High Flux Radiant Energy as a Thermal Source," Progress in Astronautics and Rocketry, Vol. I, (Academic Press, New York, 1960).

2. Kaufman, Alvin B., "Using Thermocouples with Nonstandard Reference Temperature," Instruments and Control Systems 33 (I): 106.

3. Handbook of Chemistry and Physics, 43rd ed., The Chemical Rubber Publishing Co., Cleveland, Ohio.

4. "Ignition Study Model," Operating Manual for ADL-Strong Arc Imaging Furnace, Part VIII-D, pp. 65-94.

Chapter 3

A Thermal Radiation Heat Source and Imaging System for Biomedical Research

D. L. Richardson

Arthur D. Little, Inc.
Cambridge, Massachusetts

A. INTRODUCTION

A thermal radiation imaging system has been developed which is capable of heating an area of 1 cm^2 with up to 18 cal/cm^2-sec of radiant flux with a uniformity of ±1%. The source and the optical imaging system provide a high degree of stability and reproducibility for extended periods. The radiant heat flux is achieved using a resistance-heated graphite element enclosed in an inert atmosphere. The radiant energy from the source is collected and refocused by means of a compound optical imaging system, sometimes called a "clamshell" imaging system. The high degree of flux uniformity is achieved by means of a heat flux redistributor placed at the image focus of the system.

A radiation probe has been developed for determining the uniformity of the heat flux distribution over the image plane. Radiation falling on the tip of a quartz light pipe is transmitted by multiple reflections to a photomultiplier tube where the radiation intensity is detected. Tube output is amplified and recorded with a Visicorder. The radiation probe is attached at the end of a pendulum arm and is arranged to pass through the image area at the bottom of each swing.

Studies have recently been made on thermal imaging techniques ([1]) to generate high heat fluxes and temperatures under controlled laboratory conditions. Figure 3-1 shows several of the imaging systems currently in use. The single and two-mirror ellipsoidal systems and the two-paraboloidal mirror system require that the source and its image be placed within the optical system. As a result, the source supports and experiment apparatus cast shadows which reduce the flux intensity at the image. The refractive system does not have shadowing losses because the source and image are located outside of the system. However, due to the limited pickup angle of the lens system, only a small portion of the energy radiated by the source can be used at the image. In addition, the lenses absorb portions of the spectrum and place an upper limit on the magnitude of the transmitted radiant energy.

In a compound reflective system, the radiant heat source and its image are located outside of the optical system. In this system, the 140° angle subtended by the source mirror and the collecting mirror affords very good

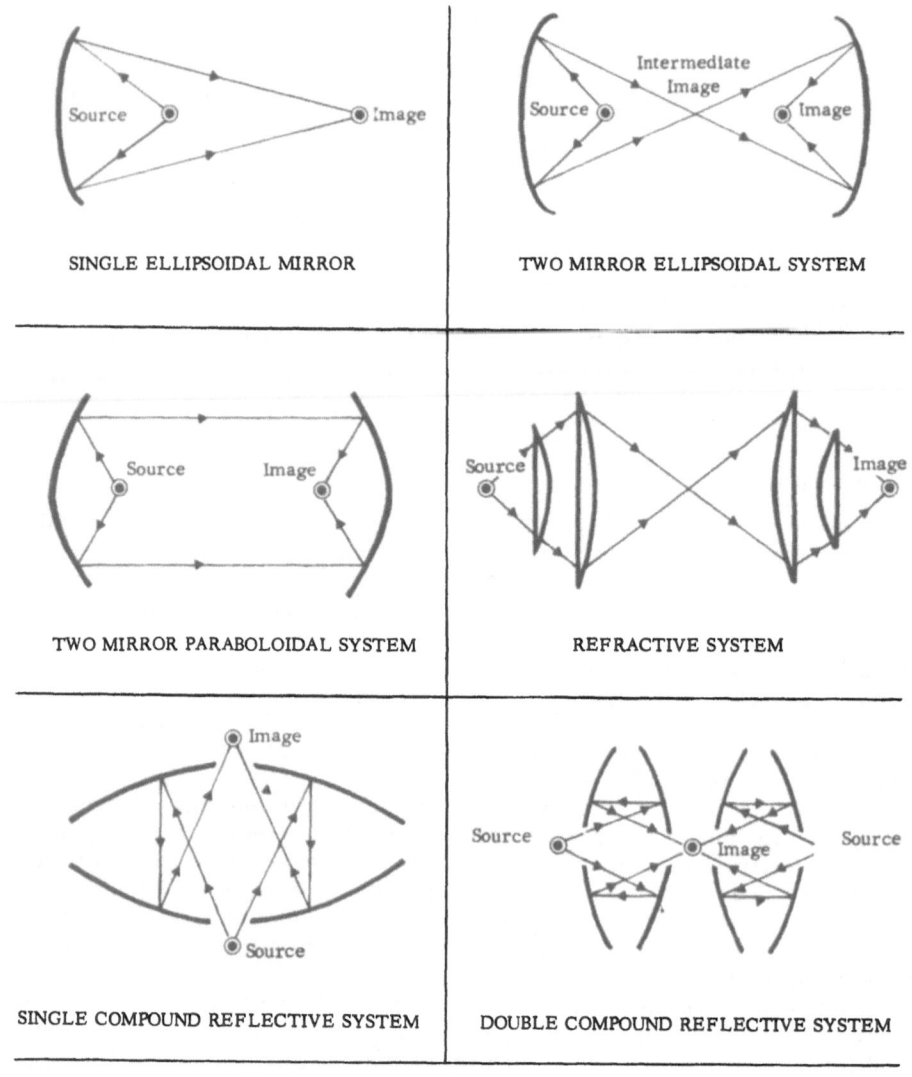

SINGLE ELLIPSOIDAL MIRROR   TWO MIRROR ELLIPSOIDAL SYSTEM

TWO MIRROR PARABOLOIDAL SYSTEM   REFRACTIVE SYSTEM

SINGLE COMPOUND REFLECTIVE SYSTEM | DOUBLE COMPOUND REFLECTIVE SYSTEM

Fig. 3-1. Thermal-radiation imaging systems.

utilization of the energy radiated by the source. By using front-surfaced mirrors, absorption losses are minimized.

A variety of radiant heat sources are currently in use with the imaging systems discussed. They fall into two categories: (1) discharge sources with radiation emanating from a volume of hot gases, and (2) heated radiating surfaces. Of the discharge sources, the air-blown carbon-arc source has the highest power rating. It provides an arc with a core temperature of 8000°K [2], which when imaged properly has produced a peak radiant flux of 450 cal/cm^2-sec [3]. Xenon and xenon-mercury short-discharge lamps [4-6] are available in sizes up to 10-kW input. No information is presently available on the operation of these lamps in radiation imaging systems.

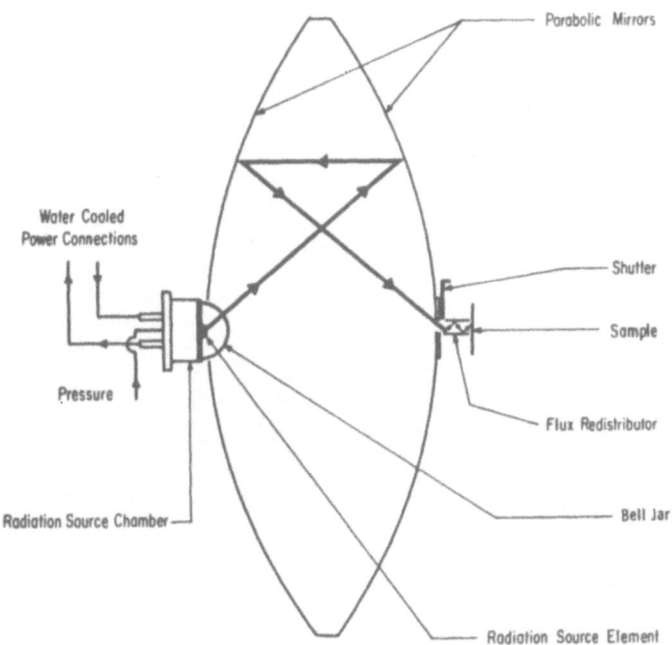

Fig. 3-2. Compound thermal-radiation imaging system and source.

The tungsten filament lamp is an example of the heated surface type of radiation source. Tungsten lamps of large energy output use coiled filaments, which when reimaged result in a nonuniform distribution of the radiant energy. A source consisting of a thin piece of graphite through which an electric current is passed has been designed to provide a more uniform irradiation [7].

B. SURVEY OF EXISTING SOURCES

A detailed survey of existing thermal radiation sources showed that the solid thermal radiation source is best suited to meet the objectives for biomedical research because the radiant energy is fairly uniformly distributed over the surface of the source. The source surface has a gray body spectral distribution of energy. Solid radiating surfaces are not limited in size, provided there is sufficient power to heat them to the desired temperature levels. Discharge sources are not suitable because the radiant energy distribution is nonuniform and concentrated in too small an area.

Of the two heated surface thermal-radiation sources investigated, a formed carbon source [7] was considered most applicable in this study. This source consists of a piece of graphite $\frac{3}{4}$ in. in diameter and approximately $\frac{1}{16}$ in. thick, through which is passed sufficient direct current to heat the element to 2500°K. By use of an ingenious concentric shielding of the source with argon and nitrogen gas it can be operated without an enclosure. This source was incorporated in a double paraboloidal imaging system and provided a heat flux of up to 21 cal/cm^2-sec with a uniformity of ±3.3%.

An alternate source considered by us consisted of an inductively heated

Fig. 3-3. Compound thermal-imaging system and power supply.

tantalum carbide disk, $5/16$ in. diameter ([8]).   When operated at 3200°K this source provided a radiant flux of 80 cal/cm^2-sec from the front face.   However, the radiating area of $5/16$ in. diameter was not sufficient for our purposes.

C. THERMAL IMAGING SYSTEM

The thermal imaging system used in this study consists of two paraboloidal mirrors mounted face to face, a radiant energy source, and a thermal flux redistributor, as shown in Fig. 3-2. This compound-reflecting optical system has the advantage that both the heat source and its image are located outside of the optical system, and the 140° subtended angle at the source mirror affords good utilization of the radiant energy output from the source. By use of front surfaced mirrors, the absorption in the optical system is minimized. Figures 3-3 and 3-4 show the complete imaging system and its associated power supply.

1. Radiation Source

The radiant heat source shown in Fig. 3-5 is a formed graphite element mounted in water-cooled holders. A water-cooled enclosure surrounds the source element.   Water-cooled power leads enter the source chamber

Fig. 3-4. Compound thermal-imaging system.

through the bottom plate. A pyrex bell jar covers the source and permits the chamber to be pressurized with dry argon at 20 psig.

2. Flux Redistributor

The flux redistributor (Fig 3-6) has a square cross section lined with four front-surfaced mirrors. The flux redistributor is installed at the image position and the radiation is directed into the opening. Multiple reflections within the redistributor result in a uniform distribution of radiant energy at the exit. A more detailed description of the flux redistributor is given by Chen et al. ([9]).

The flux redistributor used in this study (Fig. 3-6) was $\frac{1}{2}$ in. square in cross section and 2 in. long. The surfaces of the stainless steel water-cooled walls are optically flat and held within approximately two wavelengths of light. These polished surfaces were aluminized and coated with silicon monoxide to obtain optimum reflectivity. Total normal reflectivity

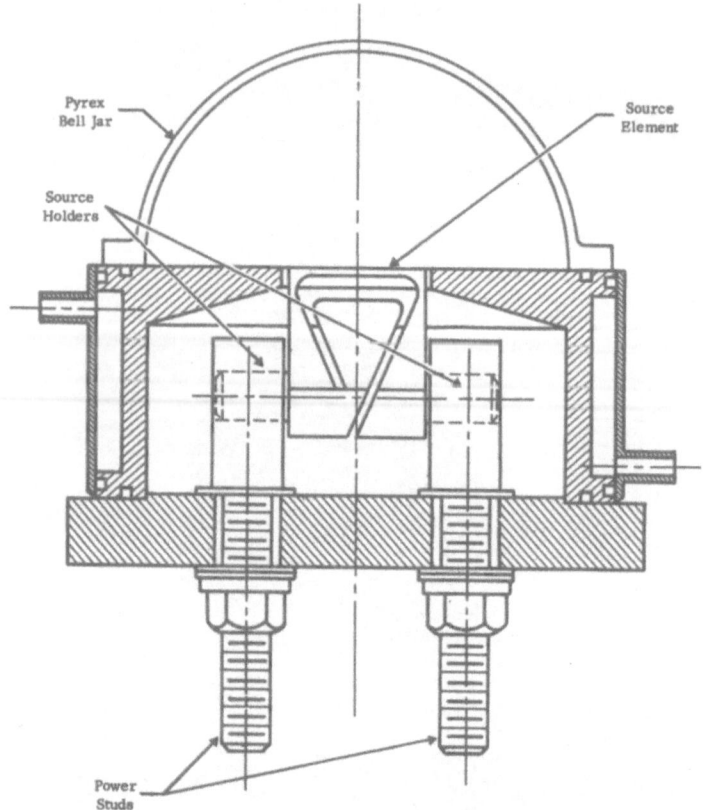

Fig. 3-5. Thermal-radiation source element and holder.

in the wavelength region of 0.5 to 3.0 μ is approximately 90% for an aluminized and coated optical surface. The measured overall flux redistributor efficiency was 50%.

D. INSTRUMENTATION

The following instruments were used to measure the thermal radiation
characteristics of the source and optical system:

1. Foil radiometer—for measurement of the absolute radiant flux at
   the image plane.
2. Optical pyrometer—for measurement of source surface temperature.
3. Radiation probe—for measurement of the uniformity of the radiant
   flux redistribution across the image plane.

1. Foil Radiometer

A foil radiometer consists of a thin, blackened circular foil of constantan
fastened around its circumference to a water-cooled block of copper. Radiant energy striking the blackened surface is absorbed by the foil and flows

Fig. 3-6. Flux redistributor.

radially to the copper block, which acts as a constant temperature heat sink.
As a result of this radial heat flow, the temperature at the center of the foil
rises above that of its circumference.   This temperature difference is a
function of the intensity of radiant flux striking the foil and is measured by
fastening a fine copper wire to the center of the foil. The differential
copper—constantan thermocouple thus obtained directly measures the temperature difference across the radius of the foil. Temperature variations
in the thermal conductivity of the constantan and in the thermal electric
power of copper—constantan thermocouples are such as to produce a linear
intensity—emf relation for circular foil radiometers made of these two
materials.   The electrical output of the instrument is not only linearly
proportional to the incident radiant flux; it is also independent of the temperature of the heat sink. Cooling water is used only to keep the temperature of the unit from rising to destructive levels. A more detailed description of this instrument and a quantitative treatment of its performance are
given by Gardon ([10]).

2. Optical Pyrometer

Surface temperatures of the source were measured with a disappearing
filament type of micro-optical pyrometer. Measurements were made at the
wavelength of 0.65 μ. Corrections were made for the monochromatic emissivity* of the graphite surface at this wavelength by the following formula:

$$1/T = 1/T_B + \lambda \ln E_\lambda / C_2 \qquad (1)$$

*See list of notation at end of this paper.

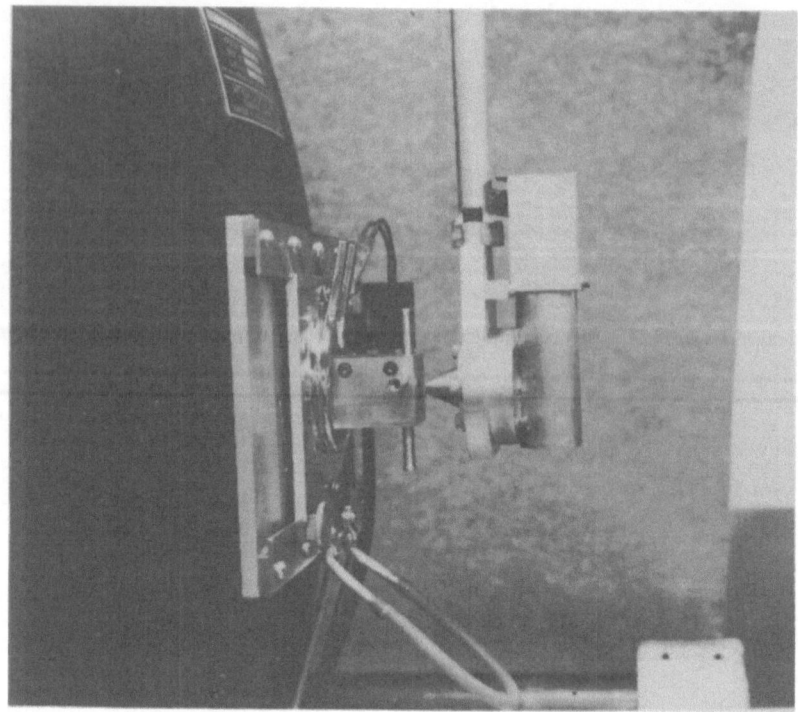

Fig. 3-7. Flux redistributor and radiation probe.

3. Radiation Probe

A swinging radiation probe, shown on the right side of Fig. 3-7, was designed to determine the uniformity of the heat flux distribution over the image plane. The probe tip contains a quartz light pipe within a platinum tube. Radiation falling on the tip of the tube is transmitted by multiple reflection in the light pipe to a series of filters (neutral density and $0.65\,\mu$), and then to a photomultiplier tube where the radiation intensity is detected and amplified. The radiation probe is attached at the end of a pendulum arm which is arranged to pass through the image area at the bottom of each swing. Variation in the velocity of traverse and differences in height due to the arc followed by the probe are negligible, since the pendulum arm is $27\frac{1}{2}$ in. long. Close control of the probe position is maintained with a compound micrometer adjustment. The power supply for the photomultiplier tube is amplified and recorded with a Minneapolis-Honeywell Model 906B Visicorder Oscillograph with a Heiland subminiature fluid-damped galvanometer, Type M-5000. This galvanometer has a flat frequency response in the range 0-3000 cps.

4. Calibration of the Foil Radiometer

The foil radiometer was calibrated in a high-temperature laboratory oven heated with "Globar" elements. The sensing element of the foil radiometer was passed very close to the Globar element and the peak output was

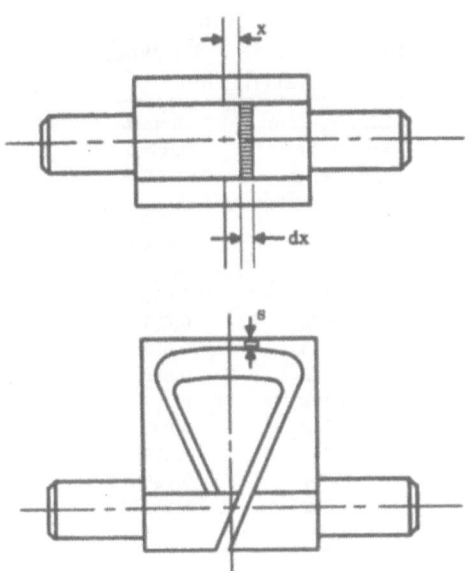

Fig. 3-8. Radiation source-element.

recorded.   The temperature of the Globar was measured with an optical pyrometer at the exact point where the radiometer passed. The sensing element of the foil radiometer is 0.9 mm in diameter, and the Globar is 13 mm in diameter.   The radiation view factor was calculated to be 0.96. The emissivity of the Globar was 0.93 (manufacturer's specification). Calibrations were made at three temperatures in this oven. In the range of radiant flux from 12 to 16.7 cal/cm^2-sec, the calibration constant was

$$C = 24.2 \text{ cal/cm}^2\text{-sec} = 102.2 \text{ W/cm}^2 \pm.6\%$$

E. SOURCE ELEMENT CONFIGURATION

Figure 3-8 shows the rectangular element used in the final configuration of the source.   The source element is 16 mm wide and 38 mm long. The thickness distribution of the radiant surface was calculated by assuming the element to be thin compared to the width and taking into account only radiation from two surfaces and conduction along the element. The heat balance for the element of Fig. 3-8 is given by

$$ks \frac{ds}{dx}\frac{dT}{dx} + kws \frac{d^2T}{dx^2} - 2\epsilon\sigma wT^4 + C_1 \frac{l^2\rho}{ws} = 0 \qquad (2)$$

The first two terms represent the conduction along the element, the third term the radiation from the top and bottom surfaces, and the last term the ohmic heating.

If the temperature in the central portion of the element is assumed to be constant, the equation reduces to the last two terms, and the thickness is calculated directly. If linear temperature distribution is assumed outside

TABLE 3-I
Source-Element Thickness
Distribution

Distance from centerline of source, in.	Element thickness, in.
0	0.078
to	constant
0.394	0.078
0.413	0.079
0.433	0.081
0.453	0.083
0.472	0.086
0.492	0.089
0.512	0.092
0.532	0.095
0.551	0.098
0.571	0.102
0.591	0.105

of this central area, the second-order term disappears and the nonlinear first-order differential equation which results can be solved stepwise for the thickness distribution. Table 3-I gives the results of this calculation.

A circular source element, $\frac{3}{4}$ in. in diameter, used in the early stages of the development had a large temperature gradient in the direction of current flow as a result of large conduction losses through the element supports. The use of the longer rectangular element minimized these temperature gradients.

F. SOURCE POWER SUPPLY

The source power supply rectifies 440-V three-phase, AC power to DC power. Output of the power supply is 4-12 V, with the current variable from 200 to 600 A, depending on the load resistance. The power supply consists of a motor-driven three-phase autotransformer, a three-phase stepdown transformer, six silicon diode rectifiers, and a fully transistorized control network that maintains the output voltage within ±1% irrespective of line or load fluctuations.

G. RESULTS OF TESTS

1. Radiant Flux

Figure 3-9 shows the radiant heat flux (cal/cm^2-sec) at the exit of the flux redistributor as a function of corrected source temperature (°K). The desired maximum flux of 15 cal/cm^2-sec was achieved when the source was operating at a corrected true temperature of 2840°K.

2. Flux Uniformity

Figure 3-10 shows a typical trace of the flux uniformity obtained at the exit of the flux redistributor. The trace was taken with the radiation probe

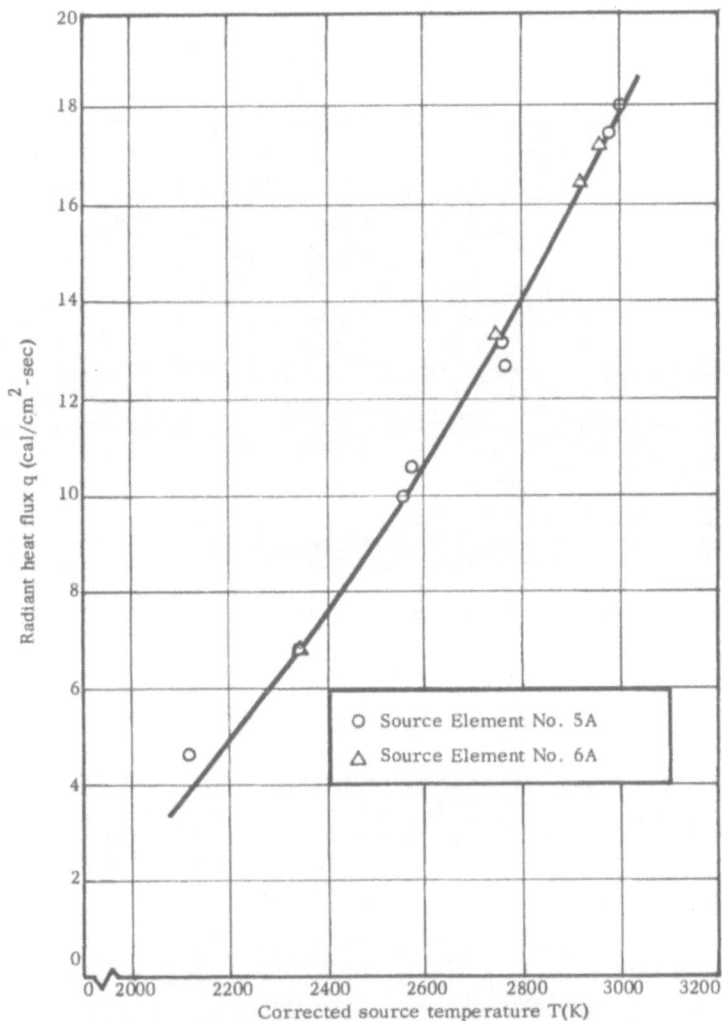

Fig. 3-9. Radiant heat flux at exit of flux redistributor vs. corrected source temperature.

and recorded on a Visicorder operating at a recording velocity of 25 in./sec. The maximum deviation over the ½-in. exit of the flux redistributor was less than ±1% of the maximum flux. The unevenness in the signal peak in this figure is due to the noise characteristic of the output of photomultiplier tubes.

Figure 3-11 shows a trace of the radiant flux uniformity without a flux redistributor. This trace is typical for the deviation in the longitudinal direction; the maximum deviation is approximately 30% from the peak to the low point. The deviation in the transverse direction is even greater—up to 40% for moderate flux intensities.

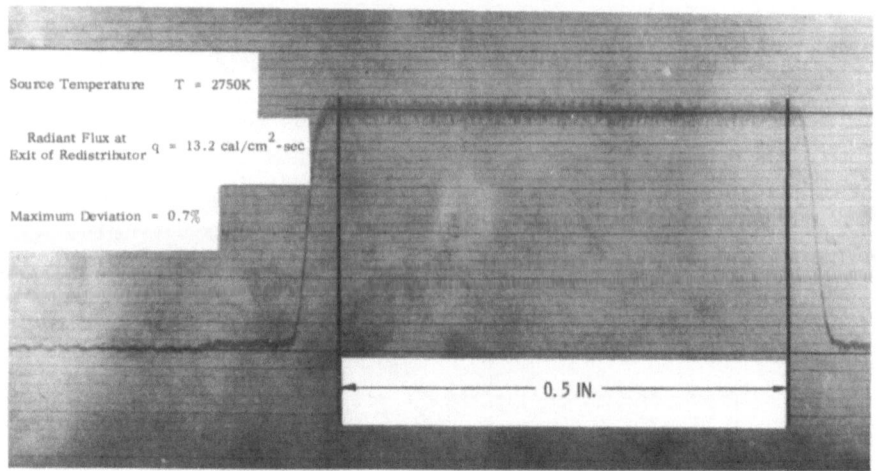

Fig. 3-10. Radiant flux profile with flux redistributor.

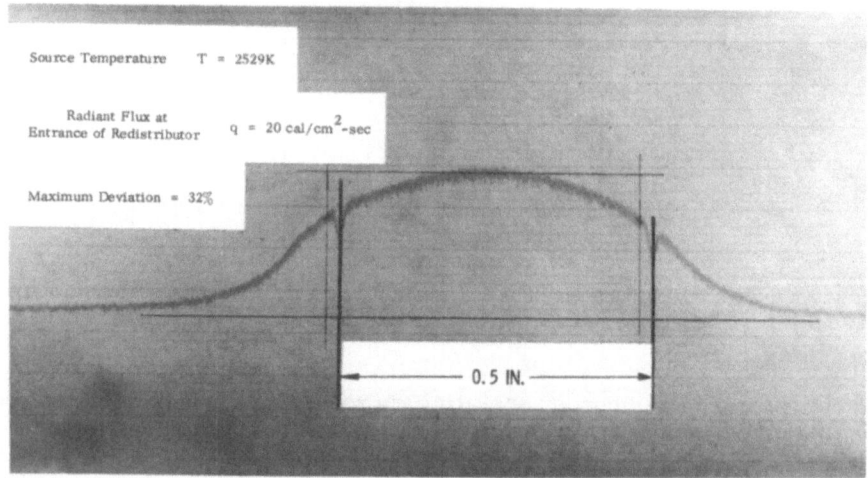

Fig. 3-11. Radiant flux profile without flux redistributor.

NOTATION

C	= Calibration constant		
C_1	= Constant	0.2389	cal/W-sec
C_2	= Radiation constant	1.438×10^4	μ-°K
I	= Total current flowing in element		A
K	= Thermal conductivity	0.075	cal/cm-sec°K
s	= Element thickness		cm
T	= Corrected temperature		°K
T_B	= Brightness temperature		°K
w	= Element width	1.6	cm
x	= Axial position		cm
ϵ	= Emissivity	0.75	—
ϵ_λ	= Monochromatic emissivity	0.82	
ρ	= Resistivity	11.6×10^{-4}	ohm-cm
σ	= Stefan–Boltzmann constant	1.354×10^{-12}	cal/cm^2-sec°K^4
λ	= Wavelength		μ

ACKNOWLEDGMENTS

This work was done for Dr. Alice M. Stoll of the Aeromedical Acceleration Laboratory of the U. S. Naval Air Development Center in Johnsville, Pa., under Contract No. N62269-1388. The author is indebted to Mr. George Ploetz of the Air Force Cambridge Research Laboratories, who developed the parabolic compound imaging system and to Dr. Edwin E. Maust of the Bureau of Mines Metallurgical Research Center, College Park, Maryland, who developed a resistance-heated graphite heating element which was the starting point in this development. The author wishes to thank those of the staff of Arthur D. Little who contributed to the success of this work: D. F. Comstock and A. L. Camus for their assistance with the radiation probe, F. F. Chellis for the design of the optical system, Dr. A. E. Wechsler for assistance in evaluating the performance of the system, and Dr. P. E. Glaser, who directed the work.

REFERENCES

1. Glaser, P. E., and Ploetz, G. P., Symposium on "Temperature—Its Measurement and Control in Science and Industry," (American Institute of Physics, New York, 1961), Paper C-5-1.
2. Gretener, E., J. Soc. Motion Picture and Television Engineers 55:391 (1955).
3. Glaser, P. E., "Progress in High-Temperature Thermal Imaging Techniques," TVF 34 (1963):3 (Sweden).
4. Osram, GMBH, Berlin, Germany.
5. Lienhard, O. E., and McInally, J. A., "New Compact Arc Lamps of High Power and High Brightness," National Technical Conference of the Illuminating Engineering Society, September 24-29, 1961, St. Louis, Mo.
6. Thouret, W. E., and Strauss, H. W., "Xenon High-Pressure Lamps," Preprint No. 15, National Technical Conference of the Illuminating Engineering Society, September 24-29, 1961, St. Louis, Mo.
7. Maust, E. E., and Warnke, W., "Performance and Operating Characteristics of a 60-in. Double Paraboloid Image Furnace," U. S. Department of the Interior, Bureau of Mines, Report of Investigation, RI No. 5946, 1962.
8. Unpublished report, Lighting Division, Sylvania, Inc., Salem, Mass.
9. Chen, M. M., Berkowitz, J. B., and Glaser, P. E., "The Use of a Kaleidoscope to Obtain Uniform Flux Over a Large Area in a Solar or Arc Furnace," J. Appl. Opt. 2(3):265 (1963).
10. Gardon, R., "An Instrument for the Direct Measurement of Intense Thermal Radiation," Rev. Sci. Instr. 24 (5):366 (1953).

Chapter 4

Investigation of Thermal Imaging Techniques

Tibor S. Laszlo* and Paul J. Sheehan†

Research and Advanced Development Division
AVCO Corporation
Wilmington, Massachusetts

A. INTRODUCTION

Image furnaces have found wide application recently in high-temperature research and testing. Because of their peculiar characteristics, however, new instruments, experimental setups, and methods have to be found for any new operation. Some new measuring and experimental techniques, as well as a new method for the fabrication of large inexpensive paraboloid mirrors for image furnaces, are presented here.

B. FABRICATION OF PARABOLOIDAL MIRRORS

Existing image furnaces use reflectors which with very few exceptions, were made for some other purpose. Large paraboloidal mirrors are available in surplus searchlights as are small ellipsoidal mirrors made for motion picture projectors. While these reflectors are generally of good quality, their greatest advantage is their low cost. The fabrication of reflectors by conventional techniques specifically for use in image furnaces is prohibitively expensive as they are not built in large numbers. Further, the cost increases greatly with a small increase in size. Irrespective of the cost, the required optical precision poses great technical difficulties for the fabrication of large reflectors.

Several attempts were made to solve the technical difficulties at a moderate cost ([1]). However, a high-precision master was already available for these attempts. If a master is not available, the fabrication of one usually involves even more difficulties and greater expense than the reflector itself. These processes, therefore, are justified only if a large number of identical reflectors are required.

A recently developed process ([2]) uses the principle that a liquid in a revolving horizontal pan takes the shape of a paraboloid. If a liquid resin, mixed with a hardening agent is thus rotated, it will harden in the shape of a precise paraboloid. A modification of this method, applied in the following process, proved to be suitable for the fabrication of large, precise, inexpensive paraboloidal mirrors.

A 36-in.-diameter stainless steel dish, which had approximately the

*Principal Staff Scientist.
†Assistant Scientist.

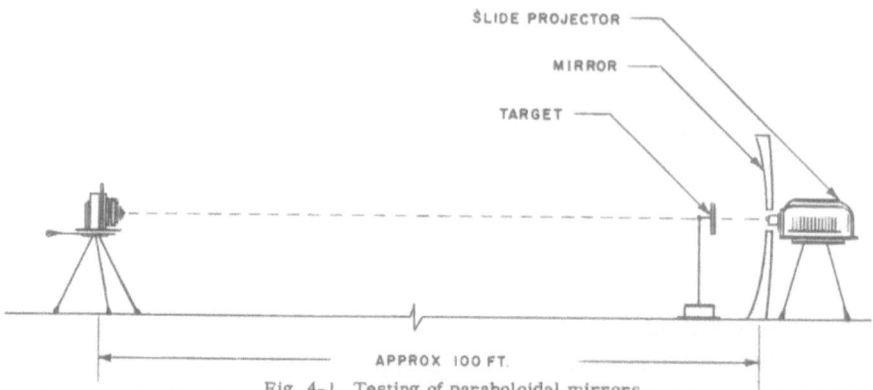

Fig. 4-1. Testing of paraboloidal mirrors.

shape of the wanted reflector was formed by spinning. The center hole in the metal dish was closed by a wooden plug that was flush with the inside surface of the metal, and the dish was mounted on a turntable which could be rotated at a very precisely controlled angular velocity. A bubble-free mixture of a clear epoxy resin and hardener was poured into the rotating dish, which was enclosed in a dust-free atmosphere. The rotating liquid mixture formed the surface of a paraboloid, and its diameter was defined by a rim on the stainless steel dish. The paraboloid had a focal length given by $f = 1474/(\text{rpm})^2$. The rotation was continued until the resin hardened (approximately 24 hr). During the latter part of the hardening period, radiant heat was applied to accelerate the reaction.

Fig. 4-2. Reflected test pattern.

Fig. 4-3. Reflected test pattern.

The resulting paraboloidal dish had a very smooth, even surface with only a few imperfections visible near the rim. The geometrical perfection of the dish was tested by the optical method illustrated in Fig. 4-1. A target pattern, consisting of a polar coordinate graph paper was mounted at the focal point of the paraboloid on three crossways in such a fashion that it could be positioned exactly on the three spatial axes. The target was illuminated with a slide projector placed behind the center hole of the dish. A wide-angle-lens camera (focal length 19 in.) was located approximately 100 ft from the dish. The camera, the target, the slide projector, and the paraboloid were all aligned on the same optical axis. Photographs were then taken of the target as reflected by the paraboloid epoxy surface. The target diameter was much larger than that of the focal area. Therefore, since the focus was well defined, it was necessary to take several photographs with the target in different positions in relation to the focus. Figures 4-2 and 4-3 show the reflection of the target in the epoxy dish photographed at different target-focus relative positions. Some optical distortion, visible on the upper-right quadrant of the dish, indicated minor local surface imperfections which could not be detected by mechanical means. This quadrant was not used during the preparation of the reflecting lining.

The conventional method used to make a reflecting epoxy resin surface consists of the vapor deposition of a metal on it. However, this process requires a vacuum chamber large enough to accommodate the entire dish, a condition which might be restrictive in the case of large dishes. In addition, it is very difficult to obtain a metal coating which adheres well to the resin and has a mirrorlike finish. Therefore, another process was used which is not limited by the size of available vacuum chambers and in which a highly reflecting metal lining is bonded to the resin with an epoxy adhesive.

First, a fiber-reinforced plastic casting was made of the most perfect section of the dish. Specially polished aluminum sheets, 35/1000 in. thick, were stretch-formed over the casting to obtain petals of the reflector. The formed sections were hand polished in order to remove stretch marks. Following this, the sections were Alzak processed in a colorless electrolyte for the formation of a corrosion/erosion resistant surface. Then the sections were trimmed to fit exactly the original epoxy dish and bound to it with an epoxy adhesive. During the hardening process, the petals were pressed into the original dish with a plastic male mold.

The completed paraboloid reflector was mounted coaxially in the 60-in. solar furnace for testing (Fig. 4-4). A fast-response radiometer was mounted in the focal area, and the maximum flux in the solar image was located by displacing the radiometer along the three spatial axes until a maximum signal was obtained. Following this, the radiometer was displaced along one of the axes until its output became zero, while keeping its position unchanged in reference to the other two axes. Then the radiometer was moved at a constant speed across the image along the same axis and its output recorded. The same procedure was followed for the two other axes. The resulting output signals are presented in Fig. 4-5. Some irregularity near the mirror periphery is indicated at the edge of the image, but the center portions show a very regular Gaussian distribution. These results are borne out by visual observations. The lines marking the edges of the individual petals are hardly perceptible in Fig. 4-4, and the continuity of the image is preserved when the image progresses from one petal to an ad-

Fig. 4-4. Thirty-six-in. aluminum-lined epoxy resin paraboloid mounted in the 60-in. solar furnace.

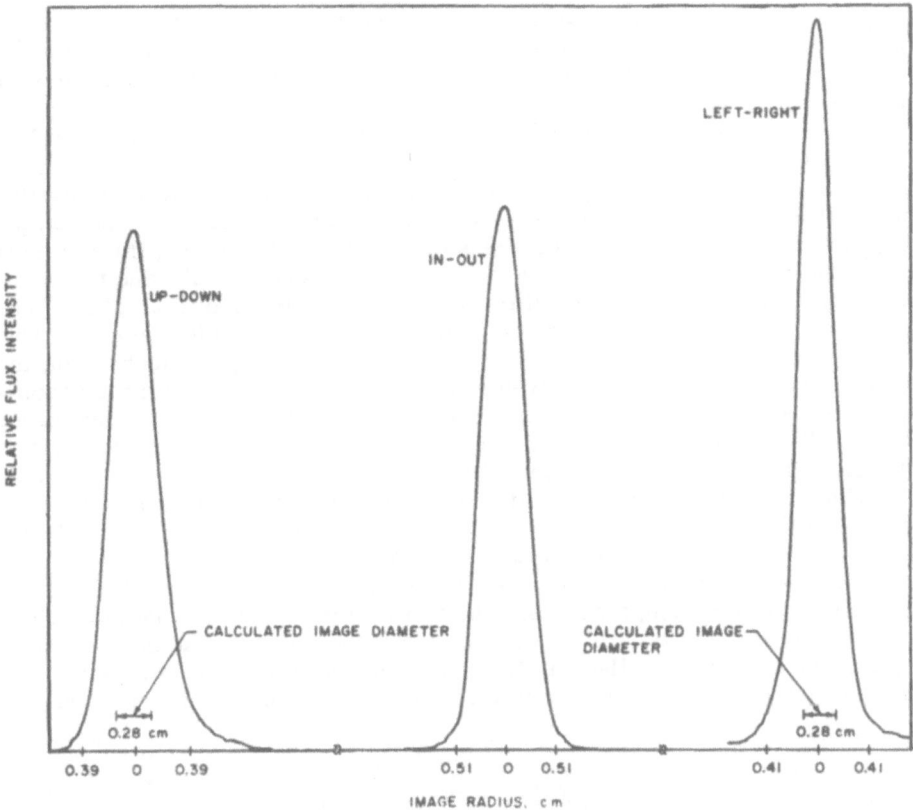

Fig. 4-5. Flux distribution in 36-in. paraboloid.

joining one. A slight distortion of the curved image line is visible at the peripheries.

This reflector was the first one made according to this new process. It is expected that the experience gained during its fabrication will make it possible to produce reflectors of even better quality. Although 36-in.-diameter reflectors of good quality are also readily available from surplus searchlights, the 36-in.-diameter size was chosen as convenient to test the procedure. Facilities are available to make paraboloidal reflectors according to this process up to approximately 30 ft in diameter.

C. SAMPLE HOLDER FOR ELECTRICAL MEASUREMENTS

One of the important advantages of image furnaces is the fact that electrical properties can be measured at high temperatures without any restrictions, since the sample is not surrounded by an electromagnetic field. Due to the peculiarities of image furnaces, however, other problems have to be solved before such measurements can be successfully performed. The greatest difficulty is caused by the fact that a sample in an image furnace is heated only on one side and, accordingly, a steep temperature gradient is established across the sample.

A sample holder has been built which permits the sample to be rotated perpendicular to the optical axis of the furnace. By rotating the sample at a sufficiently high speed, temperature uniformity around its circumference may be obtained.

Figure 4-6 is a photograph of the sample holder installed in a controlled atmosphere chamber. The "windshield wiper" mechanism for keeping the hemispherical window free of condensates has been described in an earlier publication [3]. Two hollow, stainless steel shafts are used for the sample support. A small tube inside the shafts carries the cooling fluid to their tips; the fluid then returns in the space between the tube and the shaft. Conical copper contacts are fitted tightly over the tip of each shaft. The other end of the copper contacts is made to fit the sample diameter. One of the shafts is rotated by a servo motor through a gear mechanism; the other rotates freely. The driven shaft can be adjusted laterally to accommodate samples of various length. The other shaft is spring-loaded in the lateral direction in order to make good mechanical contact with the sample. Through this contact, the shaft follows the rotation of the driven shaft. In addition, the spring-loading facilitates sample insertion and removal. Copper tubes for the cooling fluid connections are mounted on the bearing housings, and O-rings are used to make a watertight rotating seal.

This sample holder has been designed for the measurement of electrical resistivity at high temperatures. The two rotating shafts are isolated electrically by the use of Teflon gaskets and nylon-coated screws. A rotating electrical contact installed at the end of the shafts (Fig. 4-7) consists of a nickel-plated copper disk silver-soldered onto the shaft. The knife edge of

Fig. 4-6. Controlled-atmosphere, liquid-cooled sample holder.

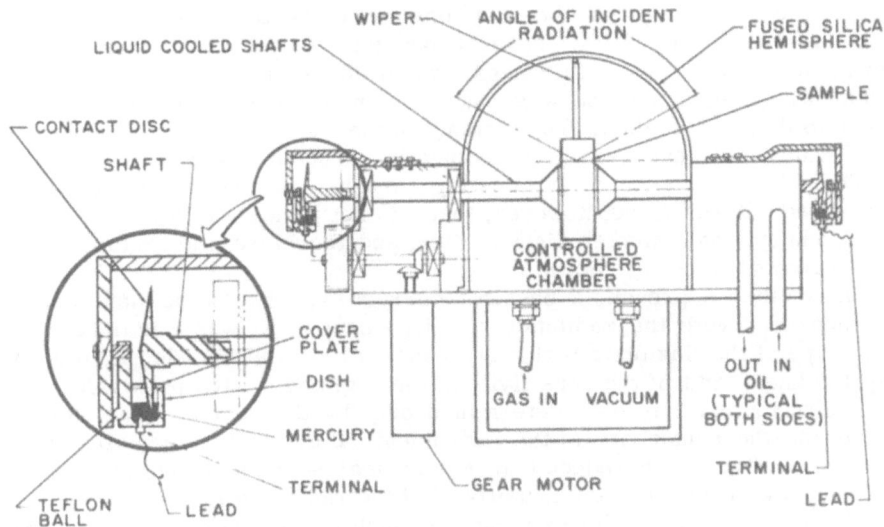

Fig. 4-7. Sample holder for electrical measurements.

the disk rotates in mercury. Electrical contact with the measuring instruments is made through a terminal at the bottom of the mercury vessel. This vessel rotates freely around a horizontal axis in order to be always in a horizontal position during the rotation of the solar furnace.

The sample holder was tested for electrical conductivity. It was found that the resistance between the terminal and the copper contacts is 2-5 ohms. The copper contacts were tested for electrical insulation from each other. The resistance between them was $>10^{10}$ ohms when the shafts were cooled by transformer oil.

D. RADIOMETER CALIBRATIONS

The use of a fine-resolution radiometer in a solar furnace has been reported earlier [4]. Observations on the failure of the carbon coating at high fluxes in an oxidizing atmosphere are listed by Laszlo [3] together with the recommendation that MgO be used as a coating material. Experimental work was carried out in order to establish the suitability of this coating.

The MgO was deposited on the radiometer sensing disk by burning magnesium ribbon in air. The assumed formation of some magnesium nitride during burning has been proven by the ammonia gas generated when the product of combustion comes into contact with water. However, the nitride is converted to the oxide shortly after the freshly coated radiometer is exposed to the high heat flux in the solar furnace. Confirmation of this was observed via the sensitivity of a freshly coated radiometer, the sensitivity decreasing during the first few minutes of exposure but becoming stable after approximately 10 min. Accordingly, before a freshly coated radiometer was used for measurements, it was "baked out" at maximum heat flux for 10 min.

Measurements were performed to find out if any correlation exists between the sensitivity of the instrument and the thickness of the coating. Such a correlation would make it possible to reduce the sensitivity determination to a coating thickness measurement. For the thickness measurement, a reference line was engraved on the housing of the radiometer close to the sensing disk. The distance of the uncoated disk from the reference line was measured microscopically. The disk was then coated with MgO, and the distance of the coating surface from the reference line was measured. It was found, however, that it is not possible to focus the microscope sharply on a smooth MgO surface, and, therefore, the precision of the measurement is unsatisfactory.

In another approach, a possible correlation between the maximum flux as measured with the radiometer and its sensitivity was examined. Since the shape of the flux distribution at the focal area is independent of the coating thickness and of the sensitivity of the radiometer, the peak value of the flux distribution curve is determined only by the normal solar incidence value and the sensitivity of the instrument. If the peak values obtained for several coatings are reduced to unit normal solar incidence, they should coincide and be equal to the sensitivity of the radiometer.

In order to prove the existence of such a correlation, the radiometer was coated with MgO and "baked out." The instrument was then positioned at the exact center of the focal area and moved across the image along the horizontal axis. Five crossings were made for each reading, and their average peak value was considered in the calculation. Five to ten readings were taken for each coating in order to establish the stability of their sensitivity.

The evaluation of the millivolt peak values in terms of flux was made according to a procedure published earlier ([4]). A refinement of this method has been made, however, by integrating the volume rather than the area defined by the output signal/image radius curve. The integral thus obtained was equated with the total energy in the same image size as determined by calorimetric measurements. From this equation, the output signal/flux conversion factor, i.e., the sensitivity of the radiometer, was obtained.

The results obtained on twelve coatings are presented in Fig. 4-8, where the peak flux values at the center of the image are plotted against the normal solar incidence. The points appear to define the linear correlation between the two parameters well. This supports the assumption that the peak flux readings depend solely on the normal solar incidence. This in turn proves that the sensitivity of the instrument is constant throughout the range investigated thus far.

Since the data presented in Fig. 4-8 were obtained with twelve separate coatings, of different thickness, the linearity of the correlation also proves that the sensitivity of the instrument is independent of the thickness provided the thickness is kept within previously defined limits.

From the twelve determinations performed so far, the sensitivity of the MgO-coated radiometer is calculated to be 0.0090 mV/cal-cm²-sec. This should be considered an approximate value, since the spread of the results points to the definite need for further work. A sensitivity of 0.037 mV/cal-cm²-sec is specified by the manufacturer for the camphor-black-coated instrument. The difference between the two values indicates that the camphor-black-coated radiometer absorbs four times more radiant energy than the MgO-coated one. This agrees well with theoretically expected values.

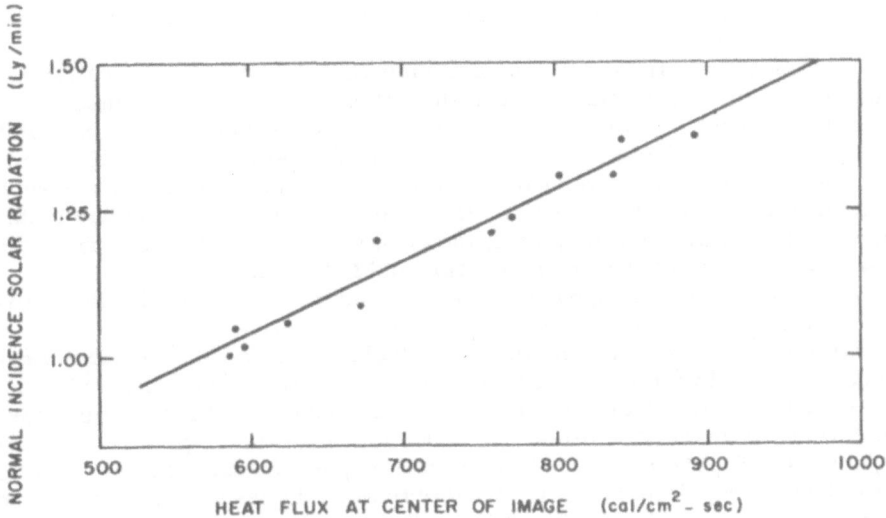

Fig. 4-8. Peak flux vs. normal incidence solar radiation.

E. FLUX MEASUREMENTS IN THE 60-IN. SOLAR FURNACE

The maximum flux measured with the radiometer was 892 cal/cm²-sec at 1.38 Ly/min normal incidence. From Fig. 4-8, a maximum flux of 975 cal/cm²-sec is obtained for 1.5 Ly/min normal incidence, a high value for the Boston area. The highest peak flux values in solar furnaces reported thus far are summarized in Table 4-I.

It is realized that the peak flux determined during this work, 975 cal/cm²-sec, is unexpectedly high, and its validity can be accepted only if proved by a method independent of the described measurement. Such a method was found in checking the melting point of several high refractory compounds in the solar furnace. If the data presented in Fig. 4-8 are true, it should be possible to melt a high-refractory compound, e.g., thoria (m.p.

TABLE 4-I
Peak Flux in Solar Furnaces

Peak flux cal/cm²-sec at 1.5 Ly/min	Method of determination	Type of furnace	Reference
573	Measurement	60-in. paraboloid	[5]
638	Calculated from measured average	60-in. paraboloid	[6]
665	Theoretical calculation	105-ft paraboloid mosaic composed of 2 ft × 2 ft spherical mirrors	[7]
975	Measurement	60-in. paraboloid	Present work

3050° C), at a flux lower than the peak flux obtainable at 1.28 Ly/min normal incidence.

Independent of calorimetric measurements and of the use of a calibrated radiometer, the correlation between flux attenuator position and modulation of peak flux has been previously determined. Accordingly, if the peak flux at a given normal incidence is obtained from Fig. 4-8, it is possible to reduce the flux to a desired value by adjusting the flux attenuator. Conversely, if the flux at a given attenuator position is determined by an absolute method, e.g., by the melting of thoria, the peak flux can be arrived at from the known relationship of attenuator position and flux modulation.

Three refractory compounds were selected for use in the checking procedure: alumina, stabilized zirconia, and thoria.

First the alumina was melted in the solar furnace with a normal incidence of 1.17 Ly/min. It was not possible to freeze the molten alumina under these conditions even when the flux was reduced to the minimum possible value with the flux attenuator. From Fig. 4-8, the peak flux at 1.17 Ly/min is 705 cal/cm^2-sec. The attenuator position reduces this value to 148 cal/cm^2-sec. If an effective emissivity of 0.5 is assumed the calculated temperature of the alumina is 2450° C. The fact that the alumina was liquid during this test proves that the temperature was definitely above 2020° C; thus, the unattenuated peak flux was higher than 341 cal/cm^2-sec.

The stabilized zirconia was melted with a normal incidence of 1.22 Ly/min with the attenuator in the maximum flux reduction position. From Fig. 4-8 the peak flux at 1.22 Ly/min is 746 cal/cm^2-sec. The attenuator position reduces this to 157 cal/cm^2-sec. Assuming an effective emissivity of 0.5, the calculated temperature of the molten sample is 2500° C. According to the supplier of the stabilized zirconia, the melting point is approximately 2590° C. Thus, the peak flux had to be at least 746 cal/cm^2-sec in order to melt the zirconia at the given attenuator position.

The thoria was melted with a normal incidence of 1.28 Ly/min with the attenuator at 9.6 on an arbitrary linear scale. The peak flux at 1.28 Ly/min normal incidence is 794 cal/cm^2-sec. The attenuator at the given position reduces this to 293 cal/cm^2-sec. Assuming an effective emissivity of 0.5, the calculated temperature of the molten thoria is 2970° C. This agrees well with the literature data of 3050° C. Thus, the peak flux had to be at least 794 cal/cm^2-sec in order to melt thoria under the experimental conditions.

In these calculations, the emissivity value of 0.5 was assumed for the various samples, because no reliable information is available. This was taken as a realistic value considering the few data reported in the literature.

A difference of ± 0.1 in emissivity does not alter the conclusions drawn from these tests.

At the present state of the flux measurement program, it can be stated already that under favorable normal solar incidence conditions in the Boston area (1.00-1.40 Ly/min) peak fluxes are available in the 60-in. solar furnace corresponding to a blackbody temperature of 4200-4800° C. Under optimum normal solar incidence conditions (1.40-1.50 Ly/min) the available flux corresponds to a blackbody temperature of 4800-4900° C. Since all known compounds melt below 4000° C, it appears possible to melt every known solid in the solar furnace.

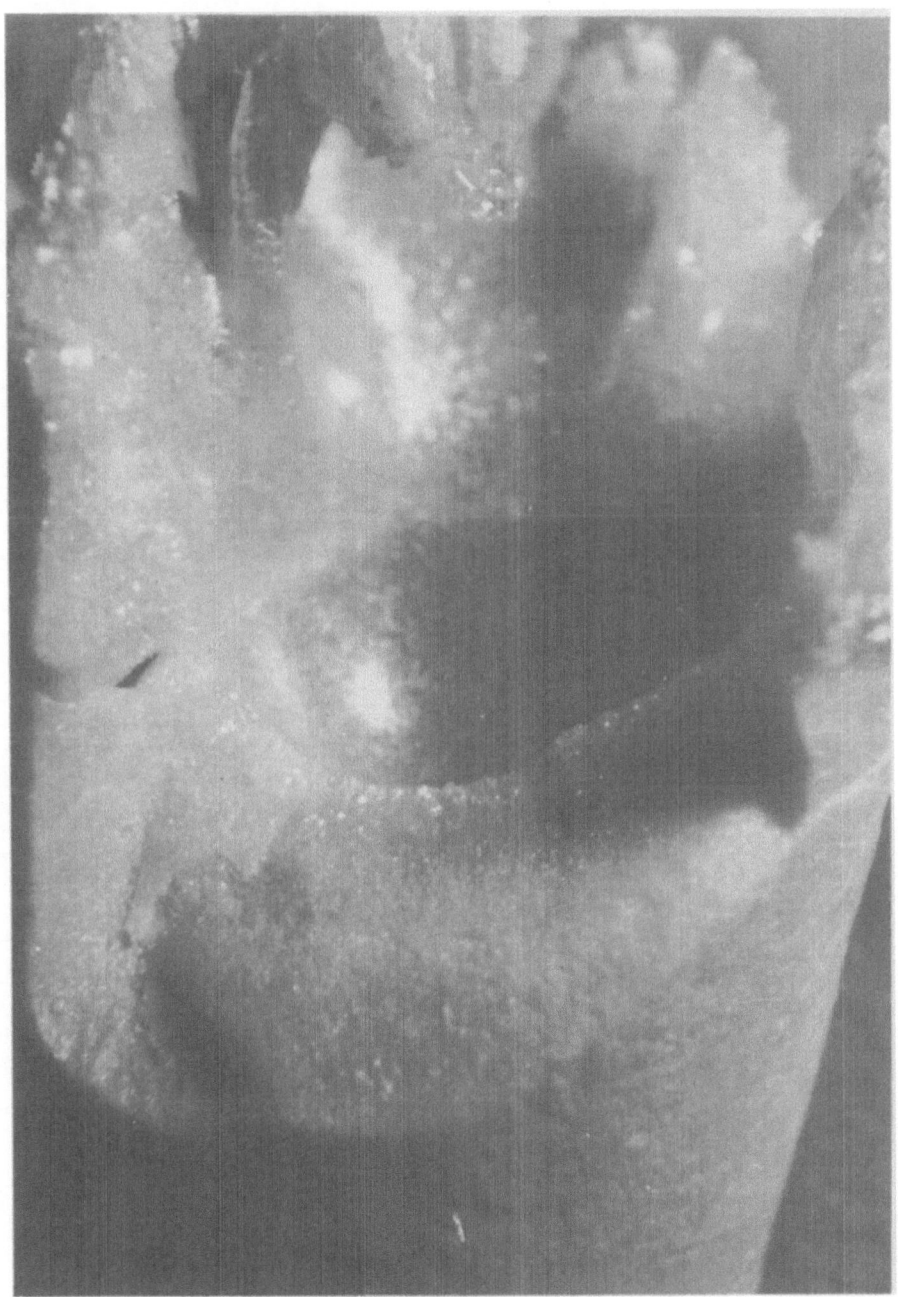

Fig. 4-9. Condensation crystals around thoria crater (150 ×).

Fig. 4-10. Rodlike single crystals of thoria (150 x).

Fig. 4-11. Octahedral single crystal of thoria (230 ×).

Fig. 4-12. Thoria crystal cluster (145 x).

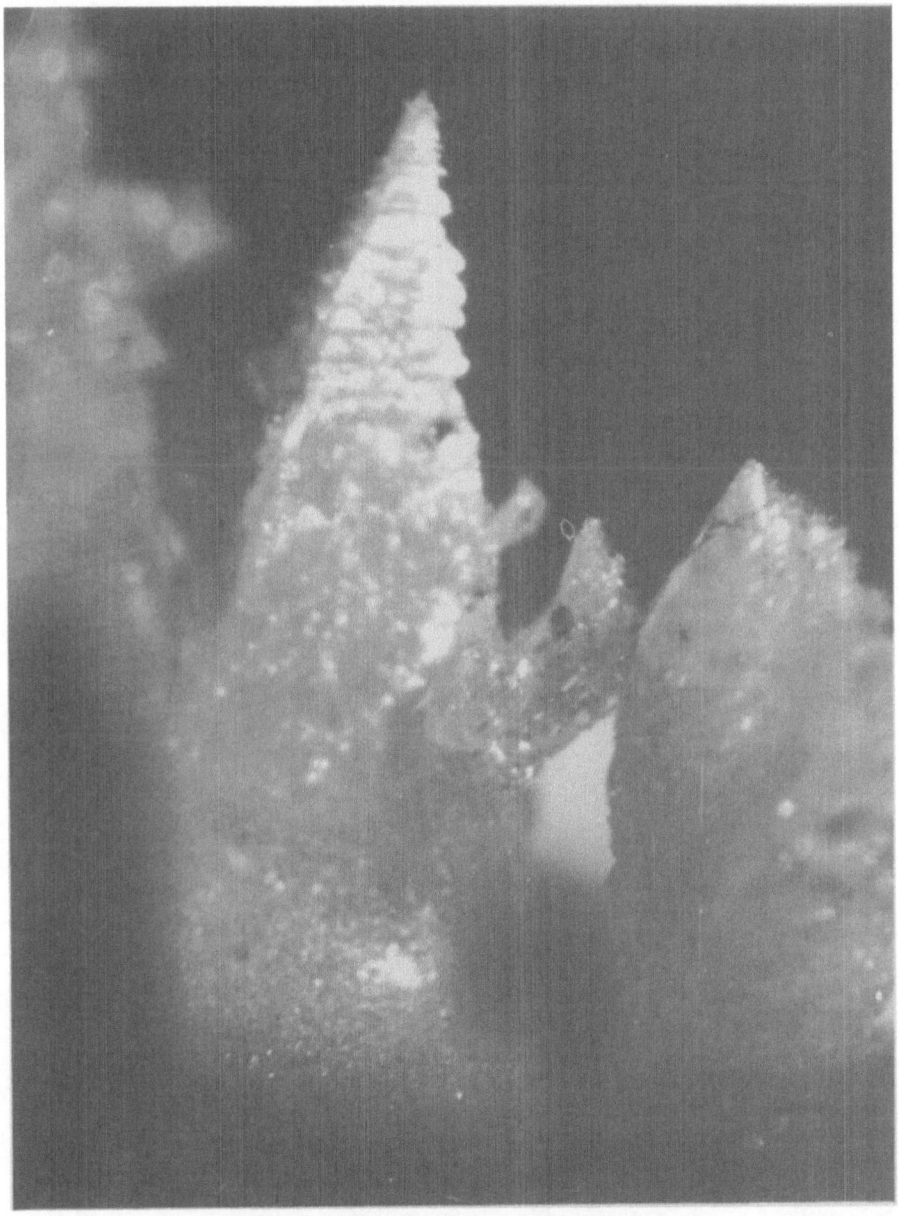

Fig. 4-13. Multiple twin thoria crystals (145 x).

Fig. 4-14. Dendritic thoria crystals (230 x).

When the thoria rods were exposed to fluxes much greater than those necessary for melting, the thoria vaporized at a very rapid rate. The vapors condensed upon reaching the cooler portion of the sample at the edge of the focal area. Figure 4-9 shows the small crystals condensed around the molten crater. Figure 4-10 is a photograph of rodlike single crystals with sharply defined faces on the top. Figure 4-11 is an octahedral single crystal attached to the rod by a threadlike formation. Figure 4-12 presents several crystal clusters composed of similar sizes. Figure 4-13 shows twinning along the C axis, the crystals decreasing in size. Figure 14 is a photograph of small crystals in dendritic growth but well-defined crystal habit.

ACKNOWLEDGMENTS

The spincasting of the epoxy paraboloidal dish was performed by the Kennedy Antenna Division, Electronic Specialty Company. The specially polished aluminum sheets were contributed by the New Kensington Research Laboratory of the Aluminum Company of America. The reflecting lining was formed and installed by The Ryan Aeronautical Company. The work was supported by contract AF19(604)-7204 from the Air Force Cambridge Research Laboratories, USAF, Bedford, Massachusetts.

REFERENCES

1. Gillette, R., Snyder, H. E., Timar, T., Solar Energy V (1):24 (1961); Bolin, J., Tenukest, C. J., Milner, C. J., Solar Energy V (3):99 (1961); Bradford, A. P., Erbe, W. W., Hass, G., J. Opt. Soc. Am. 49 (10):990 (1959); Saxton, J. H., and Kline, D. E., J. Opt. Soc. Am. 50:1103 (1960).
2. Archibald, P. B., Solar Energy I (2):102 (1957).
3. Laszlo, T. S., "New Techniques and Possibilities in Solar Furnaces," United Nations Conference on New Sources of Energy, Rome, Italy (1961).
4. Laszlo, T. S., Solar Energy VI (2):69 (1962).
5. Laszlo, T. S., Solar Energy I (2):78 (1957).
6. Farber, J., "Images of Very High Temperature Sources," Symposium on "High Temperature—A Tool for the Future," Berkeley, California (1956).
7. Allison, F., Carr, H., Hughes, G., "Notes on the Potential Performance of the Cloudcroft Solar Furnace," Solar Furnace Support Studies, AFMDC TR 58-7, AD 135014 (1958).

Chapter 5

The Carbon Vapor Lamp

G. P. Ploetz
Air Force Cambridge Research Laboratories
Office of Aerospace Research
United States Air Force, Bedford, Massachusetts

A. INTRODUCTION

A few years ago, Arthur D. Little, Inc. designed and constructed a new type of carbon arc image furnace for Air Force Cambridge Research Laboratories [1,2]. (Fig. 5-1). The axis of the furnace is vertical with an ADL-Strong blown carbon arc source mounted on top of the furnace. The thermal image of this source extends out of the double paraboloid mirror assembly and is accessible from below. This orientation is favorable for crystal

Fig. 5-1. AFCRL-ADL carbon-arc image furnace.

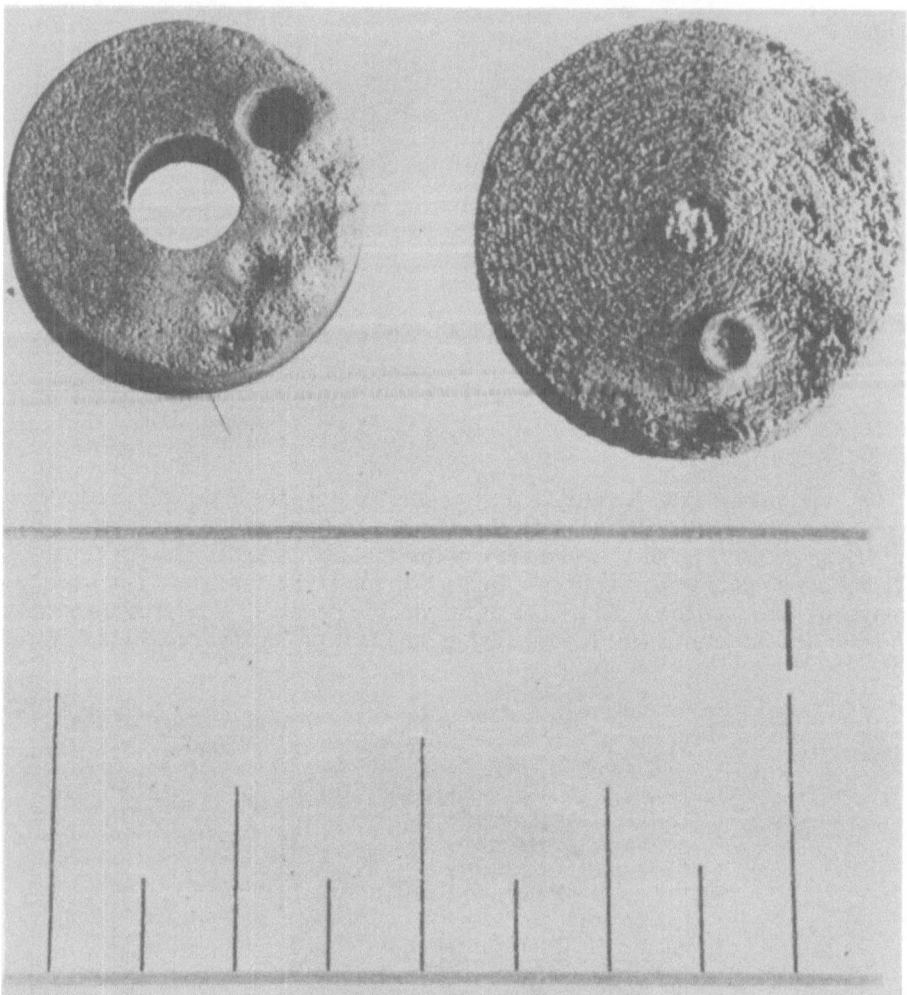

Fig. 5-2. Carbon vapor lamp.

growth and other applications. Because the carbon arc burns in air, the period of operation is limited due to consumption of the electrodes. Thus, only small crystal boules are possible. In addition, the negative electrode and air-scoop assembly produce a large illumination shadow resulting in irradiation losses. In order to fully exploit the clam-shell mirror arrangement, more appropriate sources were needed. Thus a further investigation into controllable, high-temperature sources was initiated.

The operation and history of the development of the carbon arc was studied ([3]). Then some crude experiments were performed so that the capabilities and limitations of carbon arcs in vacuum or partial gas fill could be observed. These observations led to the demonstration of a three-phase AC carbon vapor discharge in a vacuum chamber. This paper describes these experiments and the knowledge gained from these observations.

Fig. 5-3. Three-phase AC carbon vapor lamp.

B. DISCUSSION

The optimum source for the "clam-shell" mirror arrangement would be a uniformly heated solid having a variable but easily controlled temperature. Three basic geometries, a hemisphere, a thin disk, and a hemispherical hole, were considered. It was decided to investigate the thin disk first. Carbon was selected as the solid material because of its high sublimation temperature and ease of machining. Electrical or joule heating of the thin disk was considered. Such sources have been designed, constructed, and evaluated at the Bureau of Mines [4] and at Arthur D. Little, Inc. [5]. In order to heat a carbon disk more uniformly, it was decided to exploit electron bombardment methods.

Two thin, closely spaced carbon disks were positioned in the image chamber of a conventional double paraboloid mirror carbon arc image furnace. The disks were mounted so that the front disk had a central support connector surrounded by a cylinder which supports the second disk. The central support was insulated from the second disk by a central hole in the second disk. The front disk was positioned in the focal plane of the furnace and fully irradiated. By placing a positive voltage on the second disk, electron emission was drawn from the irradiated one. The anode dissipation further heated the disks resulting in more emission and greater heating. Eventually a temperature limit was reached when a vapor discharge formed between the disks. Inspection of the disks (Fig. 5-2) showed a removal of carbon material from the anode and a redeposit of carbon rim on the cathode. Thus it was concluded that the carbon vapor ionized and migrated as positive ions to the cathode. Apparently electrons streamed in the opposite direction as a core within the positive ion sheath. It also has been observed that carbon is deposited on the negative electrode of the conventional carbon arc when this electrode is positioned in the arc stream. This erosion of the positive carbon and build-up on the cathode indicated that an alternating current arc would result in less electrode consumption and therefore require less spacing adjustment.

A number of coaxial discharges were attempted, both with alternating and direct potentials. The inner electrode always eroded more rapidly due to its higher temperature.

Finally, a three-phase electrode arrangement was attempted [6]. Figure 5-3 shows the electrodes and vacuum chambers. Power was supplied directly by a three phase, ganged variac. The vacuum arc was initiated by heating the electrode tips in the carbon arc image furnace, then raising the temperature by electron bombardment, finally reaching sufficient temperature to initiate the carbon vapor discharge. The lamp operated 20 min without variac adjustment and produced a steady intense illumination.

C. CONCLUSIONS

It was concluded that a carbon vapor lamp operating coaxially or with three-phase electrode configuration could be developed. This type of source would be adaptable to the "clam-shell" mirror arrangement.

REFERENCES

1. Arthur D. Little, Inc., "Development of the 'Clam Shell' Thermal Imager," Final Report under Contract AF19(604)-6663.
2. Glaser, P. E., and Ploetz, G. P., "High-Temperature Generation and Control by Thermal Imaging Techniques," Symposium on Temperature, Its Measurement and Control in Science and Industry, Columbus, Ohio, March 27-31, 1961.
3. Finkelnburg, W., "The High-Current Carbon Arc," Fiat, Final Report 1052PB 81644 (U. S. Department of Commerce, Washington, D. C., 1947).
4. Maust, E. E., Jr., and Warnke, W. E., "The Performance and Operating Characteristics of an Image Furnace Having 60-in. Paraboloid Mirrors," RI 5946, 1962, Bureau of Mines (U. S. Department of the Interior).
5. Richardson, D. L., "A Thermal Radiation Heat Source and Imaging System for Biomedical Research," Final Report Contract N-62269-B88, March 31, 1962. (U. S. Naval Air Development Center, Johnsville, Pa.).
6. Ploetz, G. P., Cox, H. F., and Larsen, L. C., "The Carbon Vapor Lamp: A Thermal Radiation Source for Image Furnaces," AFCRL 764, September 1961, Air Force Cambridge Research Laboratories, Bedford, Mass.

Chapter 6

A Four-Hundred Kilowatt Pressurized
Arc Imaging Furnace

John C. Cook

High Temperature Research and Geophysics
Southwest Research Institute
San Antonio, Texas

A. INTRODUCTION

The pressurized arc imaging facility at Southwest Research Institute (SwRI) was designed, built, and tested in preliminary fashion during 1956–1958 under the sponsorship of the National Aeronautics and Space Administration ([1]). A subsequent objective of the NASA program was to conduct ablation studies of materials under extremely rapid heating, in connection with atmospheric re-entry problems ([2]). In 1959, other programs were begun for the Wright Air Development Division and the Sandia Corporation, which were concerned, respectively, with heat conduction in materials and spectrographic studies of ablation products. These three programs ([3–5]) were completed in 1960. Since then, additional studies of ablation have been conducted for a private sponsor and SwRI has supported a variety of short experimental studies in the facility, particularly its use as a high-intensity light source for continuously pumping ruby lasers ([6]). A program of determining dynamic thermal and thermoelastic properties of materials is currently underway for the Air Force Office of Scientific Research. Incidental to its use as a research tool, the facility has been subject to extensive modifications and improvements throughout its five years of operation. Many more improvements are possible and are needed. This paper is therefore a status report, describing the state of development of the facility as of October 1962, some of its characteristics, and its potentialities as a tool for thermal research on materials ([7]).

B. DESIGN PRINCIPLES AND GENERAL DESCRIPTION

The SwRI arc imaging furnace utilizes twin parabolic searchlight mirrors to collect thermal radiation from a DC arc and to focus it upon a test specimen held in a controlled environment remote from the arc. It was designed to provide pure thermal radiation at the highest possible flux density for at least several seconds on material specimens of the largest practical size. In accord with these design objectives, the mirror system has a magnification of unity so that the region of high thermal flux at the specimen location is about equal in volume to the volume of the "anode ball" (the region of gas just beyond the anode, which has the highest temperature in the electric arc). The optical system is normally focused upon the center of the anode ball, which is near the plane containing the rim of the anode crater.

The arc (and incidentally, the whole optical system) is operated in a pressurized environment of from 2 to 10 atm, in order to increase the density and voltage drop of the arc and decrease its size, in accordance with principles previously reported [8,9]. A few preliminary tests made at SwRI, comparing arc outputs at different pressures, indicated that radiation output does increase with pressure in this "convection stabilized" arc as it does in "wall stabilized" arcs. Air is the only medium which has been used for arc pressurization thus far.

During operation, the arc is continuously "blown" with converging streams of high-velocity air. Approximately 450 grams of air per second are injected at sonic velocity around the arc in two converging cones which are coaxial with the anode. As with other blown-arc sources [10], the air blast serves to oxidize and cool ionized matter, thus removing flame and ionized gas layers. The collecting mirror thus receives radiation directly from the "anode ball" portion of the arc, and is not obscured by absorbing matter at lower temperature. In addition, the air blast apparently "convection stabilizes" the arc and greatly increases the voltage drop across it [2].

The resulting arc, as seen through filters by means of periscopes and windows installed in the pressure vessel, is a blue-white brush about 1 in. long and $5/8$ in. in diameter. It produces a rushing sound audible above the subdued roar of the enclosed air blast. At current settings of 2500 to 3000 A, more than 400 kW of electrical power are dissipated in the arc and electrodes of which about 140 kW are probably dissipated in the arc. At present, the corresponding thermal radiation power delivered to the target region totals about 6 kW, with an average flux density up to 700 g-cal/cm^2-sec or 2600 Btu/ft^2-sec.

C. MAJOR COMPONENTS

1. Arc-Imaging Optics

Figure 6-1 shows a simplified schematic diagram of the arc-imaging system enclosed in the pressure vessel. It consists of two $37\frac{1}{4}$-in.-diameter, 15-in. focal length parabolic mirrors (from searchlights) facing one another, with the arc positioned at the focus of one paraboloid and the target, or specimen, at the focus of the other. Six-inch holes have been cut in the center of each mirror. Figure 6-2 shows the physical arrangement of various components; one mirror is shown at the right on the rolling stand used to remove it from the tank. Accurate alignment of both mirrors, the arc, and the specimen on a common axis is essential to prevent coma in the image. Both stellite and back-silvered glass mirrors have been used. Because they are heavier and thinner, the stellite mirrors require some warping adjustment, by means of shimming the clamps in the steel support frames, to reduce coma. With both types, "good" images ($1/16$ in. in diameter) have been produced from a "point" test source at the arc position.

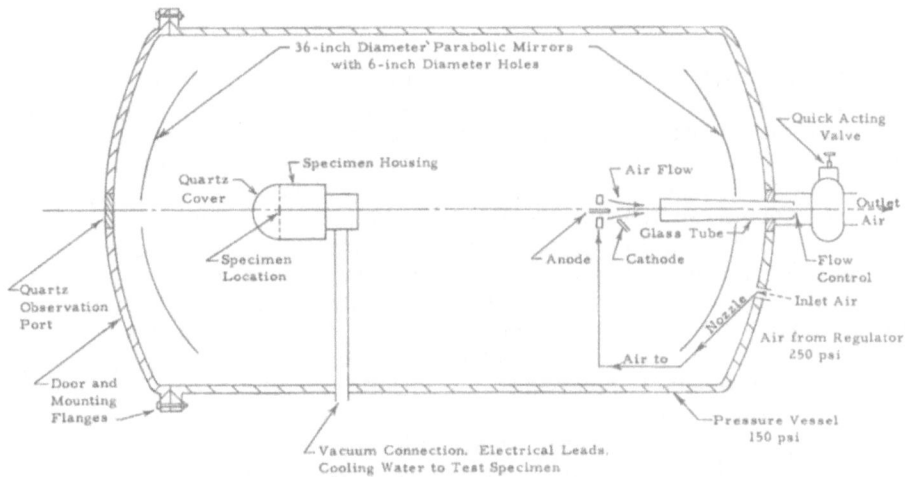

Fig. 6-1. Arc imaging system schematic showing pressure vessel and location of components.

The diameter of the image of the anode ball, or "hot spot" is approximately $7/8$ in. Consequently, the limiting diameter of a disk specimen is $1/2$ in. if fairly uniform irradiation is required. Defocusing the specimen by moving it toward the arc imaging mirror permits an increase in specimen diameter while maintaining uniformity of radiation flux density. However, the flux density falls off sharply, because the 130° angle subtended at the focal spot by the imaging mirror produces a very limited depth of field.

Fig. 6-2. View of arc imaging furnace and auxiliary components and accessories showing interior construction.

2. Specimen Environment

The specimen is contained in a pressure-and-vacuum-tight steel housing and is irradiated with thermal energy through an optical-glass dome. (Several were obtained from surplus aircraft periscopes.) The steel specimen housing is provided with connections to the outside of the arc furnace pressure vessel such that the housing can be evacuated or pressurized. The glass dome, sealed with an O-ring, can support somewhat more than the ambient pressure in the vessel; hence, the present upper limit of specimen pressure environment is approximately 11 atm. A usual low-pressure value is of the order of 0.05 mm Hg at present. Provision for a higher vacuum would require a second inner vacuum chamber free of the mechanical seals, moving parts, and electrical insulation required in the main chamber for operation of the shutters and instrumentation.

The specimen housing is so designed that it can contain alternatively a Glaser-type circulating-water, blackbody cavity calorimeter ([11]) for the measurement of thermal flux under experimental conditions similar to those used in specimen testing. Within the housing, immediately under the glass dome, are high-speed, rotary-solenoid-operated, clamshell shutters which permit the duration of specimen exposure to be controlled. Because of size limitations imposed by the optical system, the capacity of the blackbody calorimeter is limited to approximately 400 cal/cm^2-sec; therefore, the measurement of higher fluxes requires the use of a rotating hemispherical attenuator around the glass dome during calibration, which lowers the radiation incident upon the calorimeter aperture by a factor 5.5. Figure 6-3 shows a view of the specimen chamber with the shutters and calorimeter in place. The dome and rotating-sieve attenuator are shown removed.

Fig. 6-3. Specimen chamber, calorimeter, and shutters with protector dome and attenuator removed.

3. Electrodes, Holders, and Controls

The arc stand and electrodes originally used were taken from an Army surplus 60-in. searchlight. However, at operating currents above 500 A, it was found that electrode erosion was too rapid for the available motor drive to handle, and that toothed-wheel electrode drivers were unreliable [1]. A series of modifications finally produced electrode holders operable up to 3000 A; hollow, water-cooled silver anode-contact jaws [1,2] gripping the anode rod with about 5 lb of spring tension, and a remotely controlled hydraulic-cylinder anode feed; the original cathode contact jaws of heavy stainless steel were used, gripping the cathode rod with about 15 lb of auxiliary spring tension; the original toothed cathode drive was used, but was carefully insulated to prevent resistive overheating of the teeth. Graphite radiation shields, whitened by rubbing them with Tamm's silica-smoke powder, greatly prolonged the service life of these parts [3].

Continued trouble from electrolytic corrosion pitting in the silver jaws, water leaks, failures of silver-soldered joints, drive-tooth slippage, electrode spalling (presumably thermal), and high contact resistance finally caused adoption of the completely redesigned holders shown in Figs. 6-4 and 6-5, which have operated satisfactorily in over 50 tests during 1962. The anode holder is of brass assembled with bolts, having a stainless steel face plate, graphite nozzle, and graphite radiation-shield ring. The current contacts are four graphite generator brushes, each cooled and pressed with a force of about 15 lb against the anode by air pressure from the nozzle feed chamber. The anode itself forms the inside of the inner annular nozzle; hence it is air-cooled. Furthermore, since the portion between the brushes and the crater end is only an inch long, the resistive heating and spalling should be reduced during active feeding.

The anode is rotated at about 200 rpm by a small motor to ensure symmetrical erosion and crater development (although rotation does not appear to be necessary for this purpose). It is retracted or fed into the arc by the hydraulic cylinder visible at the top of Fig. 6-4. This cylinder is controlled by a two-way electric valve operated either manually, or automatically by a pair of photocells connected in a bridge circuit, viewing a projected image of the end of the anode. By these means, the anode ball is customarily maintained within 5 mm of the focus of the collecting mirror, despite consumption of the anode at rates up to 2 cm/sec.

The anodes commonly used are National Carbon Company experimental "Ultrex" electrodes 16 mm in diameter, having a 12 mm misch-metal core, rated at 325 A. Solid graphite anodes produce much less radiation. Copper- and aluminum-cored anodes unfortunately spray molten drops which adhere to the arc mirror. However, a 16-mm anode machined from solid metal (rare-earth oxides or fluorides) has given an excellent radiation output without any spalling on prolonged runs above 2000 A, and has no more electrical resistance than "Ultrex" or graphite anodes. With all anodes, the combined (cold) resistance through the anode holder, from cable to anode tip (measured by a four-probe method) is only 0.012 ohm, and the electrical power losses and resistive heating in this part of the circuit (presumably up to 100 kW) are considered relatively minor at this time.

As cathodes, 11-mm solid "H-I Experimental Projection" carbon rods from National Carbon Company have been used almost exclusively. These

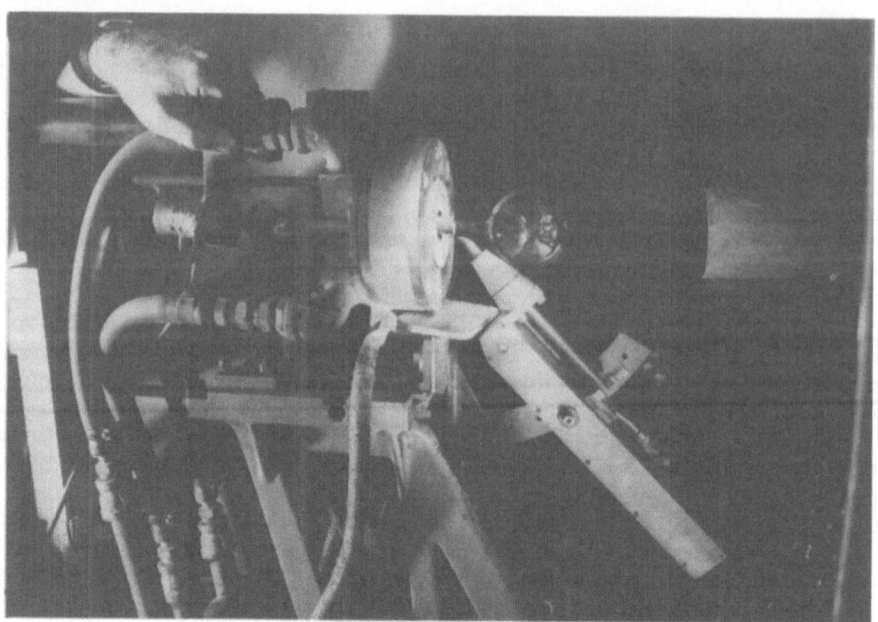

Fig. 6-4. Arc stand showing electrode relationship to air nozzle (center), air exhaust and arc mirror (right).

Fig. 6-5. Electrode holders.   Top: anode guide ring, contact brushes, and graphite air nozzle. Bottom: cathode in chain-transport, showing tungsten contact springs and graphite protector cone.

small, hard rods have a relatively high cold resistance (0.015 ohm/in.) and they glow brilliantly, erode rapidly, spall, and curl somewhat at currents above 2000 A. However, they have the advantage of casting a relatively narrow shadow on the arc mirror.

The new cathode holder employs a reliable chain drive with an insulated setscrew clamp. The cathode can now be replaced with one hand through a porthole in the pressure vessel. The angle between the cathode and the anode axis is adjustable, but 47° gives good cathode sculpturing and reasonably small mirror shadowing without as much tendency for the arc to blow out as is found at larger angles with present values of current and air-jet flow. It has been found necessary for the cathode tip to extend completely through the annular air stream into the "anode brush" to avoid frequent blowing-out of the arc.

Current to the cathode is carried as close to the arc as possible without excessive mirror shadowing, by means of twelve cantilever-spring contacts of 0.1-in. tungsten rod (Fig. 6-5). These are protected from overheating and oxidation by a replaceable graphite cap which is continuously filled and flushed with argon or dry nitrogen during operation. Unfortunately, the (cold) contact resistance is 0.020 ohm or higher, despite reasonably high contact pressures (2 lb per spring) and a large total (3 cm^2) ground and lapped contact area. Further improvement is probably needed to reduce what appear to be excessive power losses (up to 250 kW) in the present cathode and cathode holder.

Until recently, the cathode was fed by a skilled operator, who observed the arc length through a periscope and turned a large handwheel to adjust it. Cathode erosion rates range up to about 2 cm/sec at present. In order to produce more uniform test runs than are possible with a human operator, a variable-speed electric gearhead motor with an electric clutch controlled by the arc current is now in use. Upon application of the arc voltage (up to 420 V open-circuit), this mechanism advances the cathode until current begins to flow. When the electrodes burn away to the point where the current drops to 70% of its initial value (which usually occurs within 0.5 sec), the mechanism advances the cathode further in short bursts in an attempt to maintain the arc current within 5% of this value. At present, the automatic feed cannot maintain the arc current as constant as this, or as constant as is possible with manual control, but it gives more consistent and repeatable results from test to test.

4. Rectifier

The rectifier which supplies DC power to the arc is a three-phase, full-wave type using six GE Type GL5564/507 ignitrons. The continuous rating of the unit is 500 kVA corresponding approximately to the theoretical maximum DC line voltage of 560 V and an arc current of 900 A. (The actual voltage available is generally less, often only 420 V, because of rectifier malfunctions.) An overload of 2400 A is permissible for 1 min. For an arc duration of 1.4 sec, overload currents up to 9000 A are permissible. The firing times of the ignitrons can be adjusted by means of a small "synchro" three-phase transformer, whose outputs are clipped and differentiated to provide six trigger pulses in a definite phase relationship. By adjusting the angle between the rotor and stator of this synchro, the arc

Fig. 6-6. Rectifier and arc imaging furnace.

voltage and current can be varied continuously from zero to the maximum permitted by circuit resistance (about 3000 A at this time). Figure 6-6 shows a photograph of the rectifier unit and the arc imaging furnace. One panel of the rectifier has been removed to show some of the thyratron circuits which amplify the triggering pulses, two of the ignitrons, water cooling, busbars, etc. The pressure vessel which contains the arc imaging furnace is shown at the right, with a high-speed "Fastax" camera mounted to photograph specimens under irradiation, through a quartz window in the access door of the pressure vessel.

5. Primary Power Source

Because of the high power requirements for operating the arc, a separate substation (Fig. 6-7) was constructed to service the facility. The substation is powered by a 13,800-V primary transmission line. The major components of the substation are a bank of remotely-tripped oil switches in the primary circuit, and a Δy transformer bank supplying nominally 416 V (actually 380) between phase busbars to the rectifier. Ballast resistors capable of dissipating 400 kW are provided for the control of arc current. Large knife switches permit the rapid interconnection of the ballast resistors to adjust the maximum current through the arc. At present these resistors are used only for testing rectifier components, and arc stabilization is accomplished chiefly by a 12-mH, air-core choke consisting of a 600-lb spool of No. 0000 insulated copper cable in series with the arc. The resistance of this choke is 0.050 ohm, and is one of the principal factors limiting the arc current available at present.

Fig. 6-7. View of the primary power substation and compressor–accumulator facilities for the arc imaging furnace.

6. Compressed Air System

A three-stage air- and-water-cooled compressor system with an average capacity of 26 scfm and a discharge pressure of 1000 psig is used to charge a 52-ft^3 accumulator. A fast-acting pressure regulator on the discharge of the accumulator is used to reduce inlet pressure to the arc nozzles to a constant value which may be adjusted up to about 250 psig. Two fast-acting solenoid-controlled pneumatic valves operate simultaneously to turn on the air blast at the arc and to open the tank exhaust (the 3-in. steel tube at the right in Fig. 6-4), which is metered with an appropriately sized orifice (0.5 in. diameter at present). A separate 10-hp compressor with 150-psig discharge pressure is used to fill the arc pressure vessel prior to operation. The air used to charge the vessel is passed through a drier to prevent condensation on the optical components. Figure 6-7 shows a view of the primary power substation and the compressor–accumulator facilities for the furnace. Accumulator tanks for 250- and 1000-psi air are shown at the left and center; the switch gear is shown at the right, and transformers in the center surrounded by ballast resistor banks. Compressors are housed in the metal building.

The pressure vessel housing the imaging furnace has been tested to 250 psi, but is operated at half that pressure. It is equipped with two 8-in. access portholes opposite the arc and the specimen region, three 3-in. windows, and a hinged access door held by forty-four 1-in. bolts. During operation of the arc, the tank gases become heavily contaminated with brown nitrogen oxides and gray smoke consisting mostly of lanthanum and cerium carbonates coming from the anode core. These combustion products reduce the radiation flux available at the target to 30% of its initial value after 4 sec of operation and 1 or 2 min of rest, during which the smoke particles "precipitate" into an optically dense form. Hence, for each test, the tank pressure must be discharged, the portholes or door removed, the tank purged, the mirrors cleaned of gray dust, and the tank closed and refilled with dried air. In many tests, this inconvenience and the delay are unimportant, since the specimen must be inspected or changed anyway. However, more complete scavenging of the arc products would permit longer tests. Preliminary experiments indicate that an 8-in.-diameter pyrex cylinder can be installed around the arc to confine these products and channel them into the exhaust tube without seriously obscuring the light path to the arc mirror. In future pressurized furnace designs, complete arc product scavenging and easier access should be primary considerations.

7. Control System

At the extreme heating rates produced in the facility, all solid materials tested so far melt or vaporize rapidly. Consequently, many useful tests can be completed in times as short as 0.2 sec. Tests seldom need to exceed 10 sec in duration, and in any case, would be hampered by arc product accumulation, if longer. Furthermore, electrode adjustments must be made with a speed and precision seriously taxing human reaction times. A reliable test under these conditions requires complete automation of the arc-imaging furnace system and of the instruments used for measurement. For this reason, a sequence-timing and control system containing ten cam-operated microswitches is used, which automatically initiates all control and measurement functions. Operation is initiated by closing the oil switches to apply arc voltage. When the arc has been established through the automatic motions of the cathode, an arc-current-sensing relay starts the timing-cam motor. Subsequently, the shutter, cameras, spectrographs, etc. are operated in any desired sequence, and finally the arc is broken automatically by tripping the primary oil switches.

Because of the complexity of the facility, at least two operators are required at present. Check-out lists are used to prevent omission of essential steps in preparing for and operating a test. Interlocks are also built in at several points to prevent harmful malfunctions. Nevertheless, the facility remains experimental in character, and will probably always require a highly skilled and experienced operating staff.

D. MEASUREMENT AND RECORDING EQUIPMENT

The principal monitoring instrument employed is a 14-channel galvanometric recorder, which is used routinely to record on photographic paper, together with a grid of 0.01-sec and 0.1-sec timing lines, such parameters as arc current, arc voltage, electrode drive operations, tank and nozzle pressures, shutter position, radiation intensity in the parallel beam between the mirrors, and the output from specimen thermocouples and other instruments.

Two unusual instrument systems deserving special mention here are the focal-spot-scanning radiation sampler and the specimen-surface pyrometer. These and some others used in the SwRI facility are described briefly in the following paragraphs.

1. The Focal-Spot Scanner

This instrument consists of a pair of fine sapphire light-pipes, 2 in. long and enclosed in platinum tubes which are tipped with translucent white radiation-sampling spheres of a sapphire and quartz mixture, and which convey radiation to a pair of small silicon photovoltaic cells mounted in the scanner head.   In Fig. 6-3, the radiation-sampling spheres can be seen poised before the calorimeter aperture.   By means of a motor and a mechanical linkage outside the specimen chamber, the scanner head is caused to oscillate about its axis so that the sampling spheres travel in an arc passing through the focal spot nine times per second. They are exposed to the full flux for about 2 msec during each scan, but pause to cool off outside the beam for about 110 msec between scans. A detailed description of a similar instrument has been published elsewhere [12].

The output signals from the two photovoltaic cells are displayed on a dual-beam oscilloscope, where they can be photographed as radiation-intensity profiles.   Figure 6-8 shows such profiles. Note the "Gaussian" shape and fine-scale, nonrepeating irregularities.   These are believed to represent ephemeral structures in the anode ball of the arc. The signals also control a special square-wave generator in such a manner that the amplitudes of the individual square waves are proportional to the peak

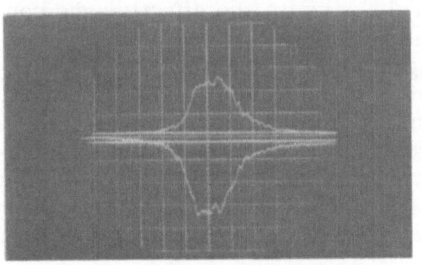

SINGLE SWEEP;
ARC IMAGE

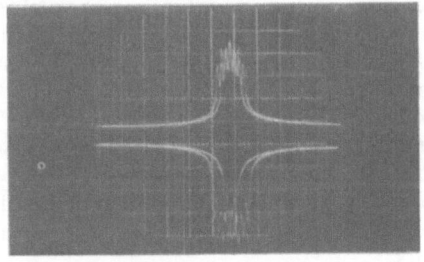

TWO SWEEPS;
6-STRAND LAMP IMAGE

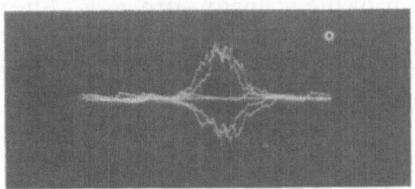

REPEAT SWEEPS ;
ARC IMAGE

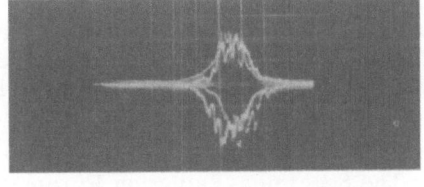

REPEAT SWEEPS;
ARC IMAGE

Fig. 6-8. Oscilloscope photos of radiation intensity profiles across the focal spot, taken with two-channel scanner.

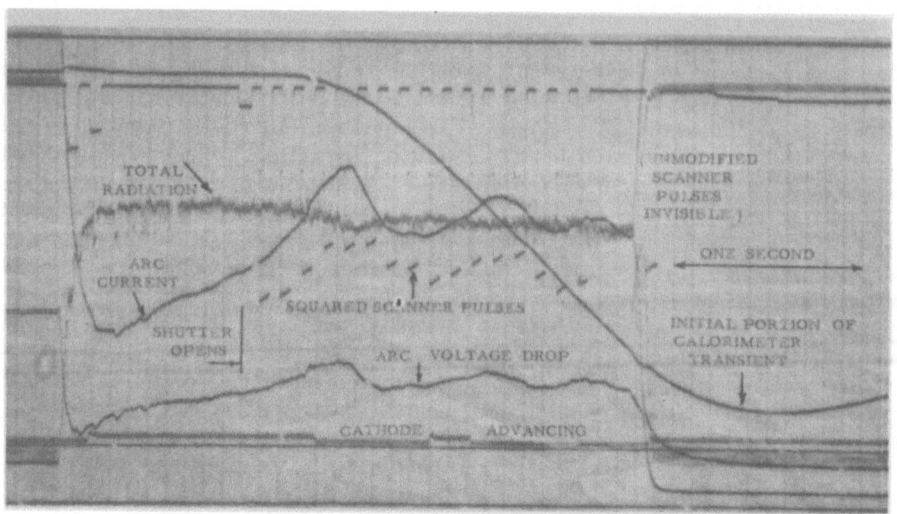

Fig. 6-9. Galvanometric oscillogram, showing squared scanner pulses monitoring specimen thermal input.

heights of individual signal profiles. This permits the effective and continuous display of the peak radiation intensities on the galvanometric oscillograph. Figure 6-9 shows an oscillogram containing such signals. They indicate rapid changes of focal-spot intensity correlated with arc current and voltage changes, both of which accompany movements of the electrodes and defocusing of the image spot. (According to the parallel-radiation monitor, the total radiation remains about constant during this process.) One scanner signal channel has accordingly been calibrated against the cavity calorimeter to provide a semicontinuous record of the incident specimen flux.

The second channel, similarly calibrated, is intended to be used to measure either the diffusely reflected or the emitted light from the specimen. For these purposes, the sampling sphere of this channel is shielded from incident radiation by means of a tiny platinum hood. For emission measurements, reflected and scattered light can be largely cut off by means of a $5/8$-in. platinum or blackened mica disk traveling with the spheres, which covers the whole specimen during transit of the scanner across it. Alternatively, the clamshell shutters can be clapped shut momentarily. Since this instrument has just recently been installed, only initial calibration and incident-light monitoring data have been obtained to date.

2. The Specimen-Radiation Pyrometer

In order to obtain surface temperatures on an ablating specimen, optical measurements are necessary. A method which has been successfully used in the SwRI arc imaging furnace consists of arranging a photocell to view the specimen continuously by red light, and recording its output transient as the incident light is interrupted momentarily by the shutters. The ap-

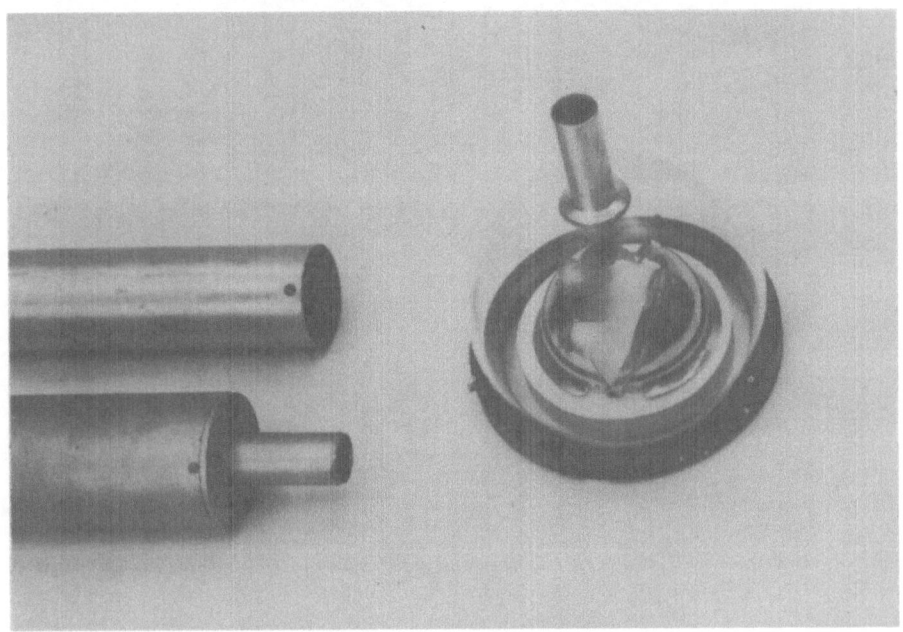

Fig. 6-10. Specimen-viewing periscope system.

paratus was calibrated by substituting a ribbon-filament lamp for the specimen in such a way that the projected image of the filament covered the
photocell, and recording the photocell outputs at various filament temperatures as determined with a standard, red-light manual optical pyrometer. The blackened specimen was assumed to radiate as a blackbody.

Figure 6-10 shows the periscope system employed to convey an image
of the specimen outside the pressure vessel through the quartz window in
the access door. The special shutters which fit around the periscope en-
trance-pupil and objective-lens housing to exclude light from the arc are
also shown. The periscope does not interfere with irradiation of the
specimen when the shutters are open, since it fits within the 25° dark central
cone blocked by the arc stand and specimen chamber (Fig. 6-1). The bundle
of rays projected by the periscope is passed through a red filter, and then
forms an image of the specimen upon a small photocell outside the pressure
vessel. By means of a beam splitter ahead of the filter, part of the light is
also directed to a color-motion-picture camera to record the behavior of
the specimen.

Figure 6-11 shows typical galvanometric oscillograms of the photocell
output signal. The final cooling-curves and those occurring during the
earlier, brief shutter closures are similar in slope. Each can be extrapolated back to the moment when the closing of the shutters began, to obtain
the specimen temperature at two different times. Such intermittent measurements will be used in future tests merely to supplement and verify data
from the scanning radiation sampler. In all probability, the camera and
possibly a second, infrared pyrometer channel to fill the gap between about
700°C and room temperature will continue to be needed.

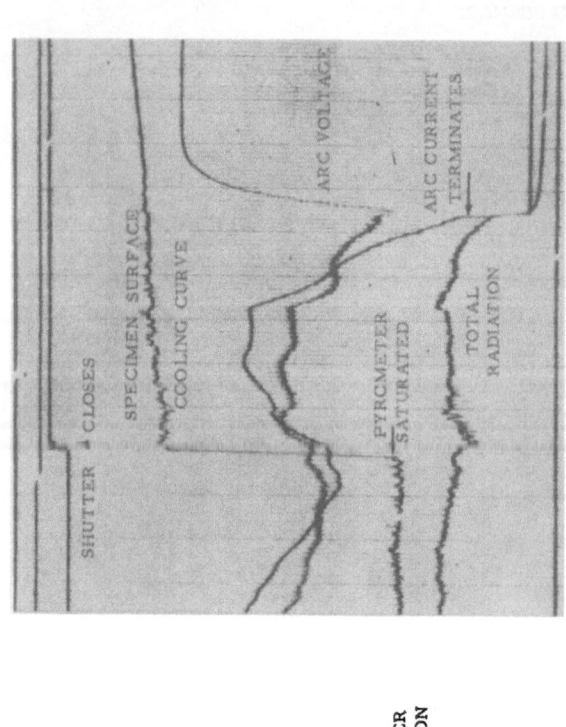

PYROMETER
CALIBRATION

Dark
1000°C –
1250 –
1500 –
1750 –
2000 –
2250°C–

METALLIC SPECIMEN

BRIEF SHUTTER CLOSURE AT MID-TEST

END-OF-TEST MEASUREMENT

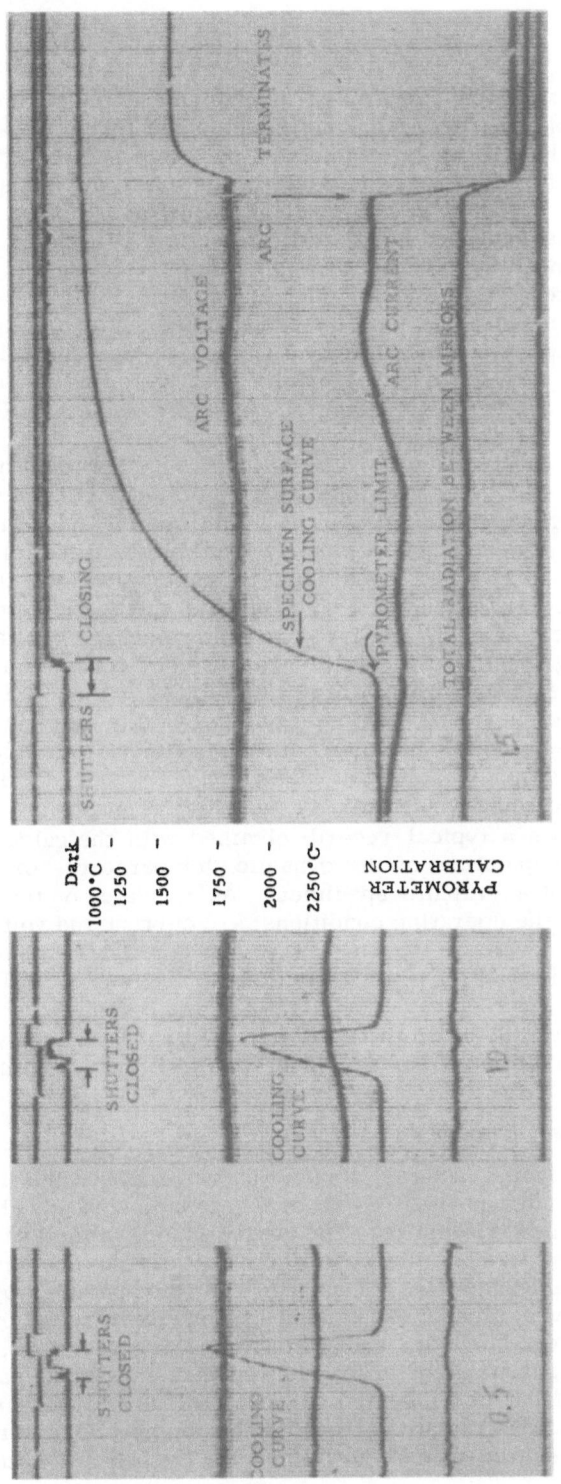

ORGANIC SPECIMEN

MEASUREMENTS AT 0.5 AND 1.0 SECOND.   FINAL MEASUREMENT AT 1.7 SECOND.

Fig. 6-11. Oscillograms of surface-radiation pyrometer signals.

3. Specimen Temperature-History Instrumentation

The thermal history of a specimen is obtained by temperature—time records obtained from calibrated thermocouples embedded in the specimen. Various techniques have been tried to obtain good thermal contact between the junction and the specimen and electrical insulation of the leads from the specimen, including spotwelding and swaging, with glass-tube or commercially made magnesia-insulated metal-tube protection.

When detailed information on heat transfer into a specimen is required, temperature—time records are obtained from several locations along the axis of the specimen perpendicular to the irradiated surface. Figure 6-12 shows a stainless steel specimen specifically designed for such determinations. A specimen of this type is typically $\frac{1}{2}$ to $\frac{5}{8}$ in. in diameter and is surrounded by an annular "guard ring" cylinder or radiation shield $\frac{5}{8}$ to $\frac{7}{8}$ in. in diameter. These concentric units are separated by a gap of 5 to 10 mils thick which is usually filled with silica flour. The radiation shield is covered with asbestos. The specimen chamber is evacuated during tests to provide additional specimen insulation. Fine metal-jacketed thermocouple wires emerge at various points and end at a terminal plate, from which thermocouple extension wires lead to an icewater junction. Both the specimen and the radiation shield are made of the same material, and the exposed surfaces of each are irradiated at comparable rates so that radial temperature gradients and resulting radiation or conduction heat loss are minimized. The faces of both are normally made intensely black with acetylene soot to reduce the incident energy lost by reflection. A whitened graphite disk is used to shield the sides of the specimen from stray radiation. Specimens of copper, stainless steel, graphite, and bakelite of this same design have been made and tested ([4]).

Figure 6-13 shows a typical record, obtained with the galvanometric oscillograph, for computation of the dynamic (temperature- or rate-dependent) diffusivity of a graphite specimen. As indicated on the figure, a record was made of the operating conditions: arc current and voltage, tank pressure, and arc nozzle pressure. The arc-current trace shows the duration of specimen irradiation, which in this record is about 2.3 sec. No shutters were used. The temperature—time records obtained from each of four thermocouples located along the specimen axis, at successively greater distances from the irradiated surface, are shown as curves A through D.

4. Spectrographic Measurements

The study of surface reactions, ablation phenomena, and the like under high thermal flux conditions requires recording of emission and absorption spectra during specimen irradiation. The quartz window in the arc imaging furnace access door permits direct observation of the specimen. The periscope now available can also be used to view the specimen by its own light with the shutters closed. Inclusion of a standard radiation source within the vessel, together with an appropriate mirror system, permits observation and measurement of absorption spectra in a plane parallel to and immediately in front of the heated surface. A lens and mirror system is used to direct emitted radiation from the heated specimen surface and from the radiation source to a 30,000-lines/in. Bausch and Lomb, dual-

Fig. 6-12. Guarded cylindrical specimen for thermal-history studies.

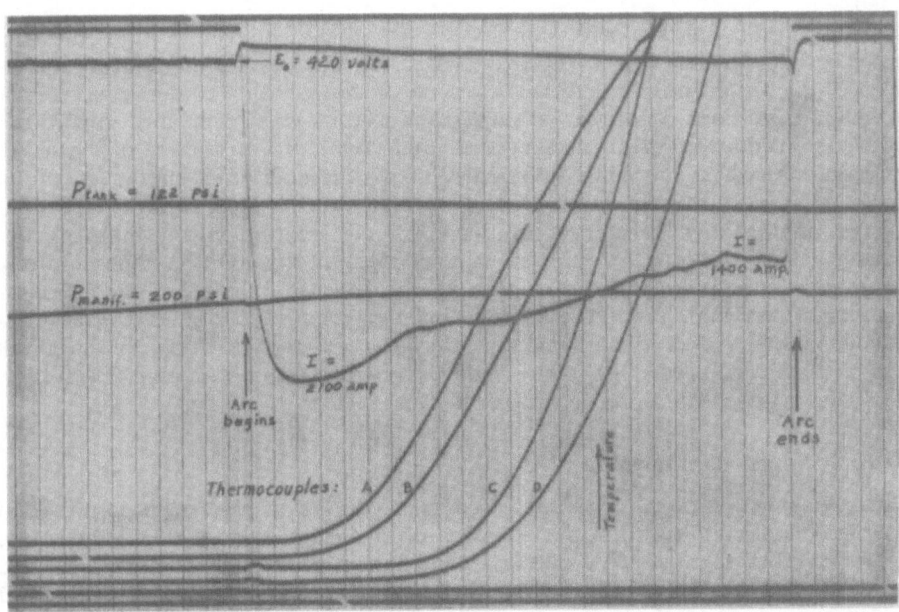

Fig. 6-13. Temperature–time data from typical measurement of thermal diffusivity under high thermal flux conditions.

grating spectrograph. The spectrograph is mounted on a rolling table so that it can be brought very close to the quartz port and yet not interfere with opening of the access door.

A Hartman slide on the spectrograph can be automatically jogged through five positions along the main slit during a test, although the test period may be as short as 0.2 sec. Hence, five pairs of emission or absorption spectra may be obtained during any one test. Two sets of five pairs, each set covering two wavelength intervals of about 1000 A each, plus standard spectra from an iron arc, can be recorded on a single glass plate for comparison. The spectra thus obtained provide information with which to interpret reactions occurring on the heated surface, to identify ablation products, etc.

Figure 6-14 shows early representative spectra obtained by a direct view of the heated specimen surface. In this case, most of the light originates in the arc, having been scattered from the graphite specimen surface. Hence, it constitutes primarily a background spectrum for ablation studies. The radiation is, of course, modified by absorption in the light path by the atmosphere in the pressure vessel. To a minor extent, for graphite at the heating rate and duration of exposure used, the arc radiation spectrum is modified by absorption by ablation products. Radiation from the heated surface and hot ablation products is also superimposed on the arc spectrum.

The top set of three spectra in Fig. 6-14 shows details of the violet, blue, and green region in which most of the light intensity occurred, with the top spectrum being a reference spectrum from a standard spectrographic iron arc source. The lower six spectra give complete spectral coverage of the emitted light from the ultraviolet to the infrared. Various exposures were used.

Absorption spectra are obtained by passing light from a high-intensity source through the vapors emitted from the surface of an ablating specimen. To date, only marginal spectra, adequate only to demonstrate the method, have been obtained. A thin graphite disk was preheated for 1.5 sec with arc radiation in an atmosphere of helium at a pressure of 1 atm. A high-intensity lamp mounted vertically above the specimen surface, a small front-surfaced mirror below the specimen, and a 2-in. quartz field lens aimed at the mirror constituted the optical system within the pressure vessel. The light path from the source passed through the vapor emitted from the specimen in a direction parallel to the specimen face about a quarter of an inch from the heated surface. One-second spectrograph exposures gave plates exhibiting several faint absorption lines in the violet and near ultraviolet. Since the only purpose was to determine feasibility, no attempt has been made to interpret these plates.

5. High-Speed Photography

A Fastax camera is sometimes used to obtain color photographs of specimens during irradiation. With camera speeds up to 4000 frames/sec, it is possible to observe rapid surface changes, material ablation, and arc image stability, and generally to assess the behavior of the test material and the experimental setup. The photographs are made through the quartz window used for obtaining spectra. They may also be made through the periscope and strikingly display the rapid cooling of white-hot specimens when the shutters are closed.

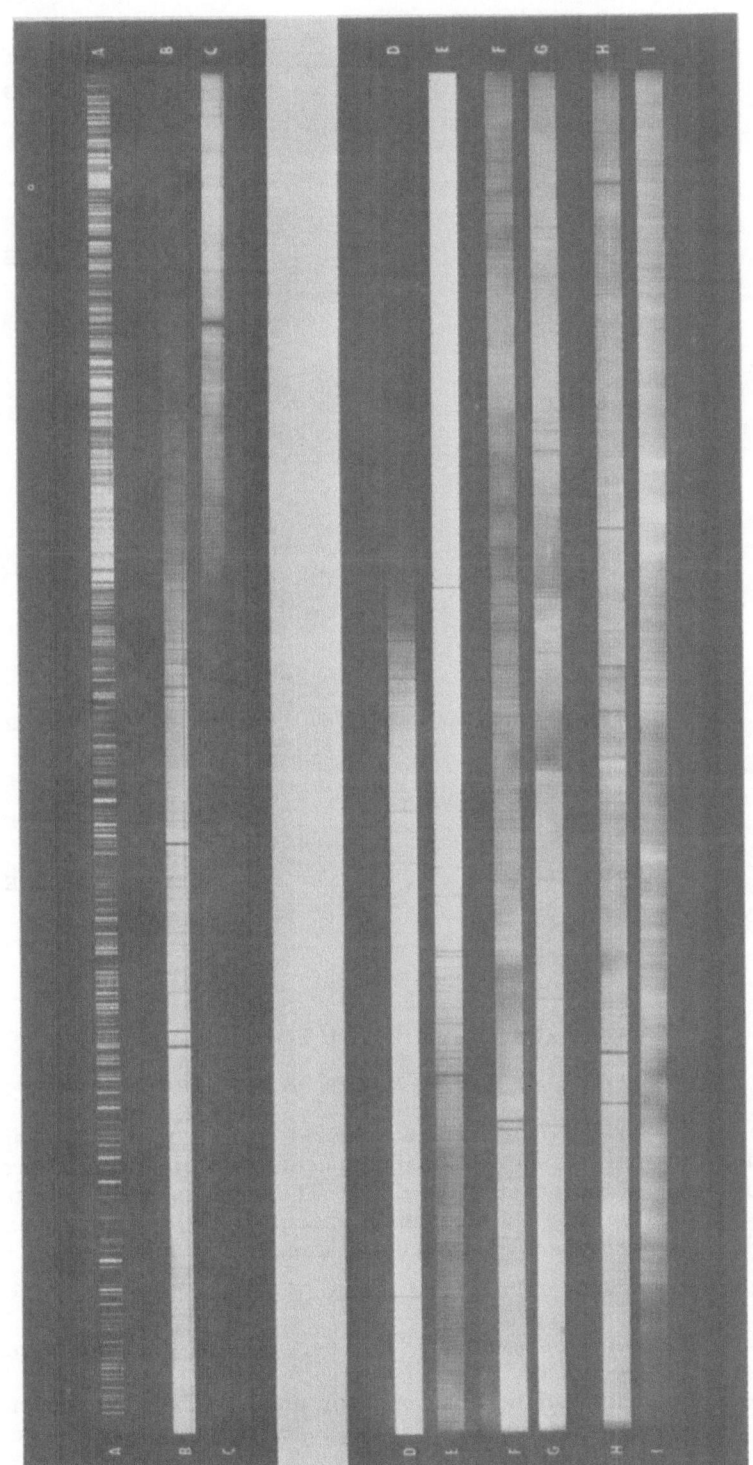

Fig. 6-14. Emission and reflection spectra obtained from a specimen of graphite irradiated at 500 cal/cm²-sec.

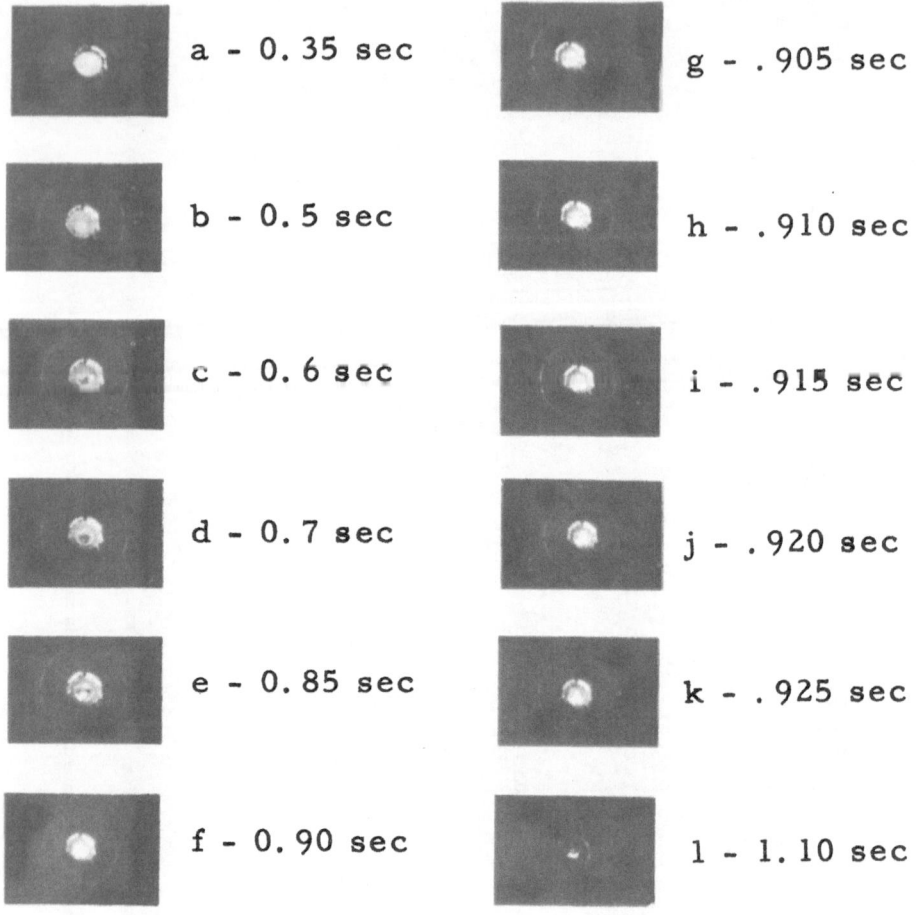

Run No. 180460-2

Fig. 6-15. High-speed movie frames of radiation-induced boiling (ablation) in stainless steel.

Figure 6-15 shows some frames selected from a reel of high-speed color motion pictures, which illustrate the usefulness of such photographs for interpretation of the quantitative data. These photographs were taken during the irradiation of a stainless steel specimen in air at a pressure of 0.07 mm Hg. Observations made on these frames are as follows:

Frame a: Reasonably uniform and steady illumination of the cold, soot-blackened specimen.

Frame b: Beginning of ablation of the soot coating near the center of the specimen.

Frame c: Extending of the ablation over much of the sample surface. "Shadow" of the thermocouple well immediately behind the surface is outlined.

Frame d: Beginning of boiling at the small point near the center of the specimen.

Frame e: A sudden "eruption" of boiling material from the hottest part of the surface.

Frames f to k: Vigorous local boiling, accompanied by the apparent transmission of glowing "waves of heated gas" away from the center of the specimen at velocities of about 10 ft/sec.

Frame l: Cessation of the arc in steps following interruption of current, with immediate cessation of boiling.

Observations such as those made from Fig. 6-15 are greatly enhanced when the original color photographs are viewed as compared with the black and white reproductions shown.

6. Ablation Smoke Scavenging

As an accessory to the measuring systems already discussed, a very successful method of removing smoke from an ablating specimen, to prevent blockage of incoming radiation, has been developed. It consists of blowing air or other gas from several nozzles distributed around the inside perimeter of the dome, toward the apex of the glass dome, and exhausting through an annular nozzle surrounding the specimen and flush with its surface. With the air moving through the dome at fairly low velocities in large volumes (estimated at 2 ft^3/min), the flow pattern is essentially laminar. The resulting pressure distribution is of such a shape as to carry any smoke, flame, and gases streaming forth from the specimen back toward it and radially outward to the exhaust nozzle. This method has successfully removed smoke and combustion products from rubberlike organic specimens, at rates estimated to be as high as 1 g/sec, without any blackening of the dome or visible obscuration of the specimen surface as seen by the motion picture camera.

E. SUMMARY AND CONCLUSIONS

This paper has described an enclosed, air-blown, arc-imaging furnace operating at up to 10 atm arc pressure, dissipating of the order of 400 kW of electric power, and delivering thermal radiation to a $5/8$-in.-diameter disk or rod specimen held in a controlled environment, at an average flux intensity up to 700 g-cal/cm^2-sec. Details of its construction and operation have been given. Several associated measuring instrument system and a few examples of typical test results have also been described.

It is suggested that this extremely versatile facility is currently or potentially suitable for the following types of scientific work:

1. Investigations of heat flow and associated phenomena (e.g., thermoelastic waves) at high heat input rates.
2. Studies of the properties of materials at high temperatures.
3. Development of new high-temperature materials through controlled melting and freezing, by vapor deposition of refractory substances, and by high-temperature chemical reactions.
4. Laser and other optical studies requiring an intense source of prolonged illumination.

It is anticipated that the versatility and available radiation intensity of this facility will continue to be improved as it is used in the research programs planned and in prospect for it.

ACKNOWLEDGMENTS

The writer is indebted for much of the material in this paper to Dr. J. M. Sharp of SwRI. Mr. Graydon Buss should receive the credit for many of the recently developed optical features and the new electrode arrangement used in the furnace. Dr. Francis Todd and Mr. Frank Wittler, who are no longer with SwRI, have contributed, respectively: the original design, construction, and testing of the facility; and two years of creditable maintenance and operation with many useful innovations. There would perhaps have been justification for including all these persons' names as co authors of this paper.

REFERENCES

1. Todd, F.C., Martinez, A., and Goland, M., "Construction of a Pilot Model of Equipment for the High Temperature Testing of Material," Summary Report on NACA Contract NAW6513 (Aug. 28, 1957).
2. Todd, F.C., "Materials Testing Program for Pilot Model of High Thermal Flux Unit," Summary Report on NACA Contract NAW6513 (Feb. 14, 1958).
3. Cook, John C., and Abramson, H.N., "Physical and Chemical Phenomena Near Surfaces Irradiated by A High Thermal Flux," Final Report on NASA Contract No. NAW6543 (June 17, 1960).
4. Abramson, H.N., Chu, Wen-Hwa, and Cook, John C., "Studies of Transient Heat Conduction at High Thermal Flux," WADD Technical Report 60-608; Final Report on Contract No. AF 33(616)-6323 (July, 1960).
5. Sharp, J.M., "A Study of High-Temperature Ablation," Final Report on Sandia Corporation Purchase Order 51-4746 (Jan., 1960).
6. Cook, John C., Proc. IRE, 50:330 (1962).
7. Romaine, Octave, (Assoc. Editor), "Arc-Imaging Furnace Tests Materials in High Thermal Flux," Space/Aeronautics (Sept., 1961).
8. Collected Works of O. Lummer, Vieweg, Braunschweig (1914).
9. Schulz, P., and Steck, B., Ann. der Physik, Series 6, 18:401 (1956).
10. Rompe, R., and Steenbeck, M., The Plasma State of Gases, see especially p. 20 of Translation by G.C. Skerlof of Mellon Institute for Project NR 223-064, Contract N90NR-83301 (1950).
11. Cook, J.C., and Levine, H.S., Rev. Sci. Instr. 31:1160 (1960).
12. Cook, John C., "A Scanning Radiation Sampler for Imaging Furnaces," in Temperature, Its Measurement and Control in Science and Industry, Vol. 3, Part 2 (Reinhold, N.Y., 1962), pp. 1051-1061.

Chapter 7

The Double Parabolic Arc Image Furnace

P. E. Evans*

Department of Metallurgy, Faculty of Technology
Manchester University
Manchester, England

A. INTRODUCTION

A description of the main features of the double parabolic arc image furnace built under the writer's direction has been given elsewhere [1,2]. The general layout of the furnace, shown in Fig. 7-1, is based on two 150-cm-diameter searchlights set on a common vertical axis. The original type of carbon electrode is used, 16 mm in diameter. Power is derived from a 22-kVA generator driven by an electric motor. Temperature is varied by varying the power dissipated in the arc and by defocusing the specimen.

B. MODIFICATIONS IN FURNACE DESIGN

The basic design, that is, the optics, of a double parabolic system are fixed but the geometry can be varied and so can the design of individual parts. This section is concerned with developments in the design of parts. Since the original report of the design and testing of this furnace was published, modifications have been made to the radiation shield surrounding the positive electrode and to the specimen support. These changes have greatly increased the flexibility and potential usefulness of the furnace.

The original radiation shield of the searchlight consisted of a thin mild steel sheet with a hole in the middle through which the positive carbon electrode projected. At a power dissipation of about 6 kVA this shield melted. It was replaced by a simple, hollow, copper disk with water inlet and outlet tubes near the periphery. The positive carbon electrode projected through a clearance hole in the center. When the power was increased above about 10 kVA, this second radiation shield also melted. Melting always occurred at the edge of the clearance hole nearest to the crater in the positive electrode. The most recent design of radiation shield (Fig. 7-2) is based on improved heat transfer. Incoming cooling water is directed to the hottest part of the disk by an inlet pipe set tangential to a tube coaxial with the disk and extending inside to within 1.5 mm of the hot surface. This design of radiation shield has allowed the use of up to about 20 kVA power dissipation and appears sound enough to allow even higher power operation were the power available.

*Senior Lecturer in Metallurgy.

Fig. 7-1. General view of double-parabolic arc image furnace.

The original form of specimen-manipulating mechanism is shown in Fig. 7-3. It allows the specimen table to be swung horizontally through about 100° of arc, to be moved 40 cm vertically, and to be moved 10 cm along the direction of the supporting tube. The horizontal stainless steel support tube also forms the means by which a bell jar placed on the specimen table is evacuated and filled with a protective atmosphere. The support tube used at present has an internal diameter of 12 mm but this is

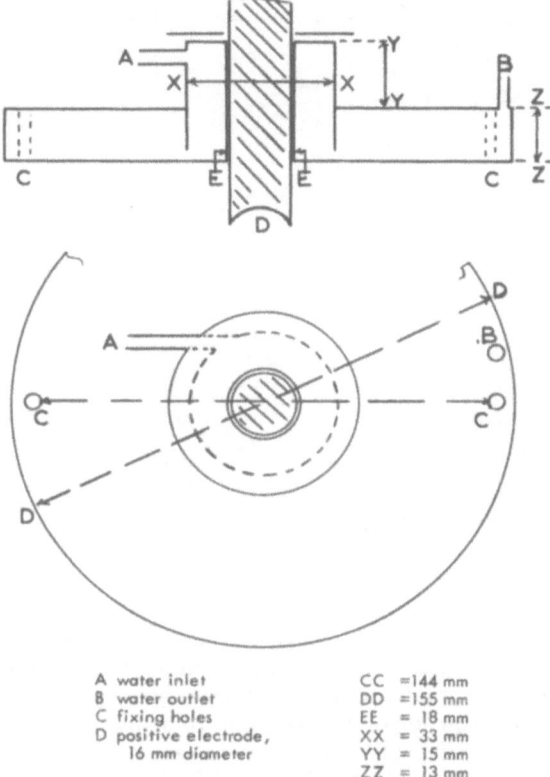

A water inlet	CC = 144 mm
B water outlet	DD = 155 mm
C fixing holes	EE = 18 mm
D positive electrode,	XX = 33 mm
16 mm diameter	YY = 15 mm
	ZZ = 13 mm

Fig. 7-2. Radiation shield for positive electrode.

shortly to be changed for one 24 mm in diameter which will have greater rigidity, and decrease the evacuation time. The most important modification that has been made here, however, is shown in Fig. 7-4. The support arm has been cut and a vacuum valve and screw seal have been inserted near the specimen table thus allowing the bell jar and its contents to be isolated atmospherically and then detached from the support arm. By this means materials can be unloaded and loaded into the specimen chamber in a dry box. This is essential for the handling of materials, such as uranium mononitride, that oxidize at room temperature.

C. EXPERIMENTAL WORK

1. Feasibility Studies

Feasibility studies have included the assessment of the extent of interaction between oxides by the simple method of melting the oxide with the lower melting point on the surface of the oxide of higher melting point and examining a cross section of the interface region. Although the technique has obvious limitations, it can be used to eliminate grossly unsuitable combinations of materials in a very short time.

In this same category are studies of the thermal stability of compounds.

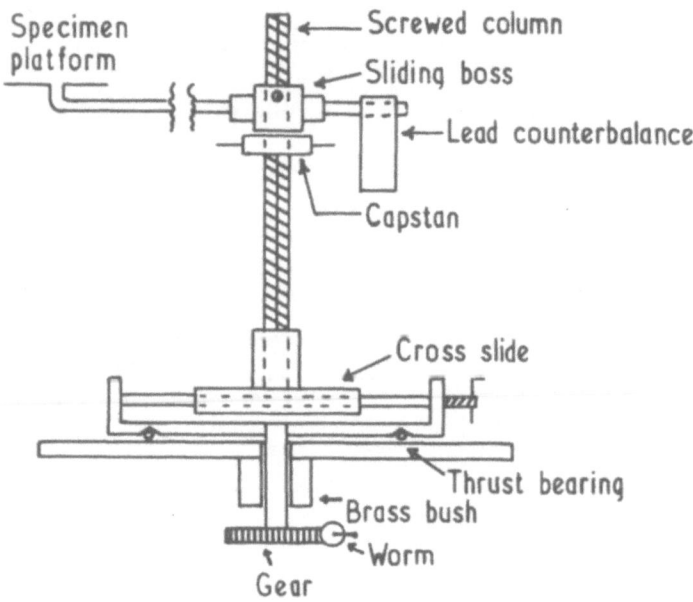

Fig. 7-3. Specimen manipulating mechanism.

In particular, the stability of certain nitrides has been qualitatively investigated by heating them in different atmospheres. For this work, the apparatus in its present form has the inherent disadvantage that the pressure within the bell jar must be kept below atmospheric pressure; therefore, the effect of high pressures on the stability of a compound cannot be investigated. However, a form of clamp for holding the bell jar to the specimen table could be readily devised, or adapted from those commercially available. Such a modification would enable some degree of positive pressure to be maintained.

2. Metallographic Studies

Metallography with this image furnace includes thermal etching studies and the study of melting and solidification phenomena in both metals and ceramics. The examples quoted are merely an indication of the kind of work being done with the furnace. It is hoped that a full account of these experiments will be published soon.

Progressive grain growth in a nickel specimen can be clearly seen in Fig. 7-5, where a number of successive positions of the grain boundaries have been clearly etched during heating in argon. This implies that the grain boundary movement was discontinuous, since only in certain places has the boundary remained stationary long enough for thermal etching to occur. In addition to the formation of grain boundary grooves, the principal effect associated with thermal etching, small surface pits have frequently been observed to form on metals heated in the image furnace. Although it was suggested independently by the writer that these pits were formed where dislocation lines met the surface of the specimen, this concept has in fact been advanced by other workers [3-5]. Such surface pits are clearly shown

Fig. 7-4. Isolation valve and uncoupling device for bell jar.

in Fig. 7-6: a specimen of electrolytic high-purity copper heated nearly to
its melting point in nitrogen. The number $(10^6-10^7/cm^2)$ and distribution of
pits is in keeping with the dislocation hypothesis. In addition to the pits,
their alignment and density, and to the different boundary etching effects
that arise by virtue of different grain boundary energies, other fields of this
specimen, e.g., Fig. 7-7, revealed an essentially nonequilibrium pattern of
highly curved grain boundaries.

When nickel specimens were superficially melted, the surface sometimes presented the appearance shown in Fig. 7-8. This suggests that
solidification of the molten surface layer has occurred around a number of
individual dislocations and that a difference in orientation exists between
the regions associated with each dislocation. By comparison alumina and

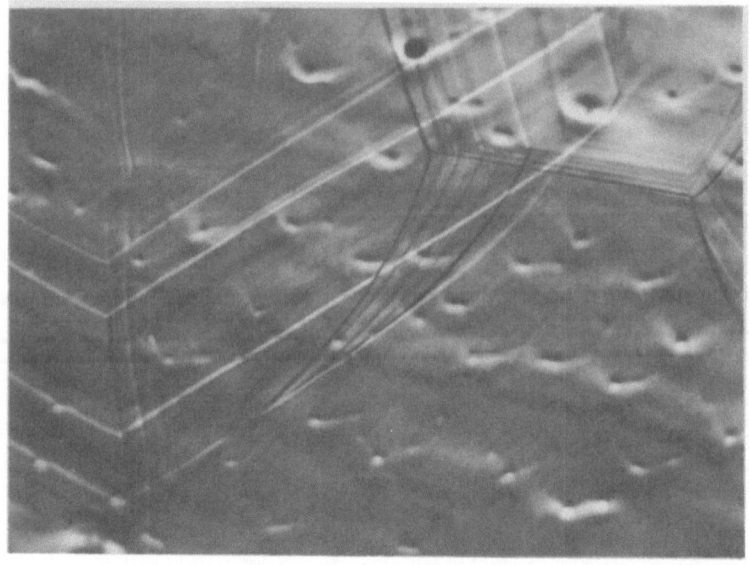

Fig. 7-5. High-purity nickel, thermally etched in argon. (250 ×)

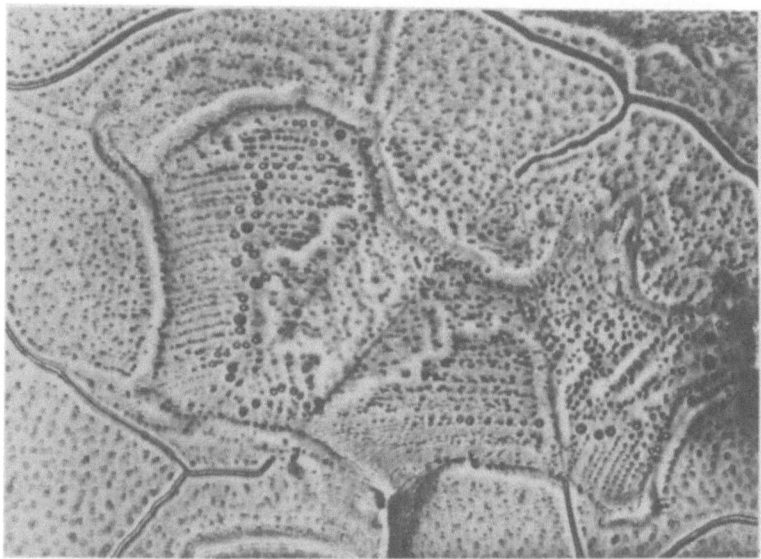

Fig. 7-6. Electrolytic high-purity copper, thermally etched in nitrogen. (250 ×)

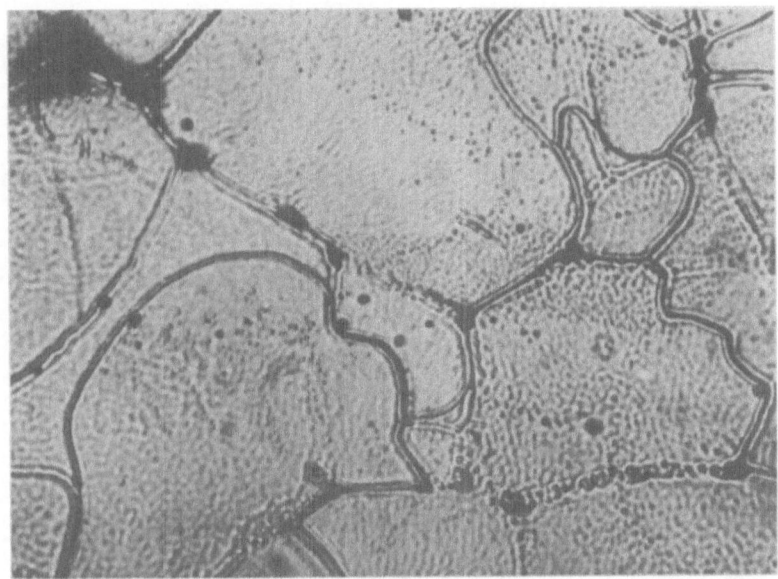

Fig. 7-7. Electrolytic high-purity copper, thermally etched in nitrogen. (250 ×)

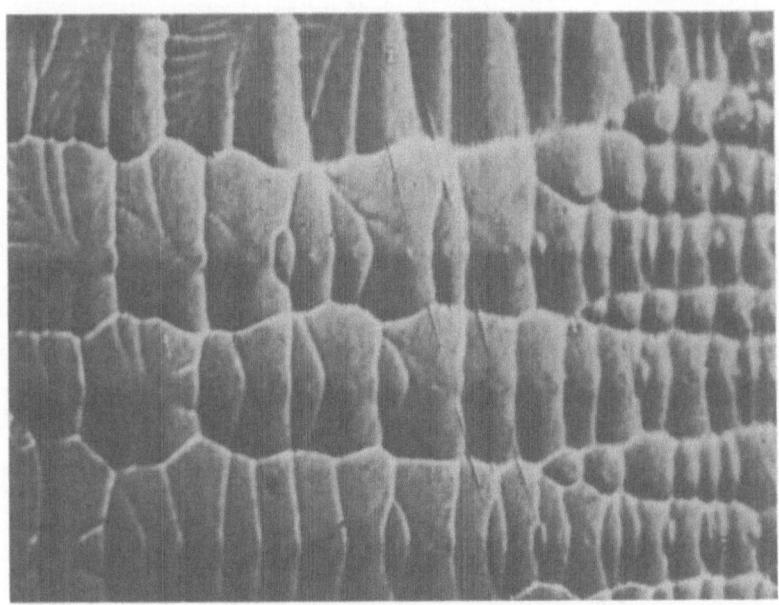

Fig. 7-8. Nickel superficially melted in nitrogen. (150 ×)

Fig. 7-9. Superficially melted alumina. (150 x)

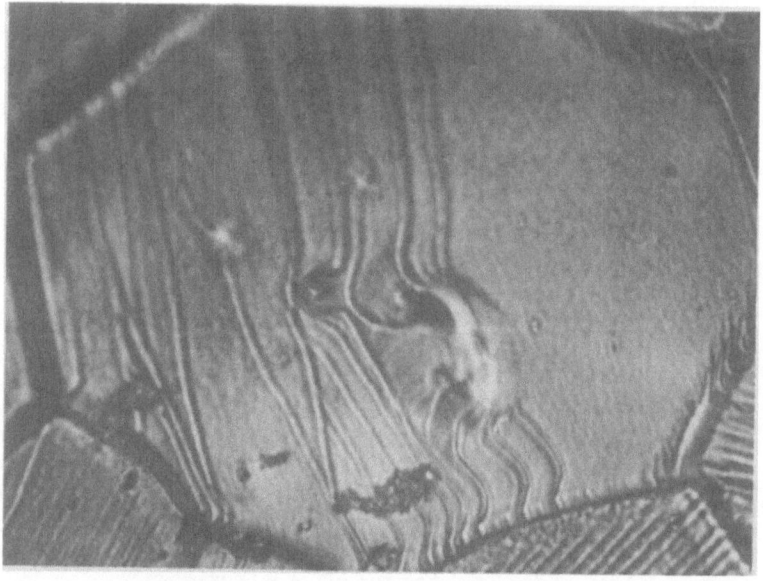

Fig. 7-10. Superficially melted magnesia. (1360 x).

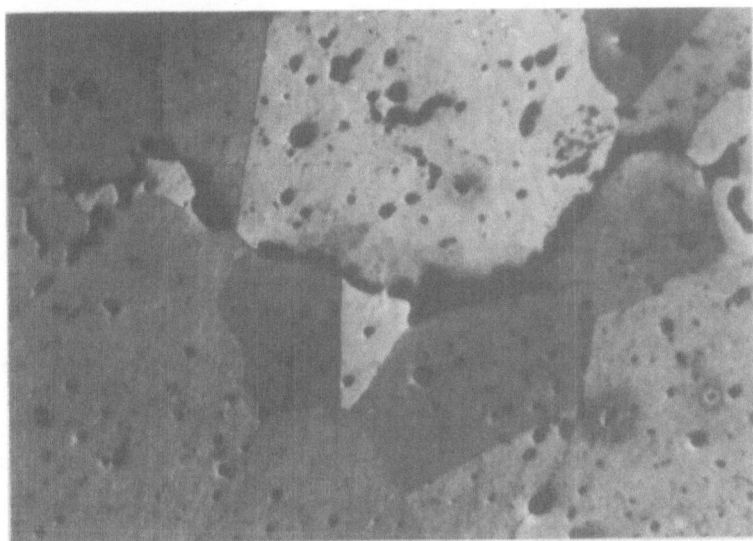

Fig. 7-11. Nickel—section perpendicular to superficially melted surface (nitrogen atmosphere). (80 ×)

magnesia presented, after superficial melting, the appearance shown in Figs. 7-9 and 7-10, respectively. It would seem reasonable to suppose that the striations across the large grain in Fig. 7-10 are not "etch-terraces" in the normal sense but represent a feature that was moving (from left to right) during heating. The curvature near grain boundaries and internal obstacles is difficult to explain otherwise.

A nickel specimen of lower purity than those referred to above showed preferential grain boundary melting. This is illustrated in Fig. 7-11, which shows a polished and chemically etched section perpendicular to the melted surface. The lower part of the photograph shows the original grain boundaries while the upper part shows that melting has begun preferentially at the original grain boundaries; the fully molten and resolidified part of the specimen is higher still, out of the field of view. The occurrence of such grain boundary melting could invalidate heat transfer calculations based on image furnace experiments.

From these and other observations it may be concluded that one of the main advantages in using an image furnace for metallographic work, its ability to present the nonequilibrium (transitional) structure, is inextricably bound up with its main disadvantage, the uncertainty of the temperature of the specimen. Apart from the difficulty of accurate temperature measurement the temperature varies across the image and this, combined with the rapid heating (and relatively rapid cooling) can be used to produce nonequilibrium structures. With a conventional furnace, on the other hand, although quenching may in general be employed to preserve high-temperature structures, those structures are usually equilibrium or near-equilibrium structures at the high temperature. The image furnace enables one to "freeze" a moving picture, particularly when short heating times are used. The high-temperature structure usually will not be an equilibrium structure. Clearly

TABLE 7-I

Percentage of Peak Intensity at Different Image Radii for 11-mm Arc Crater

Intensity	Radius, mm
Maximum	0
0.5	5.5
0.1	9.9
0.01	14.0

this attribute of the image furnace could be enhanced by incorporating a means for quenching specimens within the bell jar.

D. SOME CONSEQUENCES OF A GAUSSIAN DISTRIBUTION OF FLUX

A constant-flow calorimeter, based on an instrument described by Null and Lozier [6], has been used to measure heat flux. The receiving disk of the calorimeter had a diameter of 8 mm, and the elementary analysis that follows has been carried out to determine what proportion of the total flux was received by this calorimeter. The analysis has been extended to determine how the peak flux varies, along the optic axis, with the distance from the focal plane.

1. The Proportion of Total Flux Received by the Calorimeter

It is assumed, in accordance with experimental findings [7,8] that the flux distribution diametrally across the image is Gaussian. The equation to the Gaussian curve may be written

$$\phi = (1/\sigma\sqrt{2\pi})\, e^{-x^2/2\sigma^2} \tag{1}$$

where ϕ is the heat flux at a radial distance x from the intersection of the optic axis with the focal plane.

Let ϕ_m be the peak flux (i.e., on the optic axis) and define the image radius as the radius at which the flux has fallen to 1% of its peak value. That is, when $\phi = 0.01\ \phi_m$, let $x = R_0$. Then, if symmetry about the optic axis is assumed (which is probably reasonably valid since rotating electrodes are used) the total flux within the image will be

$$\Phi_T = \int_0^{R_0} \phi\, 2\pi x\, dx \tag{2}$$

and the fraction of total flux falling on the calorimeter receiving disk of radius r_1 will be

$$\frac{\Phi_1}{\Phi_T} = \frac{1 - e^{-R_1^2/2\sigma^2}}{1 - e^{-R_0^2/2\sigma^2}} \tag{3}$$

To evaluate σ, recourse is made to Null and Lozier's findings [8] that the radius at which the flux has fallen to half its peak value is equal to the radius of the crater in the positive carbon (5.5 mm in the present case). That is,

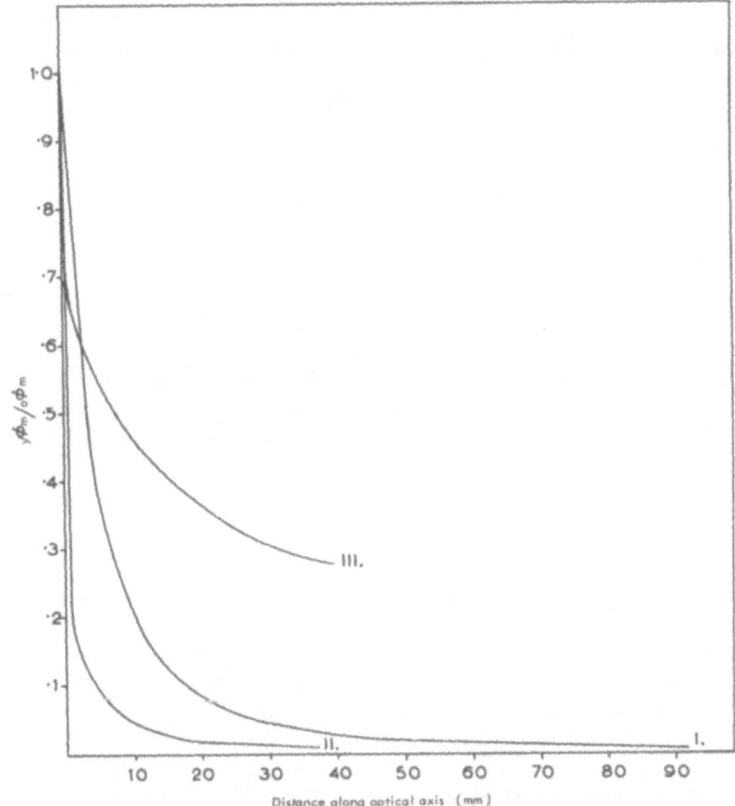

Fig. 7-12. Flux intensity along optical axis relative to flux intensity at center of image (ordinates) vs. distance along optical axis in mm (abscissa). Note: Curve I, approximation, Curve II, accurate plot. Curve III shows the corresponding variation in maximum blackbody temperature, obtained from Curve II.

when $x = 5.5$ mm, $\phi = \phi_{m/2} = 1/2\sigma\sqrt{2\pi}$. Hence substituting in equation (1) gives: $\sigma^2 = 21.79$. Furthermore, since $r_1 = 4.0$ mm and $R_0 = 3\sigma$, equation (3) becomes

$$\Phi_1/\Phi_T = 0.311$$

Thus the calorimeter, when concentric with the optic axis, receives 31% of the total flux at the image. If blackbody conditions are assumed the total flux at the image may be calculated from measurements with the calorimeter. Table 7-I gives the image radius corresponding to different percentages of peak intensity determined in the foregoing manner.

2. The Variation of Flux Along the Optic Axis

Consider the cone of radiation (semiangle 60°) converging on the image. Across any plane perpendicular to the axis of this cone the total flux Φ_T must be constant. The peak flux ϕ_m falls as the radius of the cone increases. If it is assumed that the flux distribution across a diameter of the cone remains Gaussian for small axial distances from the focus, then the variation in ϕ_m along the optic axis may be determined as follows:

Total flux across image plane is equal to total flux across circular section of cone of radiation at a distance y from image plane, i.e.,

$$_0\Phi_T = {}_y\Phi_T$$

which from equation (2) gives

$$\int_0^{R_0} 2\pi\phi x \, dx = \int_0^{R_y} 2\pi\phi x \, dx$$

which gives

$$\sigma_0 (1 - e^{-R_0^2/2\sigma_0^2}) = \sigma_y (1 - e^{-R_y^2/2\sigma_y^2}) \tag{4}$$

Since $_y\phi_m - 1/\sigma_y \sqrt{2\pi}$, the determination of the value of $_y\phi_m$ the maximum flux in a plane at distance y from the focus, requires the evaluation of σ_y. This can be done approximately by expanding $e^{-R^2/2\sigma^2}$. When this expansion is carried out, equation (4) reduces to:

$$\sigma_y/\sigma_0 = (R_y/R_0)^2 \tag{5}$$

The corresponding approximate variation of $_y\phi_m/_0\phi_m$ with distance (in mm) from the focus is plotted in Fig. 7-12, Curve I.

An accurate value can be obtained from equation (4) by inserting numerical values in the left-hand side of the equation, assigning a series of values to σ_y, evaluating the corresponding values of R_y, and from these calculating the corresponding values of y from the geometry of the cone. The variation of $_y\phi_m/_0\phi_m$ obtained in this way is plotted in Fig. 7-12, Curve II.

The corresponding variation in maximum blackbody temperature, derived from the Stefan–Boltzmann relationship is

$$T_y/T_0 = (\sigma_0/\sigma_y)^{1/4} \tag{6}$$

This variation is plotted in Fig. 7-12 as Curve III. Curve III suggests that very fine and accurate focusing is necessary if the maximum temperature is to be realized. Probably the ideal focusing arrangement would combine coarse and fine adjustments, as on a microscope.

ACKNOWLEDGMENTS

The writer is grateful to G. Wildsmith for the great energy he devoted to the construction of the furnace and the early experiments with it, to T. J. Davies for help with subsequent modifications, to M. Gould for much general assistance, and to all three for their assistance with photomicrography. Professor F. Morton generously made facilities available, and the D. S. I. R. provided a maintenance grant for one student. Figures 7-1 and 7-3 are reproduced from the writer's paper in the British Journal of Applied Physics [2], with the Editor's permission.

REFERENCES

1. Evans, P. E., The New Scientist 9:865 (1961).
2. Evans, P. E., and Wildsmith, G., Brit. J. Appl. Phys. 13:68 (1962).
3. Kitajuma, S., J. Japan. Inst. Metals 18:592 (1954).
4. Hendrickson, A. A., and Machlin, E. S., Acta Met. 3:64 (1955).
5. Fraser, M. J., Caplan, D., and Burr, A. A., Acta Met. 4:186 (1956).
6. Null, M. R., and Lozier, W. W., Rev. Sci. Instr. 29:163 (1958).
7. Swope, G. A., and Henriques, F. C., Report No. TOI 54-8, Technical Operations, Inc., Arlington, Mass.
8. McGrath, I. A., and Croft, V., Brit. J. Appl. Phys. 13:369 (1962).
9. Null, M. R., and Lozier, W. W., J. Soc. Motion Picture Television Engineers 68(2):80 (1959).

Chapter 8

High-Wattage Xenon and Mercury Vapor Compact Arc Lamps as Radiation Sources for Imaging Furnaces

Wolfgang E. Thouret and Herbert S. Strauss

Duro-Test Corporation
North Bergen, New Jersey

A. INTRODUCTION

Compact or short-arc high-pressure lamps were invented in Europe as early as 1936 ([1]). Their main characteristic is that they combine the high brightness of carbon arcs with the maintenance-free and clean operation of regular discharge or incandescent lamps. They appeared first as specialized versions of the popular mercury vapor lamps for street and industrial lighting. High-pressure mercury short-arc lamps for use in optical equipment, such as projectors or searchlights, were developed in a wide range of wattages, from 75 W to 10 kW ([2-10]).

The introduction of xenon as the discharge carrying medium ([11,12]) brought the important advantage of an essentially continuous spectrum with energy distribution similar to that of the sun in the visible and ultraviolet regions. In addition, lamps with xenon filling have undelayed light output after ignition, while pure mercury lamps generally need a warm-up time of several minutes. Xenon short-arc lamps were developed in types of up to 2000 W ([13-20]) in several countries and were first made commercially available in Europe about 10 years ago ([16]). The use of 900- and 1600-W xenon lamps as the projector light source in numerous European motion picture theaters first demonstrated on a larger scale the economic usefulness of compact arc lamps as replacement for carbon arcs. It was also proved that such lamps can operate reliably in practical service for 1000 to 2000 hr ([21]).

Higher wattage types with inputs of 2.5, 5, and 10 kW have recently become available in this country, ([22-25]) and the development of a 20 kW type is in progress ([26]). These lamps have been developed for use in military searchlights, motion picture theatres, and solar simulator installations. Such high wattage compact arc lamps, especially those with pure xenon gas filling, find increasing interest and experimental use as sources for imaging furnaces.

This paper will deal with the design principles for compact arc lamps and will place particular emphasis on the problems and limitations encountered with increasing wattage. As practical examples, the characteristics, main dimensions, and spectral emission properties of available and recently developed high-wattage types will be described.

TABLE 8-I

Principal Data of High-Wattage Xenon, Mercury, and Mercury–Xenon Compact Arc Lamps

No.	Lamp wattage, W	Type designation	Gas–Vapor filling	Max. bulb outer diam., mm	Max. over-all length, in.	Arc length in oper., mm	Int. oper. pressure, atm	Lamp oper. voltage, V	Lamp oper. current, A	Brightness data, cd/mm² Peak bright-ness	Avg. bright-ness	for Arc area of width × length (mm) (mm)	Initial lumens	Initial efficacy, lm/W	Rated life, hr
1	800 AC	SAH800C[a]	Hg	42	9½	8.5	10	70	12	180	–	–	40,000	50	600[2]
2	900 DC	XBO900w[b]	Xe	40	12 19/16	3.4	–	22	42	–	380	1.7 × 3.4	30,500	34	2000
3	1,000 DC	528B[c]	Hg–Xe	45	7	6.5	–	65	16	970	230	2.5 × 3.5	52,000	52	1000[1]
4	1,000 AC	SAH1,000A[a]	Hg	51	9½	6.5	25	65	18	475	–	–	50,000	50	300[2]
5	1,000 AC	SAHX1,000A[a]	Hg–Xe	51	9½	6.5	30	65	18	475	–	–	50,000	50	400[2]
6	1,600 DC	XBO1,600w[b]	Xe	52	14 9/16	4.2	–	26	63	–	430	2.1 × 4.2	56,000	35	2000
7	2,000 DC	UXL2,000DK[d]	Xe	53	14 9/16	6	–	28	70	–	550	2 × 6	70,000	35	–
8	2,000 DC	XE2,000[e]	Xe	54	14 13/16	3.5	–	22.5	90	4100	820	3 × 3	85,000	42.5	500[3]
9	2,200 DC	491C[c]	Xe	57	12½	4	16	20–23	100	3300	440	2.5 × 4	75,000	34	1000[1]
10	2,500 DC	929B[c]	Hg–Xe	64	12½	4	18	45–55	50	2050	540	2.5 × 4	120,000	48	1000[1]
11	2,500 AC	SAH2,500A[a]	Hg	70	13	10	15	65	45	325	–	–	125,000	50	400[2]
12	2,500 AC	SAHX2,500A[a]	Hg–Xe	70	13	10	20	65	45	325	–	–	125,000	50	200[2]
13	2,500 DC	SAHX2,500B[a]	Hg–Xe	70	13	4.5	30	50	50	2050	–	–	120,000	48	400[2]
14	2,500 DC	XBO2,500w[b]	Xe	57	16 7/8	6.2	–	30.1	83	–	450	3.1 × 6.2	100,000	40	1500
15	4,000 DC	UXL4,000DK[d]	Xe	–	15 3/4	7	–	33	120	–	600	3 × 7	120,000	30	–
16	5,000 DC[4]	932B[c]	Hg–Xe	86	13½	5	15	50–60	100	2250	780	3 × 5	230,000	46	–
17	5,000 DC	XE5,000[e]	Xe	89	19½	7	–	34.5	145	5900	870	3 × 5	275,000	55	500[3]
18	10,000 DC	XE10,000[f]	Xe	92	27½	8	10	40	250	9500	1250	5 × 5	480,000	48	500[3]
19	20,000 DC[4]	XE20,000[f]	Xe	120	33½	13.5	6	50	400	7500	1700	5 × 5	1,000,000	50	–

*Latest available published data at time of preparation of this paper.

a. Westinghouse, U.S.A.   d. Ushio Kogyo Kaisha, Japan   1. 12 hours per start.
b. Osram, West Germany   e. General Electric, U.S.A.   2. In vertical position 8 hours per start.
c. Hanovia, U.S.A.   f. Duro–Test, U.S.A.   3. Preliminary data.
4. Experimental lamp.

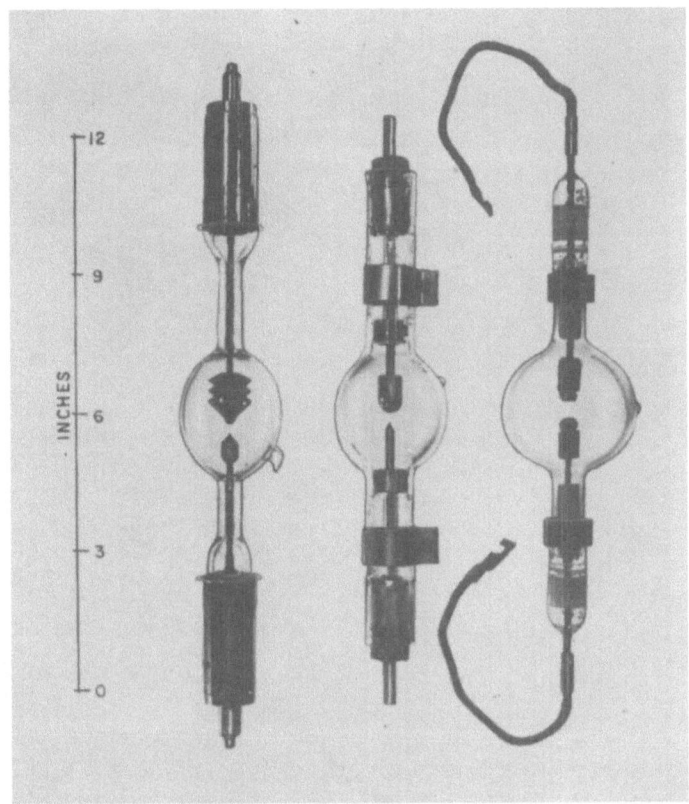

Fig. 8-1. Typical compact arc lamps in the 1600 to 2500 W range—from left to right: 1600-W xenon lamp (see Table 8-I, No. 6); 2200-W xenon lamp (Table 8-I, No. 9); 2500-W mercury–xenon lamp (Table 8-I, No. 12).

B. CHARACTERISTICS OF PRACTICAL LAMP TYPES

The design principles for high-wattage compact arc lamps can be explained best on the basis of practical lamp data. For this purpose, Table 8-I lists the main characteristics and dimensions of a number of lamp types that have been developed and made available by several manufacturers. Figures 8-1 to 8-3 give the outward appearance and relative size of some typical examples selected from these lamps. Figure 8-1 shows three lamps in the 1600 to 2500 W range, two have pure xenon filling, while the third is a mercury–xenon lamp. Figure 8-2 shows two 5000-W lamps, one with pure xenon and the other with mercury–xenon filling. Figure 8-3 presents the largest xenon compact arc lamps developed so far: a 10-kW and a 20-kW type.

The lamp pictures demonstrate that all compact arc lamps are similar in basic construction: they consist of a spherical or ellipsoid-shaped envelope, which is usually made of clear fused quartz, and has two diametrically opposed cylindrical extensions. These extensions contain metallic support and connecting rods for the two electrodes and the hermetic seals

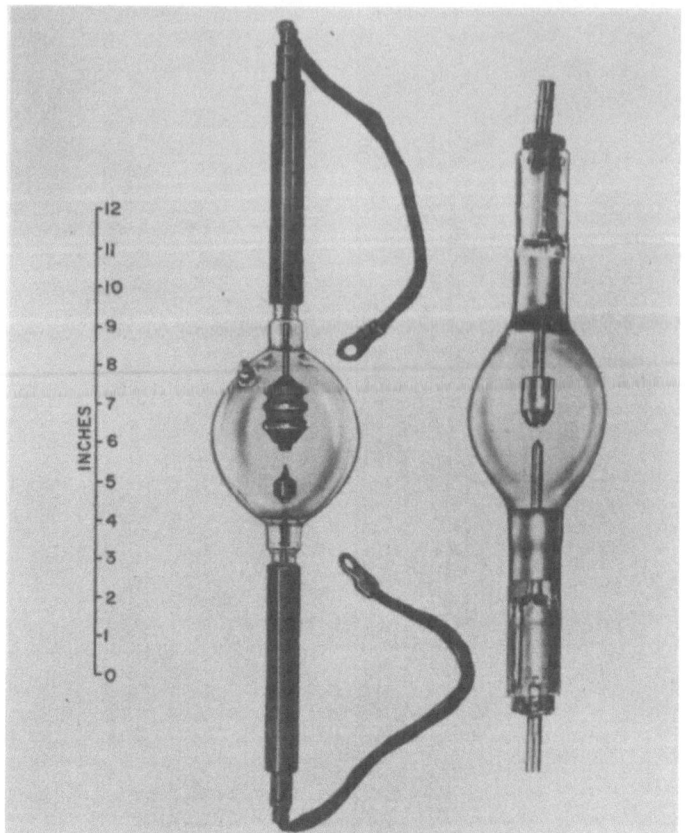

Fig. 8-2. 5000-W compact arc lamps—left: xenon lamp (see Table 8-I, No. 17);
right: mercury–xenon lamp (Table 8-I, No. 16).

and carry the outer electrical contact elements. They also serve to
mechanically support and adjust the lamp in proper relation to the optical
elements of the equipment.

In operation the lamps have internal pressures between 6 and 30 atm.
Due to the high pressure, the arc discharge fills only the small volume
directly between the tips of the solid tungsten electrodes and thus forms a
highly concentrated source. The degree of source concentration is illustrated by Figs. 8-4 to 8-7, which give brightness distributions of DC
lamps. A photograph of a typical high-wattage DC xenon arc is given in the
left part of Fig. 8-8. Xenon DC arcs typically have very uneven brightness
distribution, characterized by a "hot spot" arc constriction at the cathode
and a strongly increasing arc width toward the anode. The cathodic constriction is less pronounced in mercury arcs, as can be verified by comparing the left and right parts of Fig. 8-4. The peak brightness of the xenon
arc on the left is more than twice that of the mercury–xenon arc on the
right, in spite of a current relation of only 1.45:1 under otherwise comparable conditions. It will be explained later why mercury–xenon arcs
have virtually the same properties as arcs in pure mercury.

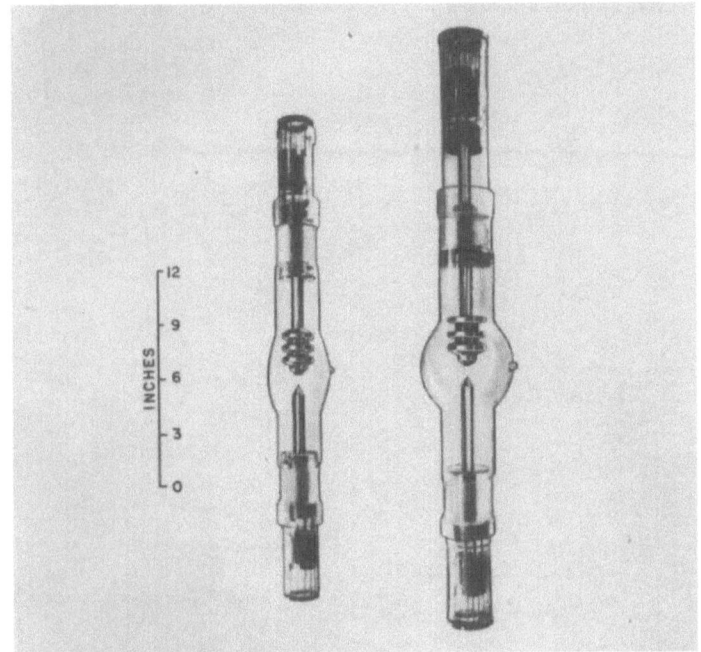

Fig. 8-3. 10-kW and 20-kW xenon compact arc lamps (see Table 8-I, Nos. 18 and 19).

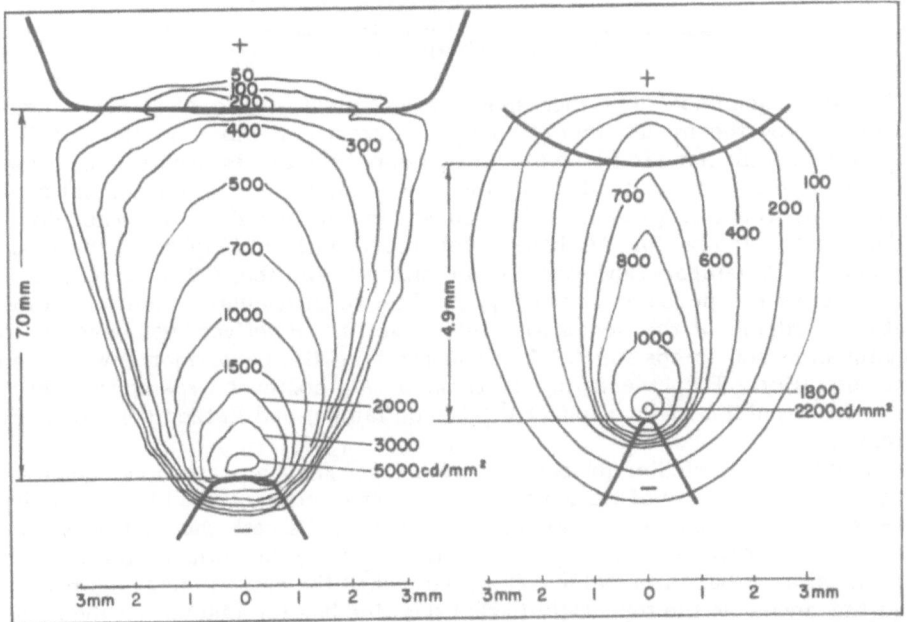

Fig. 8-4. Typical brightness distributions of compact arc lamps—left: brightness distribution of a 5-kW DC xenon lamp (No. 17 in Table 8-I); right: brightness distribution of 5-kW mercury–xenon DC lamp (No. 16 in Table 8-I).

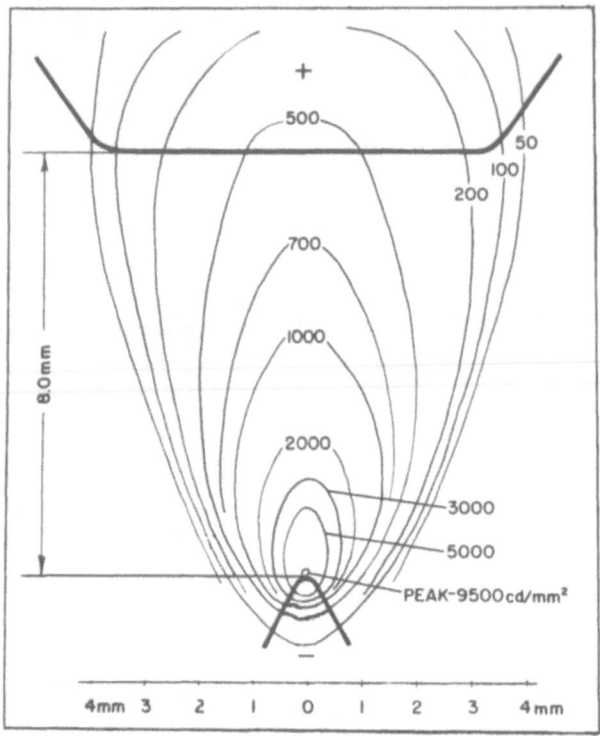

Fig. 8-5. Brightness distribution of 10-kW xenon compact arc lamp
(No. 18 in Table 8-I).

The phenomenon of a "hot spot" or "hot plasma sphere" at the cathode is caused by the mechanism of electric current transfer from the metallic cathode into the hot arc plasma. In AC lamps, the arc is usually much less constricted at the electrodes, as the anode mechanism under normal conditions prevents a pronounced increase in current density. Therefore, compact arc lamps for AC have much more uniform brightness distributions than DC lamps. However, AC operation is limited, for the time being, to mercury or mercury–xenon types. Due to particular conditions in the cathodic region of the xenon arc discharge, it has not yet been possible to obtain in xenon lamps for AC the required long-life performance with good arc stability. Brightness distributions of AC compact arcs up to 2500 W have been published by Thouret ([17]) (xenon lamps) and Retzer ([22]) (mercury lamps).

Due to the particular properties of the xenon gas, the cathode mechanism generates in the axis of DC xenon arcs a narrow concentrated high-velocity gas stream (plasma jet). This originates from the cathodic "hot spot" region and is directed toward the anode. Due to the cathodic plasma jet, xenon DC arcs burn in a stable fashion only when the cathode is at the bottom and the anode at the top. Only in this position has the plasma jet the same direction as the strong convection currents surrounding the arc column. In inverted position, xenon DC arcs become very unstable, while limited

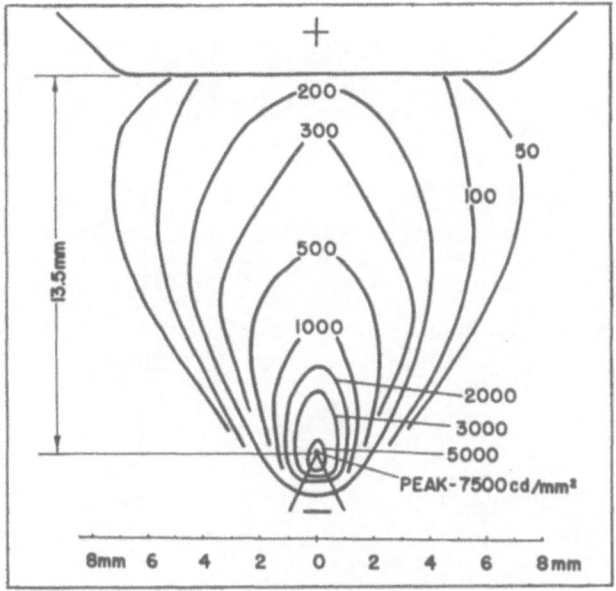

Fig. 8-6. Brightness distribution of 20-kW xenon compact arc lamp
(No. 19 in Table 8-I).

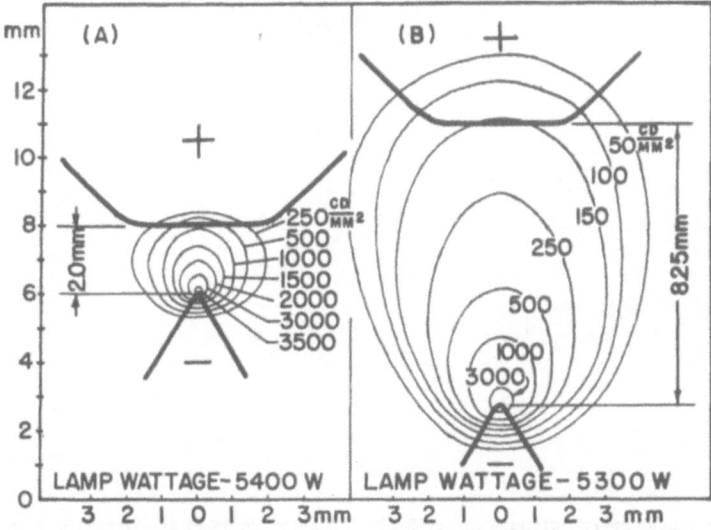

Fig. 8-7. Influence of electrode spacing on brightness distribution of compact
arc lamps. Brightness distributions of experimental 5-kW xenon lamps with
(A) 2-mm and (B) 8.25-mm electrode spacing during lamp operation.

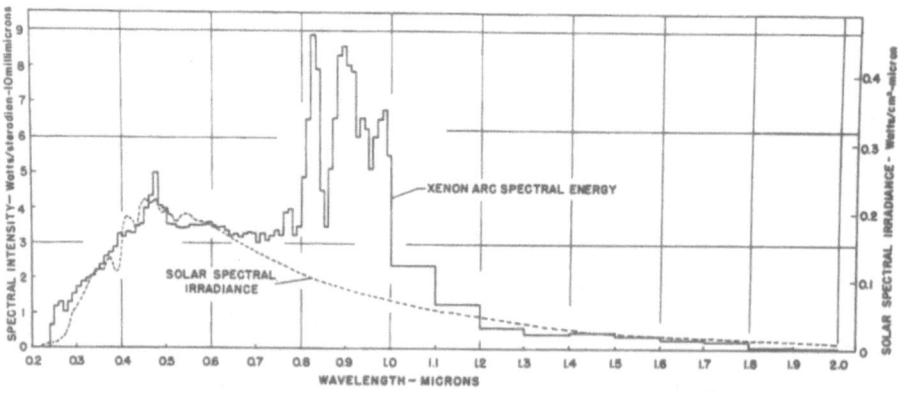

Fig. 8-8. Five-kilowatt DC xenon arcs—left: in pure xenon; right: in xenon–hydrogen mixture (64% Xe, 36% H_2).

Fig. 8-9. Spectral energy distribution of 5-kW xenon DC compact arc (lamp No. 17 in Table 8-I, the electrode and bulb radiation is excluded) with solar spectral energy distribution according to Johnson ([29]).

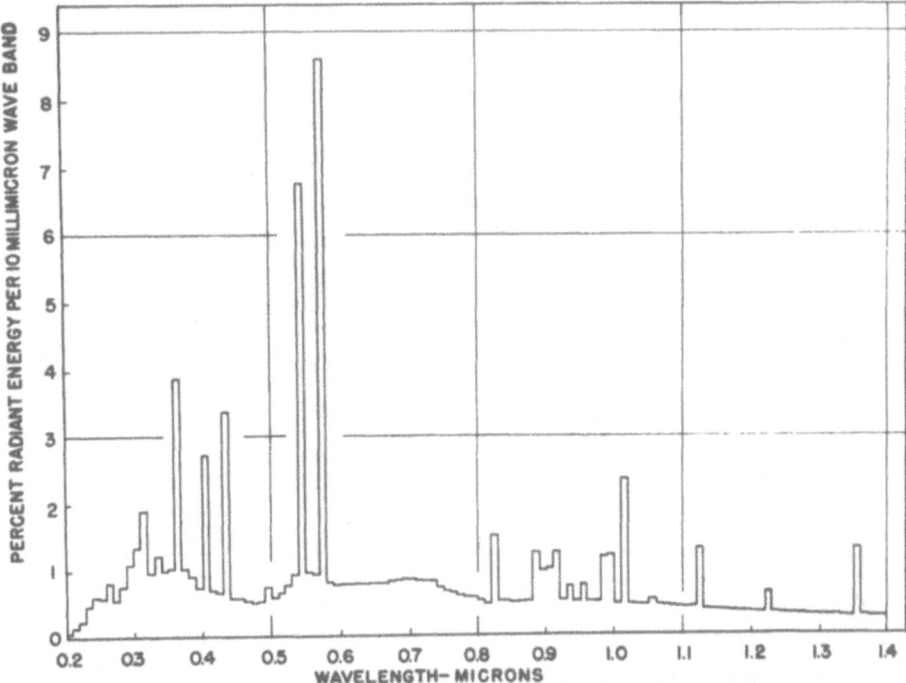

Fig. 8-10. Spectral energy distribution of 1000-W mercury-xenon compact arc lamp (lamp No. 3 in Table 8-I).

deviations from the vertical "anode-up" position do not substantially affect the arc stability. The stability of mercury and mercury–xenon compact arcs is much less critical. For example, mercury DC arcs can be operated both "anode up" and "anode down," and the desirable burning position can be determined by other design factors, such as the shortest warm-up time.

The peak and average brightness values given in Table 8-I show that the source concentration attained in compact arc lamps compares favorably with that of high-intensity carbon arcs. Some published brightness data for carbon arcs are: 950 cd/mm^2 for a 13.3 kW (180 A, 74 V) arc with 13.6 mm positive carbon diameter [27] and 1300 cd/mm^2 for a 12.7 kW (190 A, 67 V) arc with 11 mm positive carbon diameter [28]. The luminous efficacy of these carbon arcs ranges between 28 and 33 lm/W. Mercury compact arc lamps and the larger-wattage xenon lamps reach 40 to 55 lm/W. Thus, they are considerably more efficient than carbon arcs as to output of visible radiation.

The average brightness of short compact arcs is strongly influenced by the arc length or electrode spacing. When in a lamp of given wattage input the electrode spacing is reduced, the watts per millimeter arc length and the watts per cubic millimeter arc volume increase. Consequently, the arc temperature and the brightness grow. This growth becomes very pronounced in DC arcs, when the arc length approaches the arc width. The cathode "hot spot" constriction then gains decisive influence upon the arc width and thus causes a fast brightness increase to extreme values at very

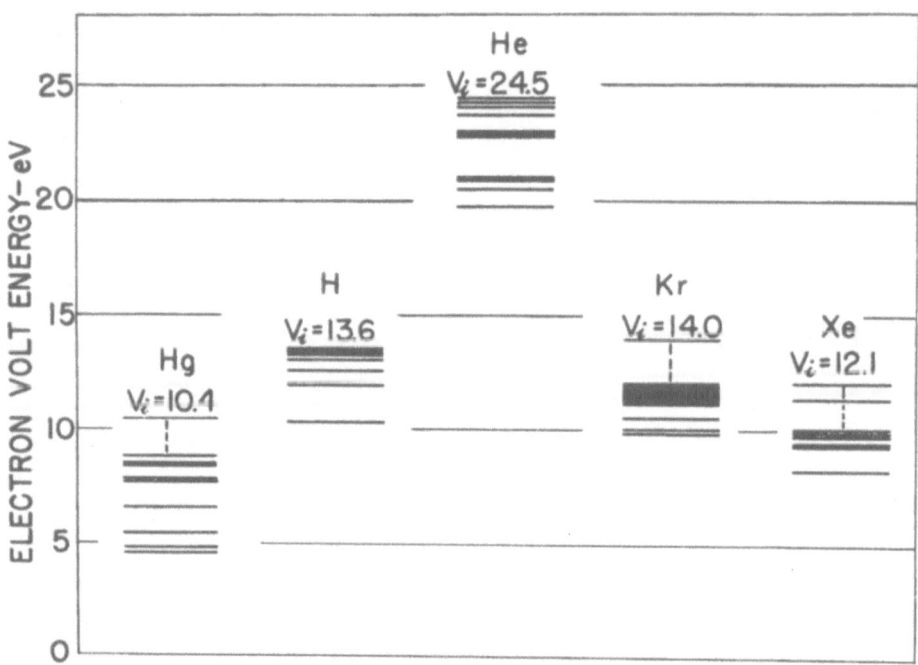

Fig. 8-11. Excitation and ionization voltages of mercury, hydrogen, helium, krypton, and xenon.

short spacing of the electrode tips. Figure 8-7 demonstrates the influence of electrode spacing on average brightness and source dimensions for the constant wattage of about 5 kW. The brightness distribution at approximately 8 mm arc length is compared with that of a 2 mm long arc. Of course, with such short spacing of relatively large electrodes, the gain in average brightness may be offset by reduced efficacy, because a large part of the luminous flux is shadowed off by the anode.

Figures 8-9 and 8-10 show the spectral energy distribution over the entire range between 2000 A and several microns for a typical xenon and a typical mercury–xenon lamp. It is well known that the xenon arc emits an essentially continuous spectrum in the visible and ultraviolet.

In the infrared, a strong line emission between 8000 and 11,000 A is superimposed on the xenon continuum. The total radiation energy emitted by xenon compact arc lamps of 5 to 10 kW amounts to between 50 and 55% of the wattage input.

The mercury arc radiates mainly the characteristic mercury lines, which are distributed between 2500 and 14,000 A. They provide a high output of radiation very efficiently. As shown by Fig. 8-10, a compact high-pressure mercury vapor arc also emits a continuous spectrum extending from the short ultraviolet to long infrared wavelengths, but the intensity level of this continuum is much lower than that of the xenon continuum [11-12].

Mercury–xenon arcs usually burn in a high-pressure mercury atmosphere to which approximately 20% xenon has been added. The excitation and ionization voltages of the mercury atom are considerably lower than those of xenon, as shown by Fig. 8-11. Also, the cross section of the mercury atom for interaction with the slow electrons of approximately 1 eV energy that prevail at the 7000°C arc plasma temperature is about one hundred times greater than that of xenon. This fact is illustrated by Fig. 8-12, which shows the atomic cross sections for electrons of mercury, xenon, and several other gases as a function of electron energy. Due to these basic atomic properties, the arc discharge in mercury–xenon lamps assumes essentially the electrical, optical, and stability characteristics of arcs in pure mercury, as soon as a minimum partial pressure of mercury vapor is established. The xenon infrared lines, however, appear at reduced intensity simultaneously with the mercury lines.

Mercury and mercury–xenon compact arcs are generally more efficient radiation sources than xenon arcs because their voltage gradient is four to five times higher under otherwise equal conditions. Due to the lower voltage drop in the arc column, xenon lamps have to be designed for higher currents than mercury or mercury–xenon lamps of the same wattage. The higher

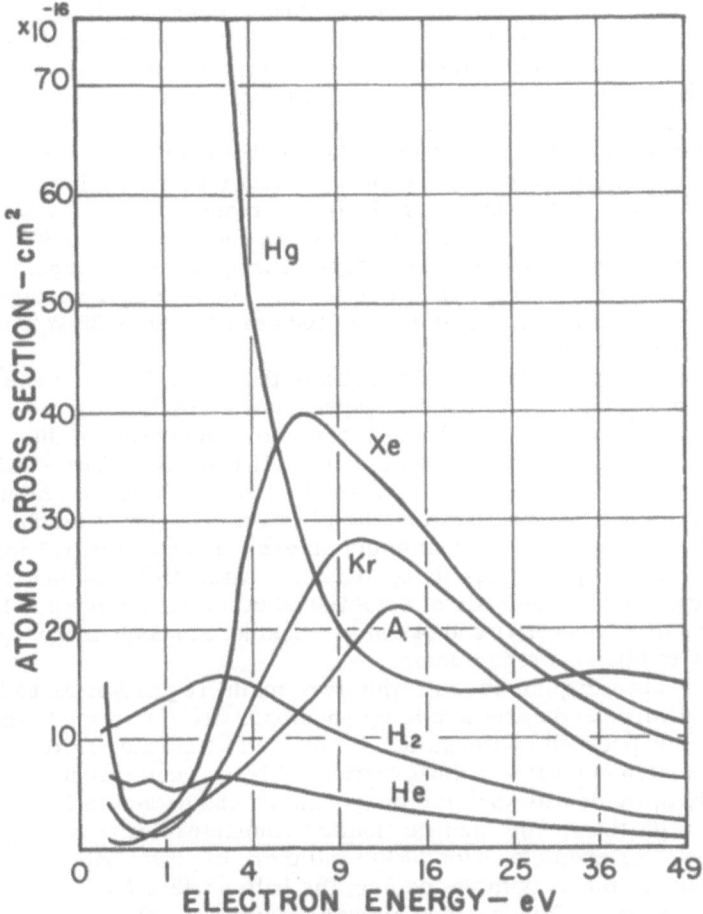

Fig. 8-12. Atomic cross section for electrons of the heavy rare gases—mercury, hydrogen, and helium—as a function of electron energy.

currents cause greater energy losses at the electrodes. Also, lamp cost is increased, because larger electrodes and seals are required. Possibilities of increasing the voltage gradient of xenon arcs without changing their radiation characteristics—through addition of other gases have been investigated recently [26] and will be discussed in the last paragraph of this paper.

C. ENVELOPE DESIGN

Because of the high internal operating pressure, the clear fused quartz envelopes of compact arc lamps have to withstand considerable tensile stresses. In addition, during lamp operation the bulb wall is subjected to substantial thermal stresses. These result from the inner and outer surface temperature difference, which is caused by the passage of energy in the form of heat conduction; there are also unavoidable temperature differences between different bulb areas. Clear fused quartz has a tendency to recrystallize locally if it is held at temperatures over 900°C for long periods. Therefore, this fact rather than the high softening point of quartz (over 1700°C) limits the upper temperature in practical lamps. The average bulb surface loading has to be limited, so that all parts of the bulb in all allowed lamp operating positions remain below the above-mentioned recrystallization temperature. The design value for the bulb surface loading, in W/cm^2 or W/in^2, must also take into account the fact that during lamp life the envelope operating temperature gradually increases, due to reduced transmittance for radiation. This reduction is caused by deposited electrode material and other inherent structural changes. An accepted maximum value for the average inner bulb surface loading of larger wattage compact arc lamps rated for a useful life of 1000 to 2000 hr is 28 W/cm^2, corresponding to 180 W/in^2.

The bulb surface loading determines the inner bulb dimensions for a given lamp wattage. The operating pressure allowable with a certain bulb wall thickness can be found by approximate computation of the tensile and thermal stress condition within the bulb wall [30]. The theoretical strength of a lamp envelope can be characterized by a safety factor n. As usual, this is defined as the relation between the ultimate strength of the material and the working stress, i.e., the maximum stress considered "safe" for reliable operation of the particular lamp over its rated life. Computation of the safety factor n for a number of established practical lamp types has shown that values of n between 10 and 14 can be considered adequate for operational reliability of higher-wattage lamps.

It has been explained that the two main requirements to be met by properly designed compact arc lamp envelopes are (1) adequate mechanical strength as expressed by the safety factor, and (2) surface loading under the accepted maximum value. Observation of both requirements becomes increasingly difficult with growing lamp wattage and seems to impose definite limitations on the design and economic manufacture of very large-wattage units, those, for example, with 30 kW and more of power input.

Because of the maximum loading, the bulb surface has to grow linearly with the wattage. This surface growth requires a corresponding increase in diameter, and consequently, a reduction in pressure to avoid an increase in the tensile stress. The latter grows in direct proportion with the diameter, if the wall thickness remains the same. Greater wall thickness reduces the tensile stress, but at the same time increases the thermal stress component. Besides this, it is still difficult to manufacture at reasonable cost large, clear quartz bulbs with a uniform wall thickness of more than 3 to 4 mm, the predominantly used values.

For the reasons given, it is accepted practice to reduce the pressure in compact arc lamps of increased wattage and bulb dimensions. This pressure reduction has the following disadvantages: (1) the arc column diameter increases, and consequently, the attained degree of source concentration (the brightness) is lower. (2) The voltage gradient in the arc decreases, which means that a correspondingly higher current is required, and this leads to reduced overall lamp efficacy, because of higher electrode losses and an increased lamp cost, due to electrodes and seals. (3) The evaporation rate of the electrode material increases, because it is dependent upon the surrounding gas or vapor pressure. This phenomenon causes shorter useful lamp life through bulb blackening, unless the electrode operating temperature is reduced sufficiently to compensate for the lower pressure. To lower the electrode operating temperature, one must use larger electrodes which becomes increasingly difficult with increasing lamp wattage, as will be discussed later in the paragraph on electrodes.

The following means of avoiding the reduced pressure in high-wattage compact arc lamps have been suggested: (1) Increase the allowable surface loading by forced cooling of the envelope with gases or liquids. (2) Increase the allowable stress by providing a second envelope that completely surrounds the first and is filled with, for example, half its pressure, thus relieving the inner envelope of a substantial part of the tensile stress. (3) Increase the strength of the envelope by hot shrinking a second envelope directly onto its surface and thus attaining uniform stress distribution within the wall in operating condition. (4) Use a bulb material which allows higher bulb surface loading and higher stress than fused quartz.

While some of these suggestions undoubtedly have practical value, none is used so far in technical compact arc lamps. The following disadvantages have hindered the practical use of the suggested measures:

1. Forced cooling can keep the bulb operating temperature below the allowable limit under increased surface loading, but does not prevent the thermal stress from increasing linearly with the loading. Liquid cooling is successfully used with long arc tubular or capillary type lamps, but their inner tube radius is in the range below 5 mm, where the thermal stress is smaller than in larger diameter envelopes. These water-cooled lamps also have a much smaller safety factor than that considered necessary for uncooled types [30].

2. A second envelope would greatly complicate the lamp manufacture and increase the cost. There would also be increased radiation losses through absorption and reflection by the second envelope wall.

3. Synthetic crystalline sapphire is at present the only bulb material that can come under consideration as improvement over fused quartz for compact arc lamps. It has a definitely higher melting point (2040°C) and can withstand higher temperatures over long periods. However, its thermal expansion coefficient and its heat conductivity, both of which strongly influence the thermal stresses, are such that the thermal stresses generated in a sapphire lamp envelope are approximately four to five times greater

than those generated in a quartz envelope under similar conditions. The ultimate tensile strength of sapphire is also four to five times greater than that of quartz. Thus, it does not seem likely that sapphire envelopes can have an adequate safety factor with much higher bulb loading than is allowable for quartz [30]. Sapphire absorbs less energy at the very short ultraviolet and the very long infrared wavelengths; this may somewhat reduce the thermal stresses. Important disadvantages of sapphire envelopes at present are the extremely high price and the limitations in shape and dimensions.

D. HERMETIC SEALS FOR QUARTZ ENVELOPES

It is well known that regular metallic lead-in conductors cannot be sealed directly into quartz bulbs, as the thermal expansion coefficient of quartz is 8 to 12 times smaller than that of the suitable metals. This characteristic has sometimes been considered to be a basic disadvantage of fused quartz as a lamp envelope material. However, in the case of high-pressure lamps with great surface loading, the low thermal expansion coefficient is actually an advantage, because the thermal stress in the envelope wall is directly proportional to the expansion coefficient. The surface loadings and operating pressures presently used in compact arc lamps would be impossible if quartz had a higher expansion coefficient without a proportionately increased thermal conductivity.

The hermetic metal-through-quartz seals for higher wattage compact arc lamps not only have to carry currents of up to several hundred amperes, but also have to withstand the high internal pressure and have to remain intact at a relatively high temperature of at least 200 to 300°C. Two different design principles are used in such seals: "molybdenum ribbon or foil" and "graded glass."

Molybdenum ribbon seals utilize the basic fact that in glass-to-metal seals, substantial expansion differences are acceptable, if the metal part is very thin, since it can balance the differences by stretching elastically. A basic requirement is, of course, that the adhesion forces between glass and metal be greater than the elastic forces within the metal. Molybdenum is widely used for such quartz-to-metal seals, because its melting point is sufficiently high, and because it can be rolled with relative ease to very thin sheets.

In modern quartz mercury vapor lamps for general illumination, molybdenum ribbon seals are used exclusively and carry currents of up to 8 A with ribbon conductors of $0.0016 \times \frac{1}{4}$ in. cross section. The current-carrying capacity of these seals is relatively high, because they operate in an oxygen-free outer bulb. The molybdenum ribbon seals of high wattage compact arc lamps have to carry much higher currents and also usually have to operate in air, where their outer ends are subject to oxidation. It has been possible to meet these requirements and to develop multiple ribbon seals for currents up to 250 A. Such seals consist of several ribbon conductors in circular arrangement between two concentric quartz tubes. The 2200-W xenon lamp in the center of Fig. 8-1, the 5000-W mercury–xenon lamp at the right in Fig. 8-2, and the 10-kW and 20-kW types of Fig. 8-3 are equipped with seals according to this method. Figure 8-13 A is a schematic

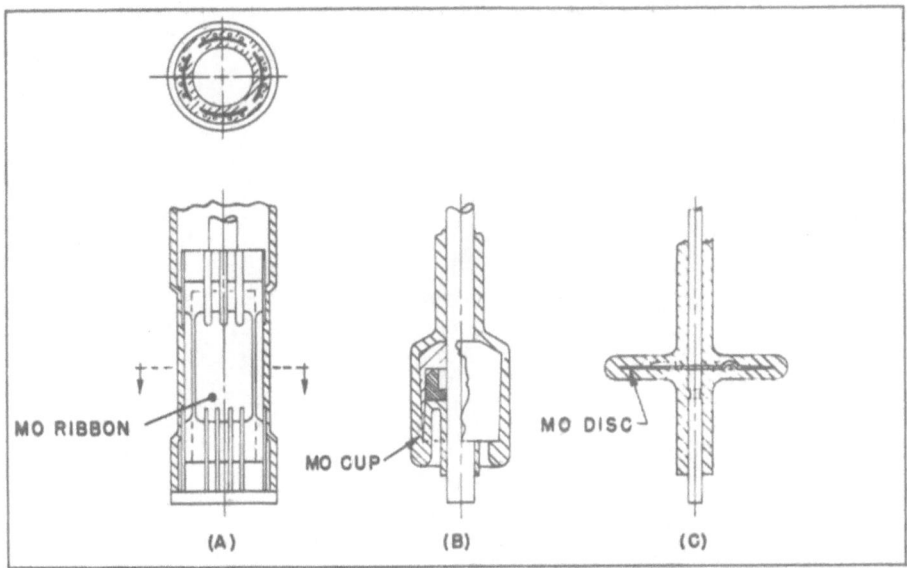

Fig. 8-13. High-current molybdenum quartz seals. (A) Multiple molybdenum-ribbon seal (used in lamps Nos. 18 and 19 of Table 8-I). (B) Molybdenum cup seal. (C) Molybdenum disk seal.

drawing of the seals used in the 10-kW lamp of Fig. 8-3. A single-ribbon conductor of this seal type is $1\frac{3}{8}$ in. wide and 0.003 in. thick.

Figures 8-13B and 8-13C show two molybdenum–quartz seal designs that are suitable for very high current and have the advantage of relatively simple construction. They have been made and tested experimentally, but are not in actual use in technical lamps. Both designs are based on the principle of separating the current-carrying and hermetically-sealing elements. This principle is well known and widely used for high current glass-to-metal seals [31]. The current flows through a straight, relatively heavy tungsten rod. In the design of Fig. 8-13B, the hermetically-sealing element is a molybdenum cup with feathered edge; the cup brazed vacuum-tight to the tungsten rod. The feather edge is embedded between two concentric cylindrical quartz surfaces [32]. The design of Fig. 8-13C uses a feather-edged molybdenum disk as the hermetically-sealing element; the disk is embedded, under pressure, between two quartz flanges [8].

In graded glass seals, the expansion difference between metal and quartz is bridged by means of the use of several glass types with graded expansion properties. Usually, a tungsten wire is beaded with a glass that has a smaller expansion coefficient than tungsten, but still produces a satisfactory bond to it. To this glass bead, several rings of special glass types with gradually decreasing expansion coefficients are sealed; the final ring is a glass that allows a satisfactory seal to quartz. The number of intermediate glass types used varies from a minimum of two up to six or even eight. With each additional glass type, the reliability of the metal-to-quartz connection usually increases, while, of course, the process of manufacture becomes more complicated.

Figure 14B shows schematically such a graded seal with relative dimensions suitable for a high-wattage compact arc lamp. Figure 14A is the sche-

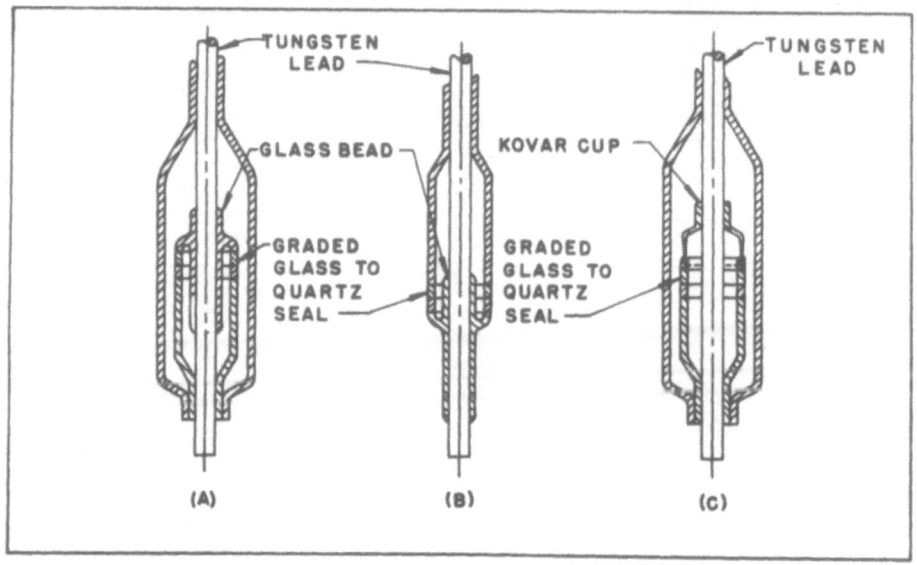

Fig. 8-14. Graded glass metal-to-quartz seals for high-wattage compact arc lamps. (A) Reversed graded glass seal (used in lamps Nos. 2, 6, and 14 of Table 8-I). (B) Direct graded glass seal. (C) Reversed graded glass cup seal.

matic drawing of a so-called reversed graded seal. This design has the advantage that the relatively weak graded glass part of the seal is only subjected to compressive stress by the pressure in the lamp, while of course the regular graded seal of Fig. 8-14B has to withstand a substantial tensile stress. Figure 8-14C is a modification of the reversed graded seal; this modified seal is suitable for extremely high currents. A Kovar cup is brazed vacuum-tight to the tungsten electrode rod, and the first glass of the graded part is sealed to the feathered edge of the cup. This seal design is also suitable for liquid-cooled electrodes (see Fig. 8-15).

Graded glass seals have the advantage of relatively simple construction and low manufacturing cost. However, they usually have limited life, if they reach a temperature higher than 250 to 300°C during lamp operation. They fail not only through oxidation of the outer lead wire end, but also because of internal structural changes. The limited operating temperature of graded glass seals makes them unsuitable for mercury or mercury–xenon high wattage compact arc lamps, since the entire inner bulb volume of such lamps must be kept above a minimum operating temperature of at least 600°C for the high mercury vapor pressure to be maintained. Therefore, only molybdenum ribbon seals are in use for lamps with mercury vapor filling.

Molybdenum-ribbon seals are not sensitive to high operating temperatures, except that the temperature at the outer end must be kept below a certain maximum value (approximately 200°C). Higher end temperature results in early oxidation of the thin molybdenum ribbon leads and eventually in seal failure. In order to protect the outer ends of the ribbon leads from oxidation, one manufacturer uses an additional graded seal (see Fig. 8-1, 2500-W mercury–xenon lamp right; No. 12 in Table 8-I). In some larger-wattage xenon lamps the seals must be forced air cooled so that their

operating temperature will remain within the proper limits. Liquid cooling
of high-current seals does not seem practical, except in conjunction with a
liquid-cooled electrode (see Fig. 8-15).

E. ELECTRODES

The electrodes of all modern high-pressure discharge lamps are made
of tungsten because this is the only metal with a sufficiently high melting
point (3380°C) and a sufficiently low vapor pressure at elevated temperatures. Compact arc lamps absolutely require solid tungsten electrodes, due
to the high wattage concentration and correspondingly high plasma temperature within the short arc gap directly between the electrodes.

In addition to the thermal energy transferred to them from the arc
plasma by conduction, convection, and diffusion of charged particles, the
electrodes are heated by the processes of electric current transfer from
the cathode into the arc column and from the arc column into the anode.
The details of the transfer mechanisms, especially those of the cathode
mechanism, are too intricate to be discussed in this paper. They have been
investigated by several authors [33-37], and it has been established that the
cathodic "hot spot" constriction, which so strongly influences the brightness
distribution of DC arcs, is closely connected with the cathode mechanism.

The heat energy supplied to the electrodes by the current transfer
mechanisms increases essentially in direct proportion to the current. The
cathode receives only a fraction (between one-fifth and one-seventh) of the
energy supplied into the anode. This great difference in received energy
between cathode and anode is explained by the fact that the same substantial
amount ("electronic work function" × "current") is withdrawn from the
cathode and transferred into the anode. With a tungsten work function of 4.5
V, the minimum difference between cathode and anode energy is nine × current. The heat transferred into the electrodes from the arc column largely

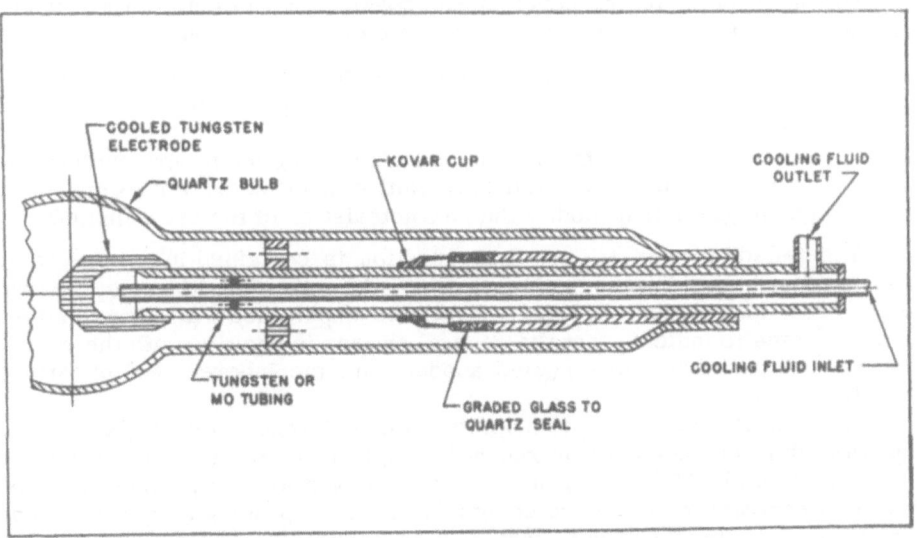

Fig. 8-15. Liquid-cooled anode for high-wattage xenon compact arc lamps.

depends upon their position: the upper electrode receives much more energy than the lower electrode, regardless of which one is the cathode, anode, or AC electrode.

The energy amounts received by the electrodes have to be removed mainly through heat conduction and radiation. Compact arc electrodes are usually so designed that they operated at a high temperature and a large part of their received energy is radiated, while only a small part is conducted away through the support rod and the surrounding gas. The operating temperature is limited, however, by the requirement of keeping the evaporation rate low in order to prevent early blackening of the lamp envelope.

Illustrations to the outlined design principles for compact arc lamp electrodes can be found in the practical lamps shown by Figs. 8-1 to 8-3. It has been mentioned in the first paragraph that stability requirements make it necessary to operate xenon DC lamps always in "anode up" position. Therefore, the anodes of such lamps have to be particularly large, and certain technological difficulties seem to appear in the manufacture of anodes for very large-wattage lamps. There is as yet insufficient practical experience with anodes of, for example, 20-kW lamps, for definite limitations to be stated. However, there are certain indications that it will be rather difficult to find suitable anodes for future lamps above this wattage category. Possibly, it will be found advantageous to use liquid-cooled anodes, as shown by Fig. 8-15. The figure presents schematically the design of a liquid-cooled anode for xenon compact arc lamps of 20, 30, or more kW of input. Such anodes have not yet been tried out in practice, but their use seems feasible, given recent innovations in the field of fabricating and connecting molybdenum and tungsten parts.

F. FUTURE DEVELOPMENTS

The interest in future development possibilities of compact arc lamps for imaging furnaces is mainly concentrated on two questions:

1. Can one develop reliable long-life lamps with useful optical characteristics and reasonable economy for wattages as high as 30, 50, or 100 kW?

2. Can one increase the brightness of existing lamp types without increasing wattage or reducing radiation efficacy, for example, by adding gases that modify the characteristics of the arc column?

A detailed answer to the first question is contained in the preceding paragraphs discussing the problems involved in obtaining larger envelopes, electrodes, and seals. Presently available experience with large-wattage lamps seems to indicate that the general answer is positive with the reservation that possibly water-cooled anodes with the inherent complications will be required.

The second question can be answered by reference to recently obtained results with compact arcs in xenon–hydrogen mixtures [26]. The presence of approximately 35% hydrogen in xenon compact arc lamps causes a pronounced contraction of the arc column, with corresponding temperature and brightness increase and with virtually no change in radiation efficacy and spectral characteristics. Figure 8-8 shows at the right the photograph of such a contracted xenon–hydrogen arc burning at approximately 5-kW input. Comparison with the pure xenon arc of the same wattage, shown in the left part of the figure, clearly demonstrates the strong contraction. The reduction in arc width is due to the particularly high thermal conductivity of dissociated hydrogen in the temperature range prevailing at the edge of the arc column. The addition of hydrogen also reduces the electrical conductivity in the xenon arc column and thus increases the voltage gradient. For example, the 5-kW xenon–hydrogen arc shown in Fig. 8-8 has a 65% higher gradient than the pure xenon arc of the same wattage. Large-wattage arcs in xenon–hydrogen mixtures definitely require a cooled anode, because the top electrode is overheated by the recombination energy of upward streaming hydrogen.

ACKNOWLEDGMENTS

Most of the authors' own results reported in this paper were obtained under contracts and in cooperation with the U. S. Army Engineer Research and Development Laboratories, Fort Belvoir, Va. The authors are greatly indebted to Messrs. S. M. Segal and S. B. Gibson for their advice and suggestions. Dr. O. Lienhard, Hanovia, and Mr. W. S. Till, Westinghouse, have kindly contributed lamp data and pictures.

REFERENCES

1. Rompe, R., and Thouret, W. E., Z. Tech. Physik 17:377 (1936); 19:352 (1938).
2. Hoepcke, O., and Thouret, W. E., Kino-Technik 20 (6) (1938).
3. Francis, V. J., and Wilson, G. H., Trans. Illum. Eng. Soc. (Great Britain) 4:59 (1939).
4. Rompe, R., and Thouret, W. E., Tech. Wiss. Abhanal. Osrem–Ges, 5:44 (1943).
5. Rompe, R., and Thouret, W. E., Licht 14:14, 73 (1944).
6. Aldington, J. N., Trans. Illum. Eng. Soc. (Great Britain) 9:11 (1946).
7. Bourne, H. K., Discharge Lamps for Photography and Projection (Chapman A. Hall, Ltd., London, 1948).
8. Thouret, W. E., Lichttechnik 2:73, 107 (1950).
9. Freeman, G. A., Illum. Eng. 45:218 (1950).
10. Beggs, E. W., Illum. Eng. 53:22 (1958).
11. Schulz, P., Reichsber Physik 1:147 (1944).
12. Schulz, P., Ann. Phys. 1:95, 107 (1947); Z. Naturforsch. 2a:583 (1947).
13. Aldington, J. N., Trans. Illum. Eng. Soc. (Great Britain) 14:19 (1949).
14. Cumming, H. W., Trans. Illum. Eng. Soc. (Great Britain) 16:129 (1951).
15. Anderson, W. T., J. Opt. Soc. Am. 41:385 (1951).
16. Ittig, K., Larche, K., and Michalk, F., Tech. Wiss. Abhanal. Osram–Ges, 6:33 (1953).
17. Thouret, W. E., and Gerung, G. W., Illum. Eng. 49:520 (1954).
18. Anderson, W. T., J. Soc. Motion Picture TV Engr. 63:96 (1954).
19. Cumming, H. W., British Kinematography 28:5 (1956).
20. Schulz, P., and Stubert, K. F., Lichttechnik 8:254 (1956).
21. Ulffers, H., Zeiss Ikon Bild und Ton, No. 36:3 (1954); Kino-Technik 10:268 (1956); Z. angew. Phot. Wiss. u. Tech. No. 10 (October, 1956); J. Soc. Motion Picture TV Engr. 68:389 (1958).
22. Retzer, T. C., and Gerung, G. W., Illum. Eng. 51:745 (Nov. 1956).
23. Lienhard, O. E., and McInally, J. A., Illum. Eng. 57(3):173 (March 1962).
24. Thouret, W. E., and Strauss, H. S., Illum. Eng. 57(3):150 (March 1962).
25. Breeding, H. A., "Control and Application of the 5kW Xenon Compact Arc Lamp," paper presented at the National Technical Conference of the Illuminating Engineering Society, Dallas, September 1962.
26. Thouret, W. E., and Strauss, H. S., "Xenon Compact Arcs with Increased Brightness through Addition of Hydrogen," paper presented at the National Technical Conference of the Illuminating Engineering Society, Dallas, September 1962.
27. I. E. S. Lighting Handbook (Third Edition, Illum. Eng. Soc., New York, 1959); Fig. 8-88, p. 8-801.
28. Segal, S. M., Illum. Eng. 50(5):259 (May 1955).
29. Johnson, F. S., J. Meteorol. 11:431 (1954).
30. Thouret, W. E., Illum. Eng. 55(5):295 (May 1960).
31. Housekeeper, W. C., J. Am. Inst. Elec. Eng. 42:954 (1923).
32. Greiner, A., U. S. Patent 2,504,521 of April 18, 1950.

33. Thouret, W. E., Weizel, W., and Guenther, P., Z. Physik 130:621 (1951).
34. Weizel, W., and Thouret, W. E., Z. Physik 131:170 (1952).
35. Bauer, A., Z. Physik 138:35 (1954).
36. Bauer, A., and Schulz, P., Z. Physik 139:197 (1954).
37. Finkelnburg, W., and Maecker, H., Encyclopedia of Physics, Vol. XXII (Springer-Verlag, Berlin) p. 403.

Instrumentation

and Measurement Techniques

Chapter 9

Spectroradiometric Instruments and Techniques for Use in Imaging Furnaces

Leonard Eisner, Donald W. Fisher, and Richard F. Leftwich
Barnes Engineering Company
Stamford, Connecticut

A. INTRODUCTION

We have built two different types of instruments to measure the temperature of materials heated in an imaging furnace. The first instrument has two channels, one to measure total radiation and the other to measure spectral radiance between 0.25 and 2.6 μ. Some preliminary measurements were made of alumina, zirconia, and other materials heated in an arc furnace.

The second system is an ultraviolet radiometer that appears capable of indicating sample temperatures with ±2% accuracy over a range of 1000 to 2500°C, despite emittance variations as great as 25% from the mean value for which the instrument is calibrated.

The range of measurable temperatures and the accuracy that can be achieved are limited by the characteristics of the furnace, the radiometers, and the measurement techniques. We shall discuss some of these limitations and indicate how they can be minimized.

B. OPERATING PRINCIPLES

The spectroradiometer employs an adjustable reflecting system to collect target radiation. Part of the time this radiation is directed to a total radiation channel, where it strikes a detector that has previously been calibrated against a blackbody. The rest of the time, the radiation goes to a monochromator. The dispersed radiation can be received by one or more detectors for the ultraviolet, visible, and infrared regions of the spectrum. By calibrating these detectors with standard sources, one can determine the absolute spectral radiance of a sample. This radiometer has been used to measure the radiance of samples heated in an arc imaging furnace.

Under certain assumptions, we may also deduce the sample's reflectance, emittance, and temperature, in the following way. At suitable intervals, the sample is replaced by a water-cooled copper block that is coated with magnesium oxide of known spectral reflectance. This procedure permits a determination of the normal sample reflectance, both total and spectral, if we assume the irradiance by the arc to be constant. For an opaque sample, the spectral emittance, e_λ, is equal to $1 - r_\lambda$, where r_λ is its spectral reflectance. In using this relationship, we make an implicit assumption, viz., that the near-hemispheric illumination of the sample and the MgO by the arc reimaging mirror yields a reflectance in the normal direction that corresponds to that called for theoretically in Kirchhoff's law. The latter relates hemispherical emittance to hemispherical reflectance. The normal values of these quantities may have a different relationship, which in turn may not be the same for the sample as for the MgO reference. Nevertheless, by making appropriate assumptions about the directivity of the reflected radiation, one can sometimes infer the value of the hemispherical reflectance from the normal reflectance value.

The radiometric principle outlined above has been applied by other investigators ([1]) to measure the temperature and reflectance of hot samples. However, they have used filters to achieve the spectral isolation required for the application of Kirchhoff's law, whereas we have used a monochromator.

C. THE SPECTRAL AND TOTAL RADIOMETERS

1. Description of Components

Figure 9-1 is a top view of the instrument layout. A pair of spherical mirrors alternately focused an image of the sample on two slits. The difference in the focal length of the two spherical mirrors resulted in an image demagnification by a factor of approximately two-thirds. The diameter of the mirrors was chosen to match the monochromator speed, which varied between f/5.5 and f/7, depending upon prism setting. A time-share chopper diverted the converging beam periodically from the monochromator entrance slit to the total channel entrance slit. This channel chopper consisted of a 150-rpm gearhead motor driving a four-point Geneva wheel, to which was

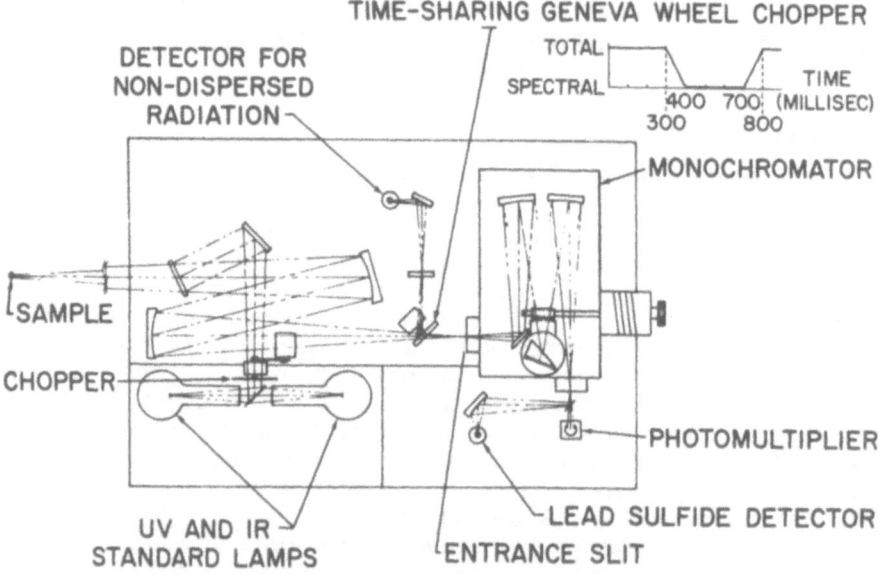

Fig. 9-1. Optical schematic of spectral radiometer.

attached a two-bladed mirror chopper. This technique resulted in a dwell time of 300 msec in each channel, with a transition time of 100 msec.

The monochromator was a Leiss instrument, for which we chose a fused quartz prism to cover the desired range from 0.25 to 3 μ. A manually operated drum was used to select desired wavelengths, but we have also built an automatic drive and wavelength pipper to scan the spectrum. Adjustable entrance and exit slits permitted compensation for the incident radiation level. For typical entrance- and exit-slit widths of 30 μ and 40 μ, respectively, $\Delta\lambda$ was 5.8 mμ at $\lambda = 0.65$ μ. These slit widths provided more than adequate signal-to-noise ratios when the reflected radiation from the arc was viewed. By still further reduction of the slit width, a limiting resolution of 1.7 mμ could be achieved.

Upon leaving the exit slit, the dispersed radiation entered a shielded photomultiplier, such as the 1P28, which we used for the 0.25 to 0.5 μ region; it could also be reflected off a small, flat mirror onto an off-axis ellipsoidal mirror, whence it fell on a lead sulfide detector.

A slit similar to the entrance slit of the monochromator was placed in the first focal plane of the total radiation channel. This slit could be adjusted so that the field-of-view of the total channel was the same as that of the spectral channel, or its width and height could be reduced to prevent saturation of the detector.

Various detectors can be used in the total radiation channel. However, it is preferable that they have a response time of approximately 1 msec or less, since the sample's emittance has to be measured during the 2.5-msec interval when a blocking shutter interrupts the arc's radiation. Since there is plenty of signal, somewhat slower detectors could be used, if there were appropriate attenuation and shaping of the frequency response of the amplifiers. We found a lead sulfide detector to be convenient, because of its rapid response. Also, the use of detectors with similar spectral characteristics in the total and spectral channels facilitated comparison of their outputs, at least in the infrared region.

The lead sulfide detectors we used had time constants of approximately 250 μsec. The bandwidth of the preamps (2 cps to 1000 cps) was adequate for use with the furnace. A 38-V bias was applied to the spectral lead sulfide detector, vs. 4 V for the total radiation detector. With 4-V bias, the lead sulfide cell saturated at a slightly higher incident radiation than the preamplifier, so the maximum available dynamic range was utilized.

2. Calibration

Although blackbody sources are available for calibration use in the infrared, their output is marginal or useless at shorter wavelengths. It was therefore decided to use a pair of tungsten ribbon filament lamps, which were calibrated against high-temperature blackbodies, in terms of absolute spectral radiance, at the National Bureau of Standards. Figure 9-2 shows such a lamp. For the 0.5 to 2.6 μ range, one lamp was used at 25 or 30 A AC. For 0.25 to 0.75 μ, the other lamp was used at 35 A. Ammeters accurate to 0.2% were used for setting of the currents. At 25, 30, and 35 A, the lamp filament temperatures are typically about 1950, 2100, and 2450°K.

The lamp radiation passes out through a quartz window. By flipping a mirror, one can select either the ultraviolet or the infrared standard lamp.

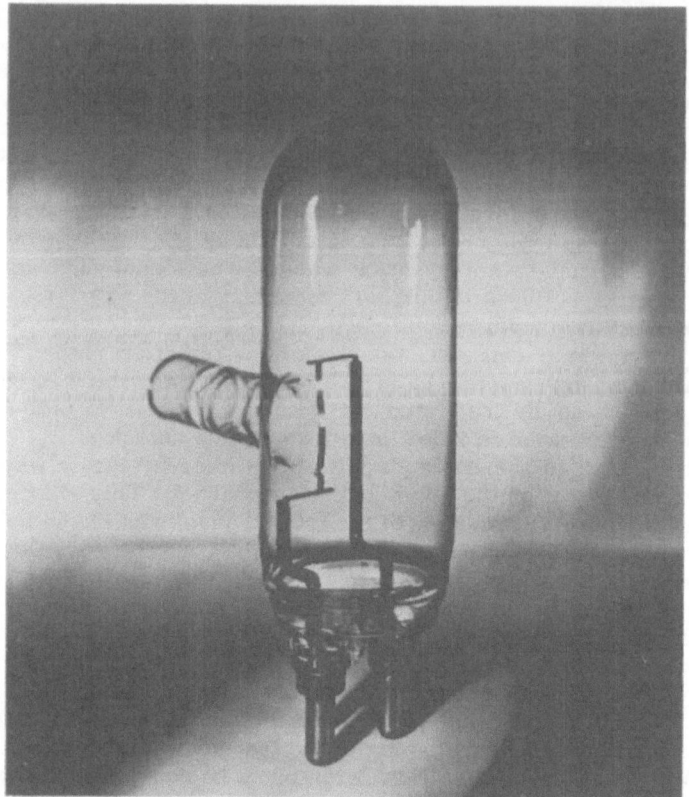

Fig. 9-2. Spectral radiance standard lamp.

Following a 30-cps chopper, the beam falls upon two flat mirrors and then onto the same pair of spherical mirrors used to focus the sample radiation. The second flat mirror can be removed when the furnace is to be operated, and is replaced by a pinning arrangement when a calibration is made.

Wavelength calibration of the monochromator was accomplished conventionally with mercury and cadmium sources. For occasional checks, it was convenient to insert special glass filters, such as didymium and holmium oxide, in the path of the radiation from the standard lamps.

While the wavelength calibration was quite stable over long periods of time, the amplitude calibration, relating incident radiation to detector output voltage, had to be repeated fairly frequently. The gain of the amplifying system following the detector could be checked by application of a precisely known electrical input to the preamplifier. We used a scope to display the radiometer output, but a fast recorder could also have been used.

3. Electrical and Mechanical Components

Figure 9-3 shows the arrangement of the radiometer system and its associated mechanical and electrical parts. The optical section discussed previously was mounted on a flat plate, provided with a cover, atop a mobile cart. Directly under this plate were mounted the preamplifiers for the

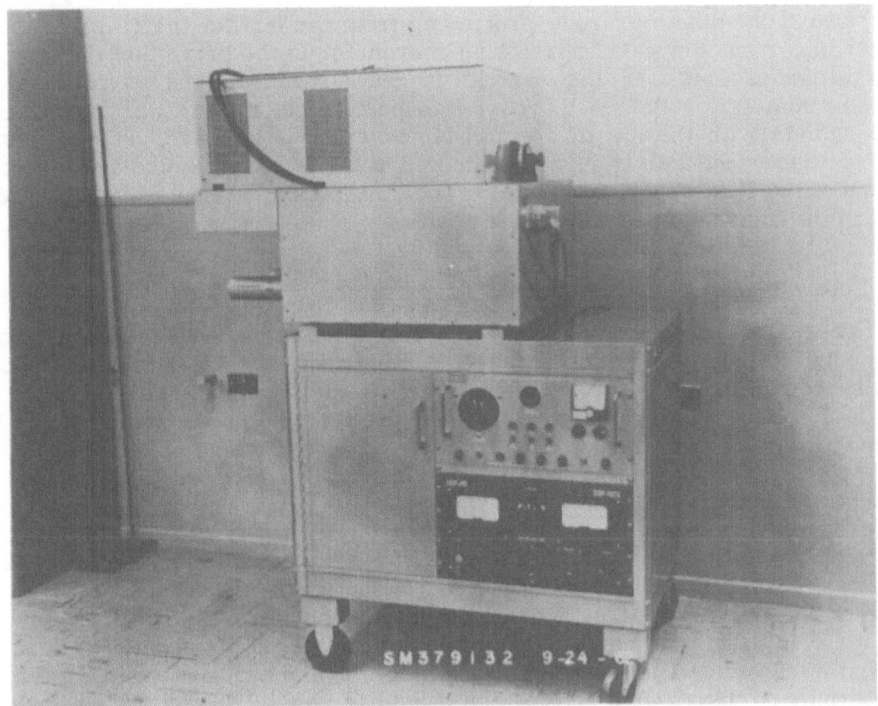

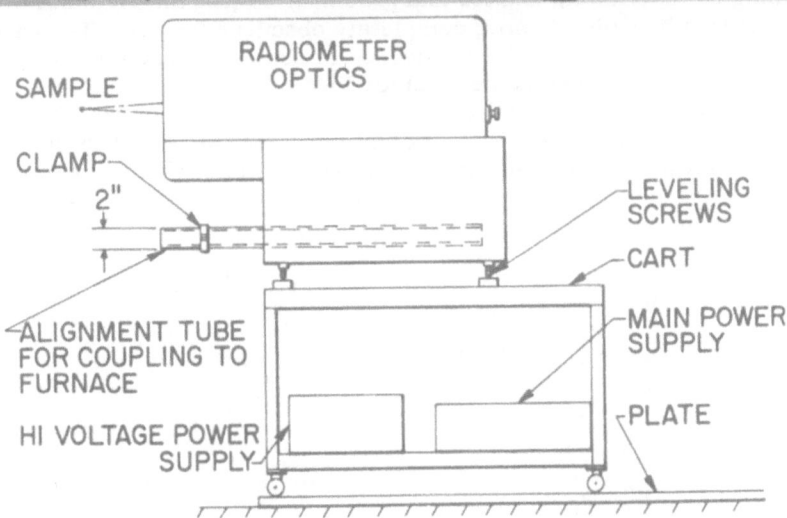

Fig. 9-3. Side view of radiometer card used with imaging furnace.

detectors.   On the bottom shelf of the cart were the power supply for the standard lamps, the plate and filament supply for the amplifier, detector bias, and amplifier calibration signal, as well as the high-voltage supply for the photomultiplier.   There was also a panel containing the various switches and controls for operating the choppers, calibration signals, and current for the standard lamps.

Two 2-in.-diameter rods protruding from the furnace mated with tubes extending from the cart, permitting movement of the cart while maintaining alignment between the sample in the furnace and the optical axis of the radiometer. Before the cart was positioned, the arc was focused on the sample with the aid of a small lamp in place of the arc. For use with the furnace, the cart was approximated until the sample was in the focal plane of the first spherical mirror. Then the rods and tubes were clamped, fixing the instrument in the proper position.

4. Determination of Emittance and Temperature

There is a chopper at the end of a rotating arm between the ellipsoidal mirrors in the ADL-Strong furnace. It blocks the incident radiation once every revolution, for about 6 msec of the 30-msec period. There is also a variable-phase viewing shutter between the sample and the radiometer. It consists of a disc with a hole near its edge, and the sample can be seen through the hole for about 2.5 msec, once each revolution. Both choppers and viewing shutter rotate at 1800 rpm.

To measure the amount of arc radiation reflected from the sample, one leaves the viewing shutter in the open position and turns on only the chopper shutter. When the detector output is presented on an oscilloscope or recorder, a pulse appears, and its amplitude indicates the reflected-plus-emitted radiation minus the emitted radiation, or simply the reflected radiation (provided linearity is maintained). One measures emitted energy by turning on the viewing shutter and phasing it so that the 2.5 msec viewing time occurs when the chopper completely obscures the arc. The base line of the resulting pulse can be considered to be zero, since the temperature of the viewing shutter can be kept rather low.

To permit measurement of the reflected radiation and the emitted radiation in rapid sequence, three more holes were cut in the viewing shutter. They were spaced 90° apart and were identical to the single, original

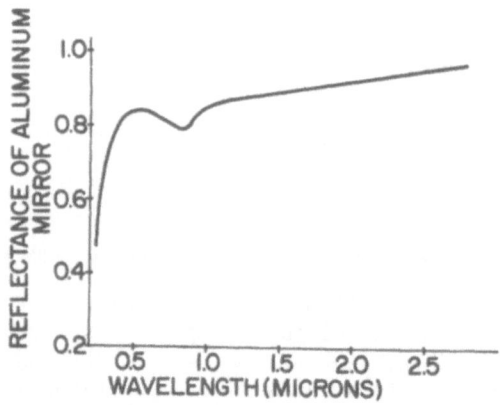

Fig. 9-4. Spectral reflectance of a typical mirror used in calibration.

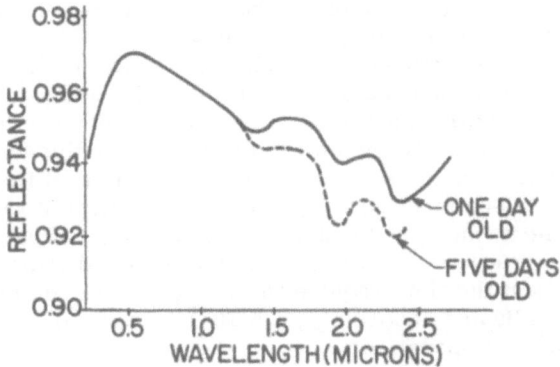

Fig. 9-5. Spectral reflectance of a MgO coating.

viewing hole. This step resulted in three pulses of reflected-plus-emitted radiation, and one pulse of emitted radiation for each revolution.

To deduce the absolute radiance of the sample by comparing its signal with that of the standard lamps, we must allow for the reflective attenuation caused by the three flat mirrors that are used in the lamp path, but not in the samples (see Fig. 9-1). Figure 9-4 shows the measured spectral reflectance of one of our aluminized mirrors. The required correction for the reflection losses of the three mirrors was obtained by cubing the ordinates of this curve.

Next it was necessary to convert the reflected-radiance data into a reflectance value for the sample. This was done by replacing the sample with a magnesium-oxide-coated, water-cooled copper block. The exchange was made as rapidly as possible, so that the effects of variations in the average output of the arc would be minimized. ADL designed a special arrangement which permitted this interchange to be made in a second or so and in such a way that there was no change in the arc obscuration.

Since the magnesium oxide was kept cool, it did not emit significantly. Its spectral reflectance has been carefully measured over the range of wavelengths of interest here, 0.25 to 2.5 μ. Among those who have made such measurements are Sanders and Middleton ([2]). A typical reflectance curve of magnesium oxide based on their data is shown in Fig. 9-5. The curve also shows the deterioration in reflectance as the film ages, especially for wavelengths longer than 1.3 μ. With the limited accuracy of our present instrumentation, such a change is not significant.

Of more significance are any differences between the sample and the magnesium oxide with respect to the geometrical distribution of the reflected radiation. Strictly speaking, the sample should be a diffuse reflector having angular reflectance characteristics the same as those of the magnesium oxide reference. For example, where there was a highly specular reflecting surface, our radiometer would see little reflected arc radiation, because of obscuration at the first ellipsoidal collecting mirror and at the hole in the reimaging mirror, both of which prevent the incidence of radiation normally on the sample. Consequently, one might assume that $r = 0$, and $e = 1$, whereas in reality e is very small.

When the reflected radiance is much greater than the emitted radiance, the latter cannot be measured readily, because its pulse amplitude is distorted by the large saturating pulses of the emitted-plus-reflected radiation that occur before and after the emitted radiance. In this case, the single-hole, viewing shutter must be used to measure emitted radiation, and the chopping shutter alone must be used to measure reflected radiation. In the latter measurement, the viewing shutter is left open, and one can follow the effects of arc variation. The output of the total radiation channel can be recorded throughout the time required for the spectral scan, and the spectral radiation can be adjusted accordingly. A 10-15% variation in arc output was typical during normal operation.

D. EXPERIMENTAL RESULTS

The main purpose of our work was to develop the instrument described above. However, a limited number of measurements were made as a check on its performance.

We irradiated the magnesium oxide sample first with the arc and then with the standard lamp, and compared the two radiances obtained. This led to the curve shown in Fig. 9-6, the normal radiance of the reflected arc radiation as a function of wavelength. The spectral radiance could not be measured reliably in the ultraviolet because of noise problems in the prototype system. The output of the UV standard lamp is quite low, and although it is chopped and the photomultiplier output could have been synchronously rectified to give a much improved signal-to-noise ratio, we were not asked to provide anything more than the raw detector output for this first model.

The arc furnace is designed to produce high temperatures. Therefore, refractory samples such as alumina and zirconia were used for our initial experiments. In principle, it is only necessary to know the emittance and radiance of the sample at one wavelength in order to determine its temperature. We made measurements at a number of wavelengths, however, to determine the consistency of the calculated temperatures. Ideally, of course, the temperatures should have all been the same. Table 9-I shows that this is not the case.

Measurements were made on both solid and molten alumina and zirconia, the temperature settings being achieved by adjustment of the size of the cone of radiation incident on the reimaging mirror, as well as by use of other furnace controls. For repeated measurements with the same furnace settings, the data were quite reproducible. We believe the calculated temperature differences are due to non-Lambertian reflection and scattering by the sample, or to differences between the sample and MgO in this respect, and perhaps to variations in the reflectance and emittance characteristics with wavelength. Melting certainly causes pronounced changes in the angular distribution of the arc radiation that is reflected and diffusely scattered by the sample.

The tabulated data have been presented graphically in Figs. 9-7 and 9-8, which show, respectively, the calculated normal spectral emittance of alumina and zirconia and the temperature dependence of the emittance at

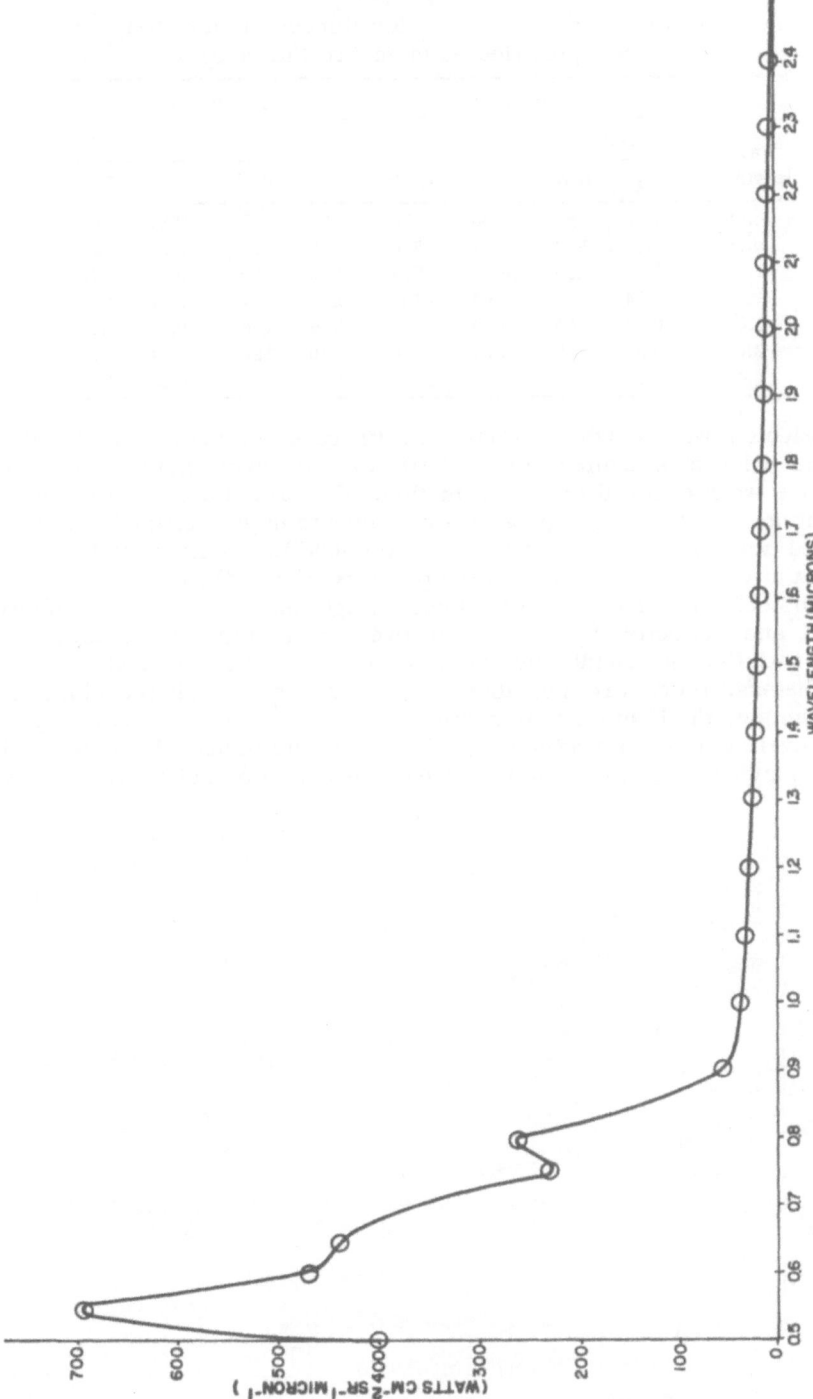

Fig. 9-6. Normal radiance of carbon arc radiation reflected from MgO.

TABLE 9-I

Temperatures and Spectral Emittances Calculated for
Samples Heated in an Arc Furnace

| Wave-length, μ | Alumina | | | | Zirconia | | | |
| | Solid | | Liquid | | Solid | | Liquid | |
	e_λ	$T,\,°K$	e_λ	$T,\,°K$	e_λ	$T,\,°K$	e_λ	$T,\,°K$
0.65	0.48	—	0.93	2800	0.82	2250	0.85	3050
0.80	0.43	900	0.91	2800	0.78	2350	0.85	3150
1.0	0.37	880	0.91	2650	0.76	2000	0.84	2750
1.2	0.34	880	0.93	2400	0.77	1975	0.82	2600
1.6	0.34	900	0.93	2250	0.75	1800	0.84	2475
2.0	0.32	850	0.91	2100	0.70	1800	0.84	2400
2.4	0.21	900	0.86	2450	0.57	1950	0.82	2600

two selected wavelengths. Comstock [3] gave an emittance of 0.82 for zirconia brick at a temperature of 2540°K and a wavelength of 0.65 μ. The emittance we measured agrees with this. For zirconia at 3200°K, he gave an emittance of 0.73. This presumably refers to molten zirconia, since the accepted melting point is approximately 3000°K. Our value for molten zirconia at 0.65 μ, at a calculated temperature of 3050°K, is 0.85.

Comstock's arrangement, which uses a light pipe to gather the radiation emitted and reflected by the sample over most of the hemisphere, more nearly satisfies the conditions for rigorous application of Kirchhoff's law. Nevertheless, there are possible causes of error in his technique also. For example, the light pipe attenuates oblique radiation more strongly than that traveling along its axis [4], and this results in nonuniform weighting of the incident radiation. It would be interesting to make measurements

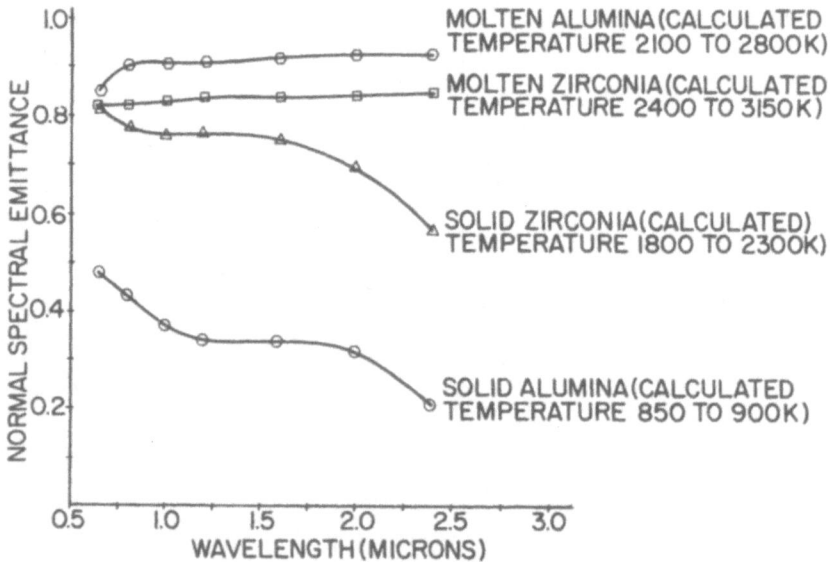

Fig. 9-7. Spectral emittance of alumina and zirconia.

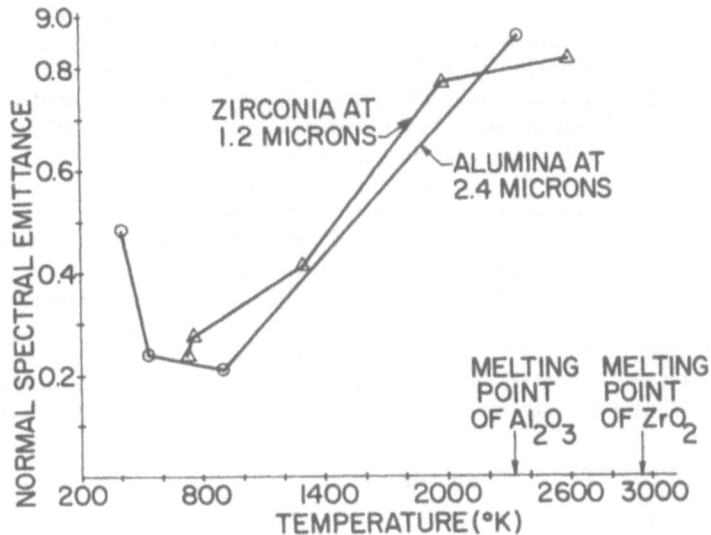

Fig. 9-8. The temperature dependence of the emittance of alumina and zirconia.

with Comstock's system, but with interference filters at considerably longer wavelengths than 0.65 μ, to see if there would be temperature differences similar to those we found.

Some measurements were made of alumina and zirconia at lower temperatures, as well as of fiberglass cloths, laminated honey-comb structures, and other structural materials. In general, the materials all exhibited emittance values which decreased with longer wavelengths. As the temperature increased, the emittance increased in the case of zirconia, but it decreased for alumina, except at melting, when it increased considerably. No general conclusions could be drawn for the temperature dependence of the emittance of other materials.

The temperature values shown in Table 9-I vary by no more than about ±15% from the mean values. At lower temperatures, however, the variation sometimes amounted to two or even three times this much. Again, this result is believed to be due to lack of a satisfactory means of reducing the incident arc energy without adversely affecting the geometry.

By way of illustrating this point, between the chopping shutter and the arc, there is a douser, which is normally used to completely block the incident radiation on the sample. When it was adjusted to let through only part of the beam, so that lower temperatures would be obtained, the illumination geometry was drastically changed. Furthermore, part of the energy reflected by the sample was found to be coming from the hot douser. This error was especially severe at the lower temperatures, when the douser intercepted a large amount of energy.

A more manageable method of attenuation was to limit the radiation cone incident on the reimaging mirror. A plate with a small circular aperture in the center was attached to a plate with a larger circular aperture, the latter being a permanent part of the furnace. By using different aperture sizes and moving the disc to or fro, we could obtain a number of different temperatures. When reflected radiation was being measured, radiation due to the hot aperture plate was not chopped, and it contributed no error; but when emitted radiation was being measured, an error resulted, because the aperture plate radiation was being chopped by the viewing shutter. This error was greatly decreased by insulation of the aperture, but it may still have been significant at lower temperatures and for low sample emittances.

Some other possible sources of errors may be mentioned here. The size of the area of the sample whose radiation is measured is determined by the entrance slit dimensions. For a slit of appreciable length, there may be an appreciable temperature gradient. However, measurements made with successively lower slit heights showed relatively little change in the calculated temperatures. In spectral regions where the available energy is less, it may be necessary to use wider slits, which will result in the use of a larger, and perhaps less uniform, sample area. Greater slit widths also mean decreased spectral resolution, but the emittance of solid materials changes rather smoothly with wavelength, so this effect is unimportant.

E. ULTRAVIOLET RADIOMETER

1. Basic Principles

For determination of the temperature of a sample by the spectroradiometric method discussed above, it was essential to know its emittance. Uncertainty in this factor was the greatest cause of error. However, if sample radiance can be measured at sufficiently short wavelengths, uncertainties in sample emittance have relatively little effect on the calculated temperature. The idea is illustrated in Fig. 9-9. Suppose that a sample shows a radiance of $1.5 \cdot 10^{-2}$ W/cm^2-sr-μ at a wavelength of 11.2 μ, but that its emittance is not known exactly. This radiance value might apply to a blackbody at a temperature of 750°K. On the other hand, if the emis-

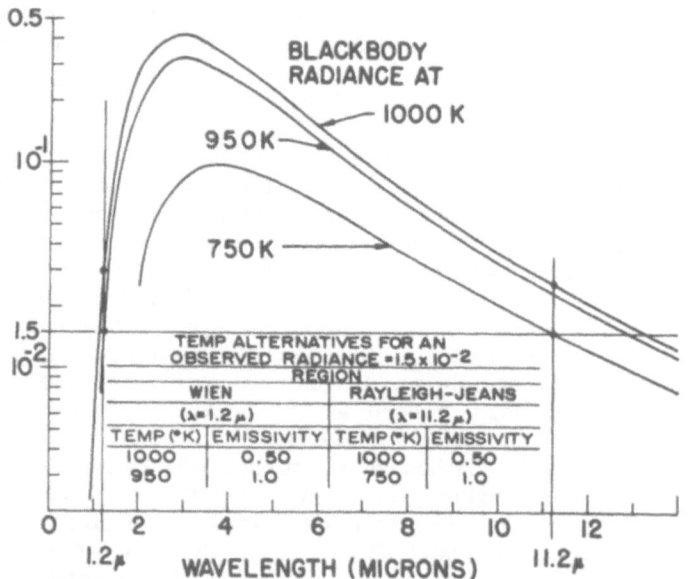

Fig. 9-9. Advantage of Wien vs. Rayleigh–Jeans radiometry.

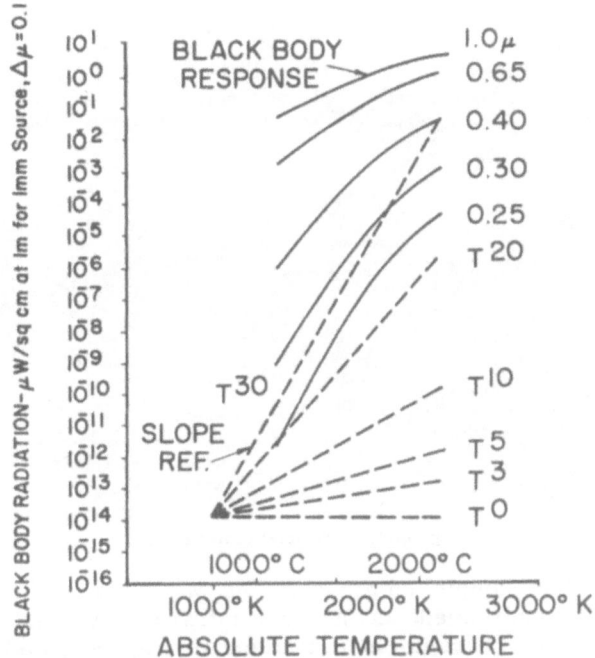

Fig. 9-10. Characteristic slopes of blackbody radiation curves.

sivity is only 0.58, the sample temperature would have to be 1000°K. Now compare these possible temperatures with those deduced for the same radiance observed at 1.2 μ. Unit emissivity now corresponds to a sample temperature of 950°K. However—and this is the important point—because of the steepness of the blackbody curve in this wavelength region, an emissivity of 0.5 would call for a temperature of 1000°K, i.e., one only 50° higher than that needed for a blackbody of the same radiance.

The same idea is shown in a different way in Fig. 9-10. At a fixed wavelength, and over a small temperature range, the radiance of a blackbody varies with temperature in a way that can be expressed as $N = A e T^n$, where A is a constant that can be expressed in terms of wavelength and the radiation constants. Using Wien's displacement law, $\lambda_{max} T$ = constant, and writing Planck's formula in the form

$$N_\lambda = (C_1/\pi) (\lambda_{max} T)^{-5} [\exp (C_2/\lambda_{max}T) - 1]^{-1} T^5$$

we see that for $\lambda = \lambda_{max}$ corresponding to some T, N_λ varies as T^5. For shorter wavelengths, the exponent increases rapidly.

The temperature range of samples heated in an arc imaging furnace is so great that λ_{max} is of the order of 1μ or less. Accordingly, to get a sufficiently high value of n, N_λ should be measured in the ultraviolet. With due regard to available signal, the shorter the wavelength, the higher the value of the exponent of T, and the less the effect of uncertainty in sample emittance.

With somewhat lower temperatures, the same principle can be used at visible wavelengths. This, in fact, is the reason that temperatures determined with an optical pyrometer are rather insensitive to the exact value

Fig. 9-11. Ultraviolet radiometer.

assumed for the sample emittance. As we go to still cooler samples, measurements can be made in the near infrared with similar advantages. To generalize the concept, as illustrated in Fig. 9-9, we have therefore used the term Wien radiometry (rather than ultraviolet) to indicate that the wavelengths of observation are sufficiently short for Wien's law to be applicable. By the same token, for wavelengths considerably longer than λ_{max}, the term used is Rayleigh–Jeans radiometry.

2. Description of Instrument

Figure 9-11 shows the ultraviolet radiometer, which consists of an optical head and an electronics package. The optical head, containing a field lens and a photomultiplier, is shown schematically in Fig. 9-12. For the 0.32 to 0.38 μ spectral region, which is isolated by the filters, the radiance varies as T^{15} at the high-temperature end of the instrument's range (2500°C) and as T^{35} at the lower end (1000°C). The instrument as shown is focused on a point about 10 cm in front of the objective, with a field-of-view approximately 3 mm in diameter.[*]

To minimize problems in reading radiance values over the large range that the detectors must handle, we used an electronic compression system, although many other arrangements are possible to meet special requirements. In the present system, a closed-loop amplifier controls the photomultiplier voltage in such a way that the anode current is kept constant. For a given sample emittance, the photomultiplier voltage is a unique function of sample temperature.

Because the radiant power falling on the phototube is very much more sensitive to changes in T than to emittance changes, relatively large variations in emittance cause only small errors in T. At 1000°C, for example,

[*]A comprehensive description of this instrument, its design and applications, has been published as a technical report by the Aeronautical Systems Division at Wright-Patterson AFB. It is available as ASTIA report AD 274-794.

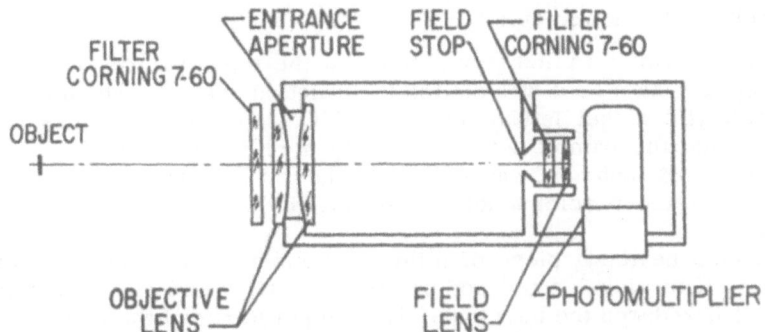

Fig. 9-12. Optical system schematic for ultraviolet radiometer.

the emittance can change by ±50% from the value for which the instrument was calibrated, with a resultant inaccuracy of only ±2% in T. At 2500°C, ±25% uncertainty in emittance corresponds to only ±2% uncertainty in T.

The anticipated performance was verified by tests with the spectral radiance standard lamps. Figure 9-13 shows the radiometer's ability to measure the temperature of tungsten filaments operated at known temperatures. The meter scale was calculated from theoretical considerations. Between 1250 and 2200°C, the temperature indications are accurate within +2 and −3%.

Temperature readings were also taken with and without the interposed metal screens that effectively change the emittance of the lamp filament without changing its temperature. For a screen that transmitted 50% of the incident radiation, the indicated lamp temperature changed by 3.5% at $T = 1400$°K and by 4.7% at $T = 2400$°K.

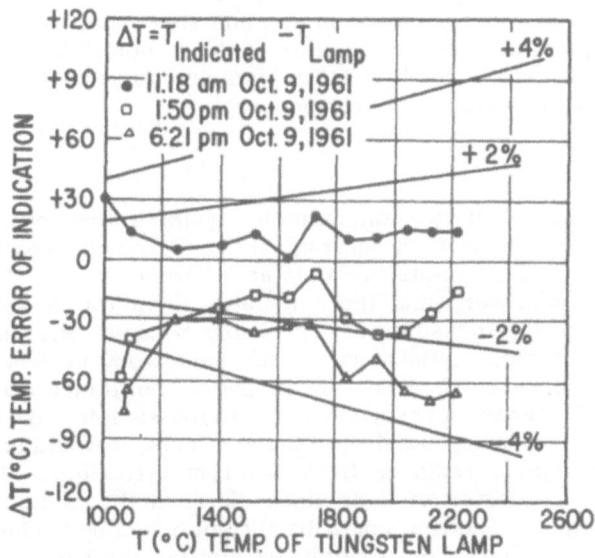

Fig. 9-13. Panel meter indicated-temperature error curve.

3. Results of Measurements

The ultraviolet radiometer monitored the temperature of a sample of electrolytic nickel as it was melted by induction heating. The melting point of electrolytic nickel is known to be 1455°C. In each of six measurements, during which the sample was alternately melted and allowed to solidify, a temperature of 1505 ± 5°C was indicated. The accuracy was therefore greater than 3%, but the precision (or reproducibility) was an order of magnitude better.

We also heated a piece of mild steel and a $\frac{1}{2}$-in. nickel rod in the arc furnace. The nickel could not be melted until it was placed in a ceramic holder that reduced the heat loss. The temperatures of the molten samples were higher than the handbook melting points, and the temperature of the samples before melting were lower than the listed melting points. Thus, for example, two different measurements on the nickel sample when it had slightly different geometry gave indicated temperatures of 1226 and 1303°C just before the sample melted vs. the accepted value of 1455°C. A temperature 150°C above the melting point of nickel was read at what appeared to be the most intense spot in the molten pool.

A small conical hole was drilled in the nickel rod, about a millimeter or two below the spot where melting occurred, in an attempt to learn the effects of increasing the sample emittance. Difficulty in controlling the furnace heating rate to maintain a stable equilibrium condition at the liquid-solid boundary made it impossible to draw any conclusions from this experiment.

For a mild steel sample, a temperature 14°C less than the melting point (1482°C) was read just outside the molten zone created by the arc furnace. Temperature readings on what appeared to be the most intense spot of the molten surface were repeatable within about 10°C of each other, but they were all about 130°C higher than the assumed melting point of the sample. These experiments were all made in air, and slag could sometimes be seen flowing from the fluid region. In view of the complex thermal effects that may occur when iron is melted in an oxidizing atmosphere, the results of the preliminary measurements are encouraging.

F. CONCLUSIONS

We have described two radiometric systems that can be used with imaging furnaces. The first instrument measured the spectral radiance and emittance of the sample, as well as its temperature. Our measurements showed, however, that the calculated temperature varied with the wavelength used for measurement, probably because of anisotropy in illumination and sample reflectance. Such anisotropy causes errors when Kirchhoff's law is used to deduce hemispheric emittance values from radiances and reflectances measured normal to the sample surface.

To gain a better understanding of these effects, it would be desirable to measure the sample's radiance from different directions. This might be accomplished by a series of light pipes distributed at a number of angles around the sample. Viewing could be accomplished by having small holes in the reimaging mirror, or possibly by slotting the latter and rotating it about its axis. Another approach would be to employ an instrument combining features used by ADL and ourselves. A hypodermic light-pipe arrangement would be used to feed a signal to our spectroradiometer. In any case, there is a need for further experimental studies of the angular relationship of reflectance, emittance, and absorptance of samples under various conditions.

The second instrument we have described is an ultraviolet radiometer which can measure temperature quite accurately, even when there is considerable uncertainty as to the samples' emittances. We have called the principle involved Wien radiometry. The same general principle can also be used for measurements in the infrared, when cooler samples are dealt with.

One of the main problems with the arc furnace is that there is no satisfactory way of varying the irradiance level. The usefulness of the furnace would be greatly enhanced if samples could be studied at, say, temperatures down to 500°C, while the illumination geometry remained fixed.

Arc furnaces have been used with samples under vacuum conditions or in a controlled atmosphere with the aid of a quartz or vycor enclosure. In some cases, it might be possible to make controlled atmosphere studies, with less chance of optical perturbations, by isolating the sample behind a high-velocity gas curtain, or even enclosing the entire furnace chamber.

When the instrumentation and techniques have been developed sufficiently, automated reduction of the data from the spectral radiometer, and therefore continuing and direct readout of sample reflectance, emittance, and temperature, will be possible. We do not believe that this effort can be justified at this time. However, despite the errors and imperfections in the instruments we have described, we believe that they are able to give useful data and that continued investigation will greatly improve their accuracy and reliability.

ACKNOWLEDGMENTS

Among the numerous Barnes Engineering Company personnel who participated in work reported here, we should like to mention especially William C. Moore, who made most of the experimental measurements with the spectral radiometer, and Morris Weiss and Guy Moffitt, who together did most of the work on the ultraviolet radiometer. Development of the ultraviolet radiometer was supported by Contract AF 33(616)-7479 with the Aeronautical Systems Division of Wright-Patterson Air Force Base.

REFERENCES

1. Comstock, Daniel F., Jr., "Methods for Temperature and Reflectance Determination in an Arc-Imaging Furnace," in Temperature—Its Measurement and Control in Science and Industry, Vol. 3, Part 2, pp. 1063-1071 (Reinhold Publishing Corp., New York, 1962). References to additional work along these lines may be found in this paper.
2. Middleton, W. E. Knowles, and Sanders, C. L., JOSA 41:419 (1951) and 43:58 (1953).
3. Comstock, op. cit., p. 1068.
4. Potter, Robert J., JOSA 51:1079 (1961).

Chapter 10

Measurement of Spectral Reflectance and Emissivity of Specular and Diffuse Surfaces in the Carbon Arc Image Furnace*

M. R. Null and W. W. Lozier

Research Laboratory, National Carbon Company
Division of Union Carbide Corporation
Parma, Ohio

A. INTRODUCTION

Basically, there are two ways of measuring the spectral emissivity of the surface of an opaque body. One widely used method involves the direct measurement of the spectral radiance of the surface along with a determination of its temperature. Combination of these data with the blackbody radiation laws yields the spectral emissivity. Another method measures the spectral reflectance of the surface. From this, the spectral absorptance is calculated and, by application of Kirchhoff's law, is equated to the spectral emissivity. Both of these methods, which we shall designate respectively as the "radiation" and "reflection" methods, have their limitations, which will receive attention in this paper.

The principal difficulty with the "radiation" method is the determination of the correct temperature of the radiating surface and strict exclusion of reflected radiation. These conditions become increasingly difficult to maintain at higher temperatures and with materials of lower thermal conductivity. The "reflection" method has the advantage that it does not require determination of the temperature; in fact, the emissivity determined under this method can be combined with the radiance temperature to provide an accurate measure of the true surface temperature. The most serious problem with the "reflection" method is the separation of reflected from emitted radiation, and it assumes ever greater proportions at higher temperatures. This paper deals with the application of a carbon arc image furnace to the "reflection" method for determination of the emissivity of opaque materials.

The use of a double ellipsoidal mirror carbon arc image furnace ([1]) to measure emissivity of various opaque materials has been previously described ([2-4]). This paper describes measurement and calibration techniques used for evaluation of surfaces which may be specular, diffuse, or a combination of both. Data are presented on the spectral emissivity of polished and roughened surfaces of both manufactured carbon and graphite. The temperature and wavelength dependence of the spectral emissivity is also included.

*This research was sponsored in part by the Air Research and Development Command and the Air Material Command, United States Air Force.

B. APPARATUS AND PROCEDURE

Figure 10-1 shows a diagram of the carbon arc image furnace and apparatus employed for the measurement of reflectance. The study sample was placed at E and heated by image furnace radiation. The emitted and reflected radiation from the sample was measured with an RCA Type 929 vacuum photocell connected across a 1000-ohm-load resistor to a Tektronix Model 502 cathode-ray oscilloscope. Traces were observed visually or were photographed with a Polaroid camera and measured with a traveling microscope print reader. Lens L and diaphragm D_2 limited the view of the photocell, C, to an area 2.3 mm in diameter at the sample position. Narrow-band interference filters, F_2, were used to limit the response of the photocell to a small spectral range near wavelengths 4305, 5010, 5545, and 6080 A. Photographic neutral density filters were also employed at F_2 to attenuate the radiation to keep within the current limitation of the photocell. The direction of observation of the sample was set at 45° from the optical axis to facilitate measurement of specular surfaces and to reduce the level of scattered radiation.

The radiation from the sample at temperatures below 1900°K consisted almost entirely of reflected radiation. Above 1900°K, the emitted component became measurable and was separated from the reflected radiation by two-blade, rotating shutters (S_1 and S_2) driven by 1800-rpm synchronous motors. Shutter S_1 interrupted the furnace radiation, and S_2 was placed in front of the photocell and filters to provide a zero signal. Diaphragm D_1 subtended 15° of rotation of shutter S_1. Three different shutter blade widths—150, 90, and 30°—having transmission values of $\frac{1}{6}$, $\frac{1}{2}$, and $\frac{5}{6}$, respectively, were employed at S_1 to produce corresponding sample temperatures of approximately 1200, 1800, and 2300°K. Shutter S_2 subtended an angle of 15°.

When a sample having a diffuse surface was observed, the photocell, C, received radiation from all portions of reflector R_2. The reflection signal on the oscilloscope was later compared with that from a standard of diffuse reflectance, a layer of freshly deposited magnesium oxide at least 1 mm thick [5-8]. For use at the high radiance levels of the arc image furnace, the magnesium oxide was deposited on a flat, water-cooled copper disc of a

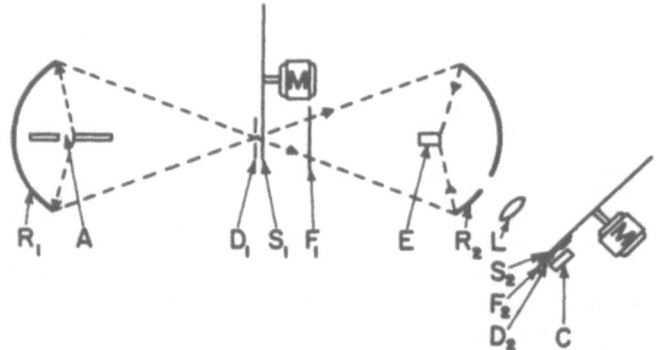

Fig. 10-1. Arrangement of the arc image furnace for reflectance measurement. A-image furnace arc; D_1 and D_2-diaphragms; S_1 and S_2-synchronous shutters; L-lens; C-photocell; F_1 and F_2-filters; E-exposed sample; M-motor; R_1 and R_2-furnace reflectors.

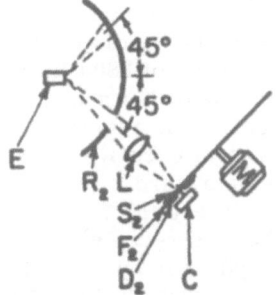

Fig. 10-2. Measurement of reflectance of specular surfaces
in the arc image furnace.

nominal 1-in. diameter. The reflectance of the surface of the study sample
was determined by use of a reflectance value of 0.98 [6,7] for the magnesium
oxide surface.

When the specimen had a specular surface, the photocell received
radiation from only one small portion of reflector R_2, namely, that part
which had the correct angle of incidence. Figure 10-2 shows schematically
this situation for a plane, specular surface at E. Location of the active
area on the mirror viewed by the surface under investigation was accomplished by positioning an opaque, nonreflecting disc of 2.5-in. diameter
over the surface of the mirror until the observed oscilloscope deflection
became a minimum. For a completely specular surface, the deflection
reduced to zero. When less perfect surfaces were measured, the remaining
deflection was contributed by the diffuse component of the surface reflectance. Degree of specularity was defined as the percentage reduction in
signal produced by the 2.5-in. opaque shield. The reflecting surface under
study was removed after location of the active mirror area and was replaced by a thin polished quartz plate. Since the index of refraction of the
quartz plate was known, the reflectance at the required angles could be
calculated. The quartz plate was then positioned so that it had the same
disposition relative to reflector R_2 as did the surface under study. This was
accomplished by placing a small incandescent bulb in the center of the opaque
disc and tilting the quartz plate until the light from the bulb fell upon the
photocell. The opaque disc and incandescent bulb were removed, the image
furnace radiation was reflected from the mirror by way of the quartz plate
into the photocell, and the corresponding oscilloscope signal was determined.
The reflectance value of the two surfaces of the quartz plate (taking into
account the multiple reflectances) at the angle of measurement then calibrated the reflected signal for this particular area of the mirror and gave
the specular component of the reflectance. The portion of scope deflection
not removed by the opaque disc was treated as the diffuse component and
was compared to the magnesium oxide calibration signal. The sum of the
two reflectance components gives the total spectral reflectance of the surface under investigation at the 45° angle.

The emitted component of the radiance of the study sample was obtained
from the oscilloscope signal during the interval when shutter S_1 was closed.
The emitted radiation component gave a measure of the radiance (brightness) temperature of the sample. Calibration of the temperature scale of

the photocell–filter–oscilloscope combination was accomplished by observation of a standard source of known spectral radiance temperature. The recently described pyrometric carbon arc [9], which has a radiance temperature of 3800°K, was used for calibration purposes in this study and was placed with its crater tip at the sample position of the carbon image furnace. The resultant spectral radiance temperature of the study sample can be combined with the measured spectral emissivity value to yield the true temperature.

The total furnace irradiance may be conveniently measured by use of the water-cooled magnesium oxide disc, which is viewed through a small diaphragm by a calibrated thermopile. Previous measurements [10] have shown that 60–70% of the arc image furnace radiation falls in the wavelength range of 4000–7000 A (where MgO has a reflectance near 0.98), that only a small fraction falls beyond 10,000 A, and that none falls beyond 25,000 A. Measurements of Sanders and Middleton [8] showed that the reflectance of MgO did not drop below 0.94 out to 24,000 A. Therefore, the mean reflectance of the magnesium oxide disc over the effective range of wavelengths incident upon it was taken as 0.96. The Eppley circular, four-junction, copper-constantan thermopile was positioned 1 meter from the magnesium oxide disc. The output of the thermopile was read with a Leeds & Northrup Style 2430A galvanometer. A diaphragm of 2.6-mm diameter was placed at a distance of 70 cm from the thermopile. These dimensions described the solid angle and sample area involved, and when combined with the galvanometer deflection and resultant thermopile irradiance, gave the level of furnace irradiance incident on the magnesium oxide disc. The galvanometer–thermopile combination was calibrated in absolute units by means of an NBS carbon filament radiation standard. Values of furnace irradiance measured in this fashion checked very closely those obtained with a calorimeter.

Two types of sample enclosures were employed to protect the carbonaceous samples from oxidation. The simplest of these was an open-end Vycor tube of 12 mm inside diameter through which argon gas was passed. The samples were placed with the exposed face $\frac{1}{16}$ of an inch inside the open end of the tube. This enclosure satisfactorily protected all materials at the 1200°K temperature, but was not adequate to preserve highly polished surfaces at the higher temperatures. A closed Pyrex cylinder of 2.5 in. inside diameter with a hemispherical end was used to enclose the samples in an atmosphere of flowing argon for the tests at 1800 and 2300°K.

The materials employed in this study were "National" Special Spectroscopic grades AGKSP graphite and L113SP lampblack-base electrodes. The samples were $\frac{3}{8}$ in. in diameter and $\frac{3}{8}$ in. long, were mounted on $\frac{1}{8}$-in.-diameter carbon rods, and were positioned so that one end would be exposed to the furnace radiation.

The ends of the graphite and carbon cylindrical samples were both given a final polish on partially used 4/0 Buehler emery polishing paper, which rested on a glass surface. With a little care, it was easy to get surfaces which were scratch-free and measured 85-90% specular by our method. The roughened samples were produced by being rubbed across fresh, 80-grit garnet paper; care was taken to avoid polishing action by particles of graphite which lodged in the grains of the abrasive paper. The AGKSP graphite samples were more prone to polishing than the L113SP lampblack ones.

C. EXPERIMENTAL RESULTS

Table 10-I gives the spectral reflectance and emissivity data as measured at about 1200°K in the open-end tube with the 150° shutter blade width for the two surface finishes of both kinds of carbon material. The reflectance data are plotted in Fig. 10-3. The polished surface for both grades of carbon shows no significant wavelength dependence for the reflectance over the visible portion of the spectrum; a horizontal straight line falls within the range of measured values. The range of values of ±10% is partly due to measurement errors, but more to the lack of reproducibility of the polished surfaces.

The roughened surfaces show a very slight increase in spectral reflectance in going from the blue to the red end of the spectrum. This increase is hardly greater than the approximate 10% spread of values at each wavelength. However, this trend toward increasing reflectance at longer wavelengths has been observed by numerous workers [11,12].

The absolute reflectance values are in reasonable agreement with those previously reported [11]. The lampblack-base material produces slightly lower spectral reflectance than the graphite specimens for both surface finishes. The reflectance observed at a 45° angle of incidence for the polished surfaces, particularly for the AGKSP graphite material, is very close to the reflectance reported by McCartney and Ergun [13] perpendicular to the basal planes of single crystal graphite. The reflectance for the roughened surfaces approaches the low values which have been observed for oxidized or sublimating graphite [3,14].

Additional data for spectral reflectance at temperatures as high as 2200 to 2400°K showed no dependence on temperature. This is most convincingly shown in Fig. 10-4, which gives an oscilloscope trace obtained with the shutter running for a roughened AGKSP sample while it was heating from room temperature to its final value over a period of 50 sec. The top trace (emitted plus reflected radiance) in each oscillogram gives the locus of the signals with the shutter at S_1 open, while the intermediate trace, when resolved, refers to shutter S_1 closed (emitted radiance) and the bottom trace is a zero signal. Figure 10-4 shows no significant change in the reflected

TABLE 10-I

Spectral Reflectance and Emissivity of Polished and Roughened Surfaces of AGKSP Graphite and L113SP Carbon Measured in the Arc Image Furnace

Material	Temperature, °K	4305 A		5010 A		5545 A		6080 A	
		r	ε	r	ε	r	ε	r	ε
AGKSP, Polished	1180	0.273	0.727	0.284	0.716	0.297	0.703	0.292	0.708
AGKSP Roughened	1220	0.050	0.950	0.054	0.946	0.053	0.947	0.059	0.941
L113SP, Polished	1230	0.251	0.749	0.243	0.757	0.251	0.749	0.243	0.757
L113SP, Roughened	1250	0.030	0.970	0.033	0.967	0.031	0.969	0.035	0.965

Note: °K = surface temperature; r = reflectance; ε = emissivity. Data measured at 45° angle in open-end tube with argon flow.

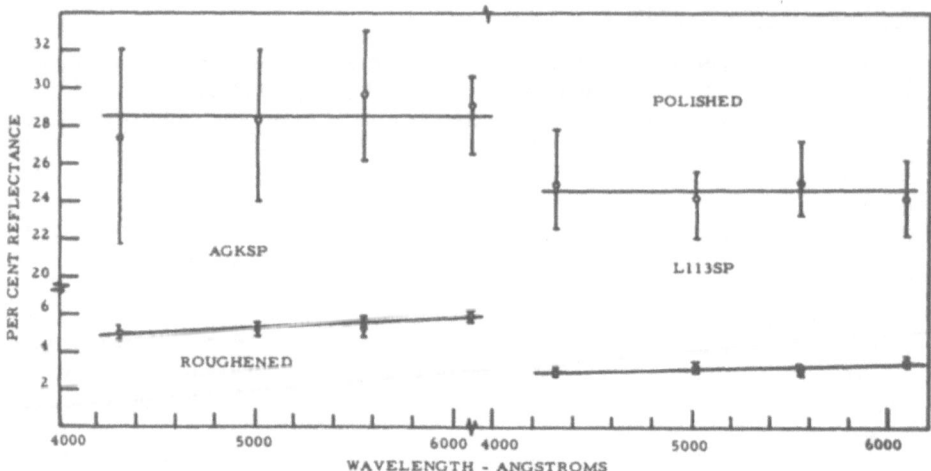

Fig. 10-3. Spectral reflectance of AGKSP graphite and L113SP carbon at 45° angle of incidence and 1200°K temperature.

component from the start at room temperature to the final temperature. The final temperature was approximately 1200°K in the right oscillogram ($S_1 = 150°$) and approximately 2200°K in the left one ($S_1 = 30°$). Emitted light was evident only at the highest temperature, which was reached after approximately 15-20 sec. These and many other traces for both polished and roughened surfaces, for either grade of carbon, show clearly that the reflectance and emissivity are unchanged up to temperatures of 2200-2400°K.

Figure 10-5 shows comparable oscillograph traces obtained with the 30° shutter width (approximately 2300°K) at S_1, with the two kinds of sample material and the two surface finishes. A separate retrace of the zero is also shown. The reflected signals for the polished specimens were several times off scale and have been clipped electronically. The 30° shutter, as it passed diaphragm D_1 of Fig. 10-1, had a completely closed interval of approximately 2 msec. During this time, the test samples underwent some cooling, clearly evident in Fig. 10-5. The cooling during shutter closure and the subsequent heating after the shutter reopened were most pronounced for the L113SP lampblack sample with a roughened surface. The temperature values associated with the minimum temperature and the maximum at

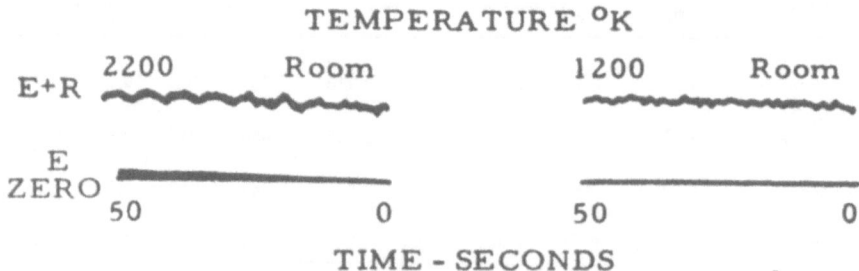

Fig. 10-4. Radiance signal vs. heating time for roughened AGKSP graphite.

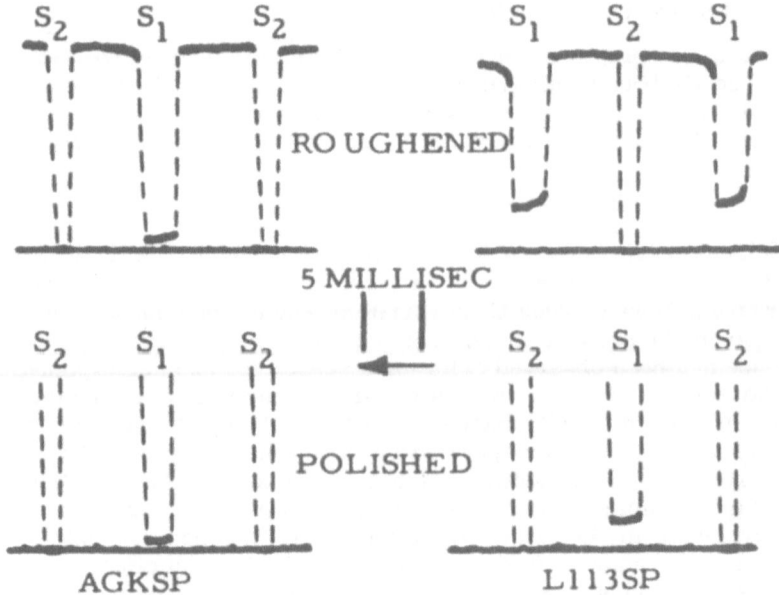

Fig. 10-5. Radiance signal showing cooling and heating of graphite and carbon surfaces during transit of shutter blades at S_1.

the tail of the shutter cutoff of the reflected light are given in Table 10-II. The roughened L113SP material reached the highest temperature and showed the greatest cooling (59°K) during the shutter closure. Lesser amounts of cooling are found for the other material and surface finish. These differences are understandable in terms of lower energy absorption by the polished surfaces and lower thermal conductivity for the lampblack material and for the roughened surfaces. As a matter of fact, quantitative studies of the cooling can give useful information on the thermal conductivity of the surface material. However, the data presented in Table 10-II are not sufficiently precise for this purpose.

TABLE 10-II
Cooling of Graphite and Carbon Surfaces During Closed Interval of 30° Shutter at Intermediate Focus

Material	Temperature, °K		ΔT
	Maximum*	Minimum*	
AGKSP, Polished	2092	2075	17
AGKSP, Roughened	2208	2173	35
L113SP, Polished	2335	2305	30
L113SP, Roughened	2474	2415	59

*Values of radiance temperature (5545 A) at beginning and end of interval.

D. DISCUSSION OF RESULTS

Results of previous workers on the emissivity of carbon and graphite have recently been summarized in excellent fashion by Plunkett and Kingery ([11]). [A subsequent paper by Grenis and Levitt ([15]) should also be mentioned.] There have been wide differences in the results obtained by the various previous observers. With a few notable exceptions, however, the spectral emissivity of carbon and graphite (generally determined near 6500 A) has shown a marked tendency to decrease with increasing temperature. The trends have been such that the corresponding spectral reflectance would have shown increases of at least as much as 50% in the temperature range from 1000 to 2000°C, a variation which would have been easily observable in the present study. The fact that no significant temperature dependence has been observed calls for a search for another explanation.

The results reported by other workers cited above on the temperature dependence of emissivity have been obtained by the "radiation" method. A spurious decrease in the apparent spectral emissivity could have occurred if differences in temperature existed in the sample between the locations where the sample temperature and its spectral radiance were observed. Such gradients might arise from the heat flow through the sample required to supply the radiated flux. The magnitude of such an effect would depend, among other things, upon the thermal conductivity of the material near the surface and would become more marked at higher temperatures.

The outstanding exceptions to the findings of decreasing spectral emissivity with increasing temperature are the results for highly polished materials: For these materials, Thorn and Simpson ([16]) and Plunkett and Kingery ([11]) have obtained spectral emissivity values almost independent of temperature in the 1000-2000°C range. It is probably also significant that these two groups of workers took the greatest precaution to reduce the errors due to thermal gradients.

Plunkett and Kingery ([11]) also reported a careful series of tests on one graphite material given various degrees of surface roughness. This series showed that the rougher the surface, the greater the spectral emissivity and the greater its decrease with increasing temperature. These data can be understood on the basis that the rougher surface has a lower effective thermal conductivity and increases the thermal gradient. Such an observation is in agreement with the findings in Table 10-II, where the rough surface was observed to increase the cooling rate during shutter interruption of the image furnace radiation.

E. CONCLUSIONS

1. The results reported in this paper have shown conclusively that the spectral reflectance and emissivity of carbon and graphite materials, either rough or smooth, are essentially independent of temperature from room temperature to approximately 2300°K. It has been suggested that the apparently discordant results of other workers can be explained in terms of thermal gradients.

2. There is a slight tendency, more pronounced for the roughened samples, for the spectral reflectance to increase with increasing wavelength over the visible portion of the spectrum.

3. Carbon and graphite samples can have a spectral emissivity ranging from as low as 0.70-0.75 for polished surfaces to values close to unity for roughened surfaces.

REFERENCES

1. Null, M. R., and Lozier, W. W., Rev. Sci. Inst. 29:163 (1958).
2. Null, M. R., and Lozier, W. W., J. Appl. Phys. 29:1605 (1958).
3. WADC Technical Report 59-789, "Development of Graphite and Graphite Base Multicomponent Materials for High-Temperature Service."
4. WADD Technical Note 61-18, "Research and Development on Advanced Graphite Materials."
5. ASTM Method D 986-50.
6. Benford, F., Lloyd, Gwen P., and Schwarz, Sally, J. Opt. Soc. Am. 38:445 (1948).
7. Middleton, W. E., and Sanders, C. L., J. Opt. Soc. Am. 41:419 (1951).
8. Sanders, C. L., and Middleton, W. E., J. Opt. Soc. Am. 43:58 (1953).
9. Null, M. R., and Lozier, W. W., J. Opt. Soc. Am. 51:1470 (1961).
10. Null, M. R., and Lozier, W. W., J. Soc. Motion Picture TV Engr. 68:80 (1959).
11. Plunkett, J. D., and Kingery, W. D., Proceedings of the Fourth Conference on Carbon (Pergamon Press, New York, 1960), pp. 457-472.
12. Betz, H. T., Olson, O. H., Schurin, B. D., and Morris, J. C., WADC Technical Report 56-222, Part I (1956).
13. McCartney, J. T., and Ergun, S., Fuel 37:272 (1958).
14. Null, M. R., and Lozier, W. W., "Measurement of Reflectance and Emissivity at High Temperature with a Carbon Arc Image Furnace," presented at the Symposium on Measurement of Thermal Radiation Properties of Solids in Dayton, Ohio, on September 5-7, 1962.
15. Grenis, A. F., and Levitt, A. P., Watertown Arsenal Laboratories Technical Report No. WAL TR 397.1/2, November (1959).
16. Thorn, R. J., and Simpson, O. C., J. Appl. Phys. 24:633 (1953).

Chapter 11

Concerning Several Devices for the Utilization of Imaging Furnaces

Marc Foex
Solar Energy Laboratory
National Center of Scientific Research
Montlouis, France

A. COMPARISON OF SOLAR FURNACES AND CLASSICAL IMAGING FURNACES

Solar furnaces are actually a special type of imaging furnace, although the name imaging furnace is reserved for the type of apparatus that uses an electric arc as the source of radiation.* While electric arcs can be used at a fixed point inside the laboratory, the sun is available as a radiating source practically at infinity, undergoing constant displacement, and available only during a portion of the day. Further, its reception is susceptible to the vagaries of the climate.

We shall not consider at present the different mirror devices employed in various types of imaging or solar furnaces, but only the general characteristics of such apparatus and their possible applications.

The radiation emitted by the sun resembles that of a blackbody at a temperature in the neighborhood of 6000°K (Fig. 11-1). However, this is primarily the case if one considers this radiation outside the atmosphere, since the latter absorbs strongly a portion of the ultraviolet (ozone) and a portion of the infrared (water vapor); moreover, the short wavelengths are more strongly diffused than the long ones. The global energy received normal to the direction of the sun is approximately 1.4 kW/m^2 in the stratosphere, and only 0.9 to 1 kW/m^2 at the surface of the earth.

Imaging furnaces proper generally use intensive arcs (Fig. 11-2), which make it possible to obtain radiation quite similar to that of a blackbody at 5000 or 6000°K. This radiation is similar to that received from the sun outside the terrestrial atmosphere. Other types of radiation sources have been used more recently, in particular xenon lamps capable of furnishing 5 kW arcs of very small dimensions (5 mm in width, 8 mm in length, for example), but the observed energy distribution is very different (Fig. 11-2), since the observed energy is almost uniformly distributed over all visible wavelengths and possesses strong maxima in the very near infrared. It is very important to have available new sources furnishing radiation that is very different from that of an arc or the sun, not only for photochemistry,

---

*At the end of this article, we present a brief bibliography relating to the characteristics of application of imaging furnaces and solar furnaces.

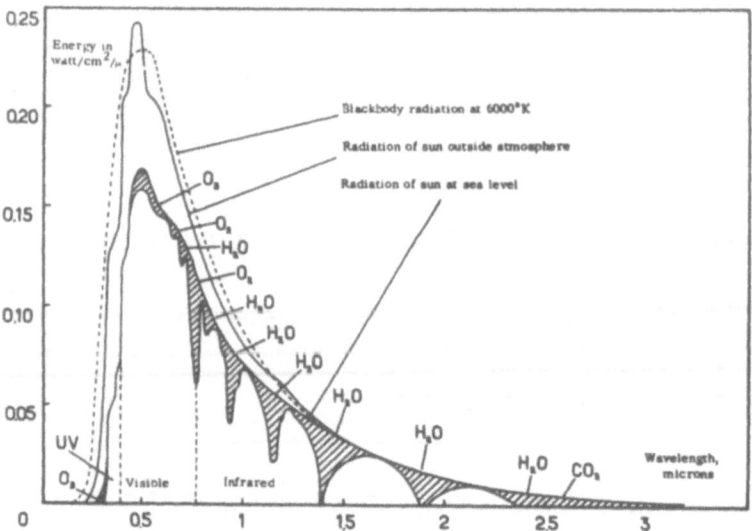

Fig. 11-1. Energy distribution of solar radiation outside the atmosphere and on earth (mean value at sea level).

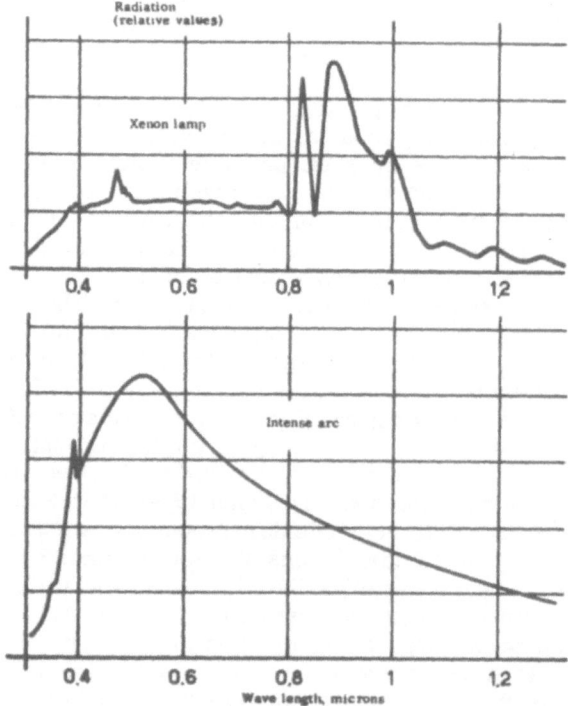

Fig. 11-2. Energy distribution of intensive arc radiation and radiation from a xenon lamp.

but also for various applications such as measurement of the temperature of substances treated in a furnace.

Let us begin by comparing the efficiency of imaging and solar furnaces.

A very good imaging furnace presently in use ([18]) possesses two conjugate elliptic mirrors and an energy source consisting of a 25-kW arc. The image has a useful dimension of 9 to 10 mm in diameter; the illumination energy at the focal plane reaches a mean value of 950 W/cm^2, while the maximum value observed at the center of the target is 1360 W/cm^2. The useful energy represents 3-4% of the electric energy supplied. This relatively low efficiency is due, among other things, to the difficulty involved in effectively utilizing all of the radiation emitted by the arc.

It is frequently easier to capture with suitable efficiency the radiation emitted by a source at infinity, such as the sun. Thus, a small solar furnace of the type used at Montlouis, ([6]) comprising a fixed parabolic mirror 2 m in diameter and 85 cm in focal length, captures approximately 3 m^2 of radiation normal to the direction of the sun, i.e., approximately 3 kW. The useful dimension of the image (obtained by eliminating the peripheral zones receiving illumination energy of less than 500 W/cm^2) is 13 to 14 mm in diameter, with a mean illumination energy of 900 W/cm^2 and a maximum of 1500 W/cm^2 at the center of the target. The efficiency as a measure of the solar radiation captured is 40%.

The above figures are for mirrors silvered on the rear surface. The values could be increased about 20% with mirrors covered with an aluminum layer on the front surface (one would then obtain about 5% and 50% efficiency, respectively). It is even possible to obtain efficiencies in the vicinity of 80% with an aluminated parabolic mirror aimed at the sun.

As for the energy distribution at the focal plane, one observes greater uniformity at the center of the spot in the case of solar furnaces than in the case of arc imaging furnaces.

The location in space of images obtained at the focus of solar furnaces can be very stable, even without the use of astronomic devices, which are too complex and costly. Certain hydraulic devices used at Montlouis made it possible to achieve accuracy of the order of 1%. Results obtained with the arcs of imaging furnaces are generally less satisfactory, despite certain stabilization devices.

The rapid variations of the order of 10% of energy flux at the focus of an arc imaging furnace may be very inconvenient. The intensity of solar radiation does not suffer any substantial rapid variation in regions of clear sky, the variations observed being frequently less than 0.1%. Rather, the intensity of solar radiation varies slowly during the course of a day, generally passing through a maximum slightly before the solar noon. On a clear day at Montlouis (1600 m altitude, 42°5' latitude), one observes the following values for the duration of the period in which the variation of intensity of solar radiation does not exceed 5%: 4 hr (winter solstice), 5 hr (equinox), 6 hr (summer solstice).

The power of solar furnaces depends directly upon the dimensions of the mirror used, since it is necessary to collect the radiation received at the surface of the earth. In contrast, in the case of imaging furnaces, it is possible to use reflecting surfaces of small dimension, and these surfaces are less cumbersome and less costly. This is a notable advantage, but only within certain limits. In fact, if the relative size of these surfaces is greatly reduced, shadowing can occur.

The techniques applicable to imaging furnaces and those used with solar furnaces are essentially identical, except in the case of experiments under pressure. It is, in fact, extremely difficult to construct walls which are very thick and simultaneously transparent to the incident radiation of short wavelength, and which are also sufficiently resistant mechanically. The difficulty can be avoided in the case of imaging furnaces by placing the entire assembly of radiation source and device for its utilization in the same metal envelope [13].

The use of transparent envelopes of pyrex glass, silicon glass, or even plastic makes it possible to work in a conditioned atmosphere sheltered from air, both in the case of imaging furnaces and in the case of solar furnaces. Vacuum techniques are of particular interest, although one frequently observes obscuration of the wall by deposit of vaporized products, which limits the energy admitted into the envelope.

There is a domain of application particularly adapted to solar furnaces. The domain in question is space, where one can use high vacua, as well as intensive and uniform solar illumination (1.4 kW/m^2 outside the earth's atmosphere), for the high-temperature treatment of certain products of high quality. The moon could also receive such apparatus.

In the absence of atmosphere (and consequently of wind), and sometimes even in the absence of gravity, the concentration apparatus used (parabolic, etc.) could be extremely light, easily oriented, and devoid of intermediate mobile reflectors (plane heliostats). Such devices should be very inexpensive for a given power. Moreover, in the absence of the corrosive effects of the atmosphere, it should be possible to use a great variety of materials.

We might recall by way of indication that parabolic mirrors 1 to 3 m in diameter and approximately 5 kg/m^2 (rigid models) or 0.5 kg/m^2 (non-rigid models) in weight have been developed by NASA for the heating of solar thermopiles under the above conditions [14].

The use of centrifugal furnaces would make it possible to maintain substances in place conveniently and still have cavity treatments with good absorption of radiation.

Working in high vacuum in the absence of pumping devices and envelopes no longer poses out-gassing problems, and numerous products can thus be heated to high temperatures without contamination by the atmosphere, or can be evaporated.

The achievement of a given atmosphere other than vacuum, on the other hand, requires the use of receptacles which possess walls that are transparent to solar radiation. Plastic balloons inflated with various gases are one example of such devices.

Work in the high vacuum of space does not necessarily require the use of apparatus for concentrating the energy. In certain cases it is possible, as indicated by Gillette [15] and Hibbard [16], to use the solar radiation directly, by receiving it on a surface that selectively captures the radiation of short wavelength and strongly reflects the radiation of longer wavelength.

We have indicated in the tabulation below the values of wavelength beyond which the absorption of surfaces should present a sharp transition threshold in order that certain temperatures can be reached. The same tabulation indicates the fraction of solar radiation thus utilized. We have assumed that only the insulated surface radiates, the other surface being, for example, thermally isolated with the aid of reflectors.

Equilibrium temperature, °C	Absorption limit, μ	Efficiency, % (for 1.4 kW/m² total)	Energy absorbed, kW/m²
500	2.6	98	1.37
1000	1.09	80	1.12
1500	0.63	43	0.6
2000	0.37	7	0.1

The practically available energy is very small and could suffice only for extended refraction near 1000°C of substances kept in vacuum in the interior of envelopes whose walls consist of selective materials, the exact nature of which remains to be determined.

Another method, according to a scheme described by Francia ([17]) would use black receptor surfaces covered with a network of small channels of, for example, honeycomb design. These blackened channels, located normal to the surface are so oriented that the solar radiation will strike the surface to be heated directly while the radiation emitted by the surface will be captured almost entirely by the honeycomb structure located nearby.

Finally, it would seem desirable in many cases to use a very light paraboloid, even one of mediocre quality, such as and inflatable apparatus ([14]), in combination with selective surfaces of type described, in order to increase the available energy per unit of surface.

It is also interesting to compare the present and possible future use of imaging furnaces and solar furnaces in apparatus of high power.

Classical imaging furnaces have a very low efficiency, and industrial use appears unlikely. On the other hand, imaging furnaces of low power (1 to 20 kW available at the target) are capable of supplying great services in the domain of scientific research, and this will be particularly true when radiation sources more practical than the classic electric arc are developed, as appears to be partly the case (xenon lamps, etc.) already. We can foresee the day when a great variety of sources, each having its domain of application, will be constantly available inside a laboratory.

On the other hand, solar furnaces of high power appear to have certain possibilities of development in different domains on the surface of the earth (arid and sunny zones) or in space.

Solar energy can, in fact, be captured quite effectively. Moreover, the efficiency of operation "in cavity," particularly for centrifugal furnaces, increases considerably with power. The efficiency almost doubles when the power changes from 50 to 1000 kW. Energy loss by thermal conductivity through the product is, in effect, lower in a large furnace than in a small one, the relative surfaces of the crucibles being smaller in the former case. Only the loss of energy through the orifice for radiation access is proportional to the power of the installation. Thus, for example, the quantity of fused or fritted zirconium oxide (which is about 60 kg per day with the 50-kW apparatus at Montlouis) could reach 2 to 3 tons per day with the 1000-kW apparatus currently being constructed at Odeillo under the direction of Professor Trombe. The efficiency obtained (20% in one case, 50% in the other) could be even further improved with more powerful furnaces. However, the construction of such units would be relatively more problematic, in view of the considerable dimensions of the parabolic mirror and its support. On the other hand, the alternation of day and night does not always permit the most effective utilization of such apparatus.

For powerful solar furnaces, it would be necessary to envisage fabrication processes of discontinuous character, not involving much thermal inertia at the beginning, and capable of being discontinued abruptly at the end of a day.

B. MEASUREMENT, CONTROL, AND REGULATION OF ENERGY AND TEMPERATURE

Work with radiation furnaces requires rigorous control of both the energy available at the focus or in the zone of utilization and the temperature obtained under each particular operating condition ([11, 19-23, 26]).

1. Control of Energy

It is easy to measure the global intensity of energy flux radiated to the focus by means of metal calorimeters possessing a blackened cavity into which the radiation penetrates and is captured; one measures the heating of a stream of cooling water at a known rate of flow.

The distribution of energy in the focal region or in a neighboring region can also be easily determined by measuring the brilliance of a diffusing surface suitably located.   Surfaces formed of a white refractory oxide frequently give good results.   Since high-energy radiation may cause deterioration of the diffusing products, one expediently utilizes sand-blasted matte metal surfaces refrigerated by a stream of water.   The brilliance is measured at different points of the surface by means of a monochromatic pyrometer of the disappearing filament type.   In order to establish clearly the existing relation between the measured brilliance and the energy, it is necessary to associate the above method with calorimetric tests.   Devices involving the use of photographs are also capable of giving good results.

The distribution of energy flux in the focal plane is not homogeneous, which is inconvenient in certain direct treatments (for example, during thermal shock tests).   One can eliminate this inconvenience (at a sacrifice in usable illumination energy) by realizing a homogenation by successive reflections from refrigerated cavities whose internal surfaces have been made diffusing by application of coatings of white pulverizing refractory oxides.   These cavities comprise two apertures, the first facing the furnace to admit radiation, the second serving for utilization.

The energy can be regulated easily by means of screens or filters located either in the trajectory of incident rays, or in the vicinity of the focus in the zone of convergent radiation.   At Montlouis, workers developed a device consisting of a shutter curtain controlled by a DC motor whose speed is regulated by controlling the excitation.   A given heating program can be realized by means of a cam, for example, calibration being performed during a preliminary operation in which a calorimeter, rather than the treated product, is placed at the focus.

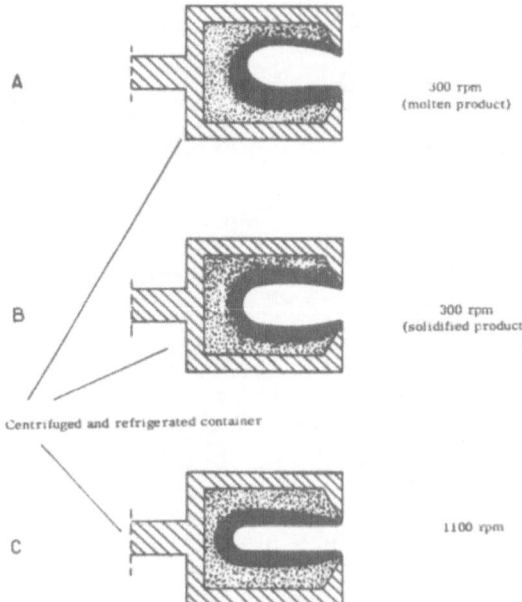

Fig. 11-3. Sections of products treated in a centrifugal
furnace to measure solidification temperatures.

2. Realization of Blackbody

Measurement of the temperature of products treated by means of imaging furnaces requires either knowledge of the substances' emission factors (emissivity) under the given measurement conditions, or knowledge of the realization of suitable blackbodies.

Knowledge of the emission factor exhibited at high temperatures by solid pulverized or agglomerated substances is very unreliable; this factor varies considerably with the development of fritting and the dimensions of crystals.   On the other hand, it is easy to determine almost exactly the emission factor (monochromatic or total) of fused masses by measuring the amounts of reflected radiation.

The realization of small cavities, which are relatively deep and of very small diameter, makes it possible, in principle, to perform correct measurements of temperature in superficially irradiated solid substances but, in practice, the results are almost always falsified by the importance of thermal gradients existing in such masses.

In order to perform correct measurements of the solidification points of various refractory substances, it is advantageous to centrifuge the molten material in such a way as to produce within it a cavity possessing essentially the characteristics of a blackbody, this can be done by means of a process developed by F. Trombe and the author ([9]). For this purpose, the product is placed in the pulverized state* into a refrigerated metallic cylindrical envelope, pierced by an axial orifice for the access of radiation (Figs. 11-3 and 11-4).   When the speed of rotation is suitably regulated, the molten

*It is convenient to use a product which has been previously melted and crushed so as to limit the volume of the cavity formed.

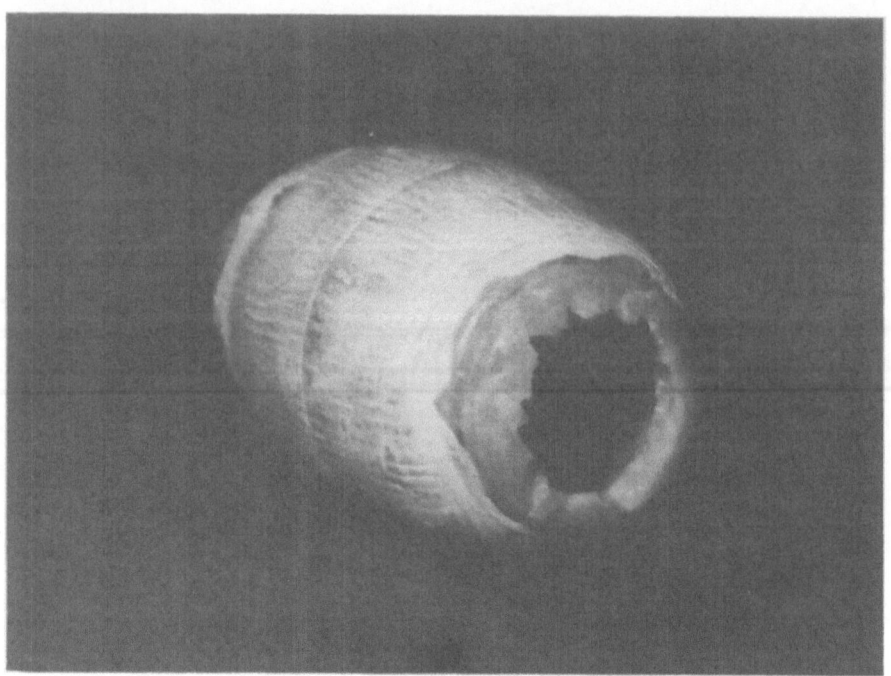

Fig. 11-4. Photograph of a "crucible" of refractory oxide fused in a centrifugal furnace.

product is forced against the walls in regions where the cavity has a sufficient diameter, while the liquid mass located near the center escapes
centrifugation and undergoes continuous mixing, which favors homogeniza-
tion of its temperature; periodic variations of the speed of rotation enhance
this phenomenon. When the action of radiation ceases, the product solidifies
progressively, forming a regular cavity centered about the axis of rotation
of the furnace. The vessels used in our last experiments* are of the order
of 20 to 50 cm³, their length being equal to, or greater than, their diameter.
The orifice for radiation access possesses a diameter of the order of 10 to
11 mm. Given these conditions, we were able to obtain with certain products
and with speeds of rotation near 1100 rpm, cavities reaching 35 mm in
depth whose inside diameter was almost uniform and did not exceed 7 to 8
mm (outside diameter 15 mm). Sighting the bottom of the cavity makes it
possible to perform correct measurements under conditions which very
nearly represent those of a blackbody. Notable progress has been made in
this area at Montlouis since the first tests were performed ([21]).

When the diameter of the cavity is relatively large (Fig. 11-3, A and B)
one observes a strong parasitic influence of the incident radiation which
perturbs the measurements by successive reflections. This is not the case
when the cavity has a great length and a small diameter (Fig. 11-3 C). It
is then possible to perform a correct measurement of the temperature by
sighting the bottom with a disappearing filament pyrometer. The parasitic
action of incident radiation can be suppressed in practice by neutralization
of the central portion of the parabolic mirror by means of a mask.

*Experiments performed with a solar furnace consisting of a fixed parabolic mirror with axis hori-
zontal, 2 m in diameter and 85 cm in focal length, served by a heliostat.

3. Devices for Temperature Measurement which Permit Elimination of the Parasitic Action of the Radiation Source

Parasitic reflections of the incident radiation (sun, arc, various lamps) are very disturbing and produce errors in excess; these are frequently much greater than the errors in efficiency caused by observations without blackbody conditions. As we have seen, the use of cavities does not always permit protection against these inconveniences, since the radiation is frequently capable of reaching the pyrometer by successive reflections.

Errors in excess are particularly important if one uses monochromatic disappearing filament pyrometers operating in the visible (red: 0.65 μ), since solar radiation and arc radiation are considerably richer in short-wavelength components than radiation from the treated substance, which is heated only to 2000 or 3000°C.

To begin with, in the regions subjected to low-energy illumination, it is possible to interpose a filter between the source of radiation and the focus of utilization. This filter must possess sufficiently intense and definite absorption for a certain domain of wavelength, while allowing passage to the larger portion of incident energy. One then measures the temperature of the treated substance by means of an optical pyrometer. This pyrometer is provided with a filter opaque to solar radiation except in a transmission band corresponding to the absorption band of the first filter.

At Montlouis, we have tried various types of large-area filters (2 m by 2 m) located in the trajectory of the radiation reflected by the heliostats serving the small, 2-kW solar furnaces of vertical axis. Certain Kodak Wratten-type filters absorbing radiation in the vicinity of 0.65 μ can be associated with disappearing filament pyrometers; however, they possess the inconvenient property of being very fragile. The same is true of colored Plexiglas. Filters of the "Manganal" type capture all of the visible, but the transmitted global energy (in the infrared) does not exceed 20%.

Far better results have been obtained with the aid of uncolored Plexiglas sheets (Fig. 11-5), which transmit the largest part of solar radiation, while possessing certain absorption bands in the infrared near 1.7, or slightly above 1.9 μ. A sheet of 15 mm thickness transmits three-quarters of the solar radiation. In the case of sunlight, it is expedient to combine the action of Plexiglas with that of the atmosphere (Fig. 11-5), which, due to water vapor, strongly absorps radiation in the vicinity of 1.9 μ. This combination provides in this region of the spectrum an absorption band which is simultaneously more intense and wider, and which makes it possible to cover almost entirely the foot of the transmission band of interference filters used on pyrometers. Measurements near 1.7 or 1.9 μ must be performed with pyrometers sensitive to this radiation. Apparatus with lead sulfide cells yields good results, provided one uses a reference lamp with which the cell can be constantly compared (optical null device). It should be remarked that, contrary to thermopiles, these cells are not sensitive to the action of long-wavelength radiation emitted by the screens and filters, which heat slightly under the action of incident radiation and radiation from the product being treated.

In the case of imaging furnaces, it is generally possible to reduce the dimensions of filters, particularly if one uses devices with two conjugate hyperbolic mirrors. Nevertheless, it is expedient to avoid excessive heat-

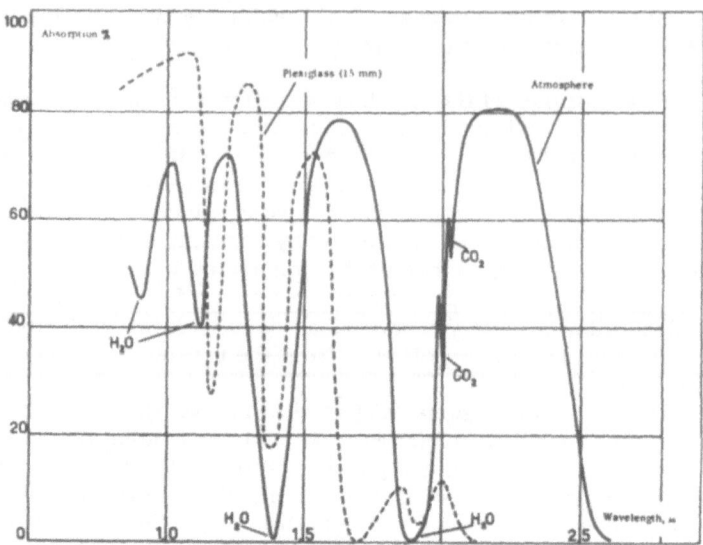

Fig. 11-5. Absorption curves of the atmosphere and Plexiglas as a function of wavelength.

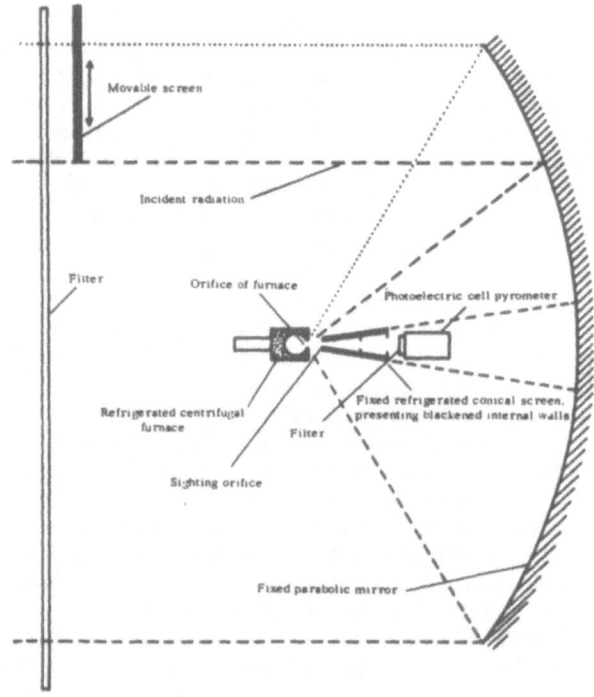

Fig. 11-6. Device for temperature measurement consisting of a photoelectric cell pyrometer, filters, and a fixed conical screen blackened inside.

ing of the filters. This can be accomplished by producing selective absorption by means of refrigerated solutions circulating in vessels with parallel surfaces.

Moreover, certain radiation sources, such as xenon lamps, exhibit very weak emission in certain regions of wavelength (above $1.1\,\mu$, for example), which makes it possible to avoid filtering the incident radiation.

Figures 11-6 and 11-7 present schematic diagrams of parametric installations incorporating filters. The dimensions and the position of the surface to be measured must be rigorously defined by means of a fixed conical screen located on the axis of the installation and separated from the focus. The conical screen aimed at the substance whose temperature one wishes to measure may be blackened inside (Fig. 11-6) so that one only uses the radiation passing through the sighting orifice in the direction of the pyrometer. Alternatively, the walls furthest from the furnace may be of a diffusing nature (see Fig. 11-7).

Another simple process used for a number of years at Montlouis involves obscuring the incident radiation by means of a shutter at the instant the temperature is measured (Figs. 11-8, 11-9, and 11-10).[*] A screen which shuts out the incident radiation is made to pass very rapidly in front of the substance being examined (in 0.1 sec, for example), and one proceeds with the temperature measurement during this short period by using an optical pyrometer with a photoelectric cesium cell, sensitive essentially between 0.6 and $1\,\mu$ and possessing good fidelity during the time required for the measurement.

It is also possible to stop the shutter screen in front of the substance heated to a high temperature and to measure the substance's temperature during cooling.

4. Thermal Analysis of Substances Treated

A very convenient technique developed at Montlouis for measuring the melting and solidification points of oxides and conducting thermal analysis of different systems consisting of refractory products involves the use of cavities formed by means of centrifugal furnaces.

For example, the system consisting of the oxides of chrome and of lanthanum has been studied in this way. The thermal analysis indicates that these two oxides, whose melting point slightly exceeds 2300°C, combine to form a lanthanum chromite, $CrLaO_3$, which is very refractory. A study of the diagram in Fig. 11-11 indicates two eutectics flanking the above compound and melting at slightly above 2200 and 1850°C, respectively. There is no important solid solution.

Figures 11-12 and 11-13 represent temperature–time curves observed during cooling of lanthanum oxide, on the one hand, and a mixture formed of equal parts of the oxides of chrome and chromite on the other. The first of these curves (Fig. 11-12) displays, outside the solidification point of lanthanum oxide, two anomalies near 2080 and 2020°C; these anomalies are similar to those one finds with most of the sesquioxides of the rare earths. The second curve (Fig. 11-13) illustrates the importance of the phenomenon of superfusion preceding the solidification of the eutectic.

---

[*]In regard to pyrometric devices involving the use of shutters for screening out incident radiation, it is advantageous to consult the works of Conn and Braught ([22]) and those of Glaser and Blau ([19]).

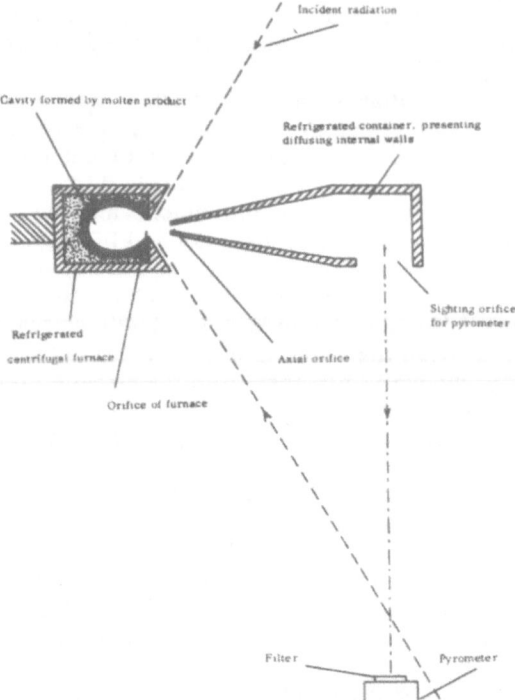

Fig. 11-7. Device for the measurement of temperature consisting of a photoelectric cell pyrometer, filters, and a fixed screen provided with internal walls which diffuse the radiation.

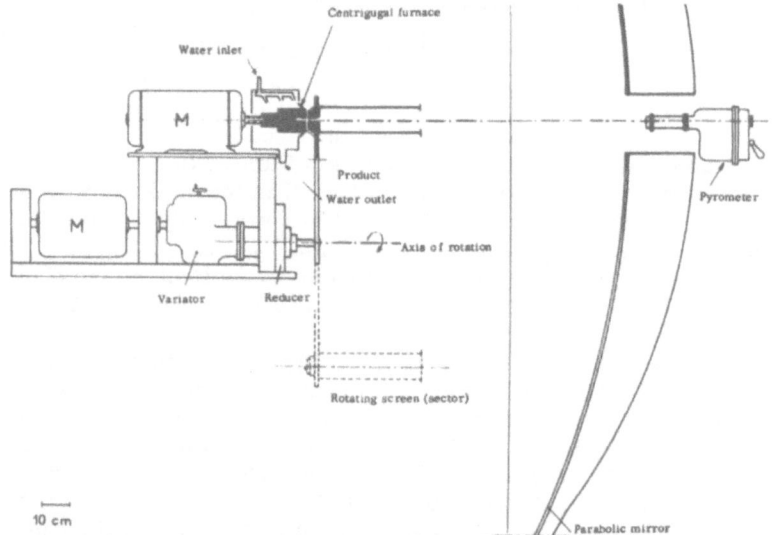

Fig. 11-8. Device with a rotating screen (sector) which permits rapid obscuration of solar radiation to permit performance of temperature measurements by means of a fast-response pyrometer.

Fig. 11-9. Detail of a solar installation for the measurement and regulation of temperature and energy.

Thermal analysis of products during the course of cooling can also be performed on products whose surface is placed directly at the focus (Fig. 11-14). The conditions realized are no longer those of a blackbody, and the measurements performed after interposition of the screen thus depend upon the value of the emission factor of the product. The emission of "white" products, in particular, decreases strongly at the instant of solidification, a circumstance which gives clear evidence of superfusion.

In the presence of a certain quantity of incident radiation, there is a strong increase in the radiation reflected by "white" products during their solidification, at least when the dimensions of the crystals do not exceed a certain value.

In many cases, it is interesting to perform a thermal analysis of products during less rapid cooling or during heating; these conditions can be obtained by progressive shielding of, or progressive admission of radiation on a product. Thermal phenomena are then sometimes more difficult to display, but it is easy to perform microscopic examinations which, for

Fig. 11-10. A solar installation for the measurement and regulation of temperature and energy.

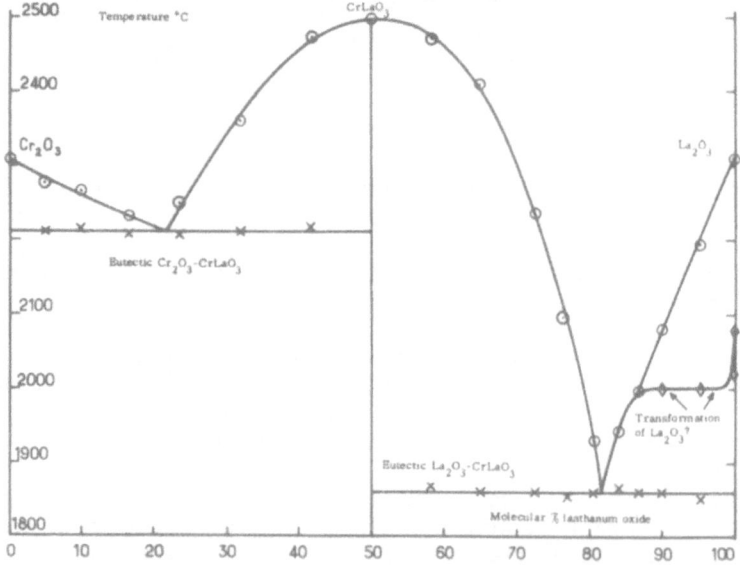

Fig. 11-11. Solidification diagram of the chrome oxide-lanthanum oxide system.

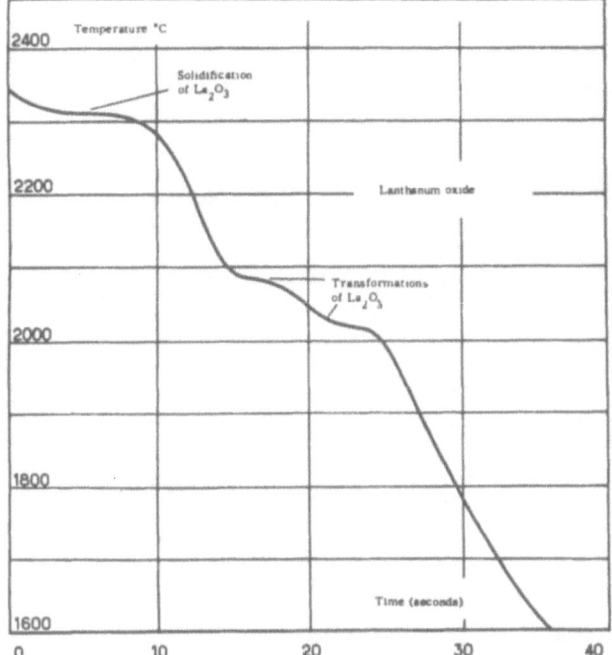

Fig. 11-12. Cooling curve of lanthanum oxide fused in a centrifugal furnace.

example, permit detection of the appearance of crystals within the molten mass at the instant of solidification.

5. Realization of Homogeneous Temperatures and Regulation of Temperatures

Obtaining directly irradiated surfaces at a homogeneous temperature has to do with the problem of homogeneous energy distribution. The realization of cavities at homogeneous temperature is made easier by: (a) displacement (rotation) of refractory pieces inside a cavity; (b) mixing of the fluid, powdered, or molten product treated in the cavity; (c) use of walls which suitably reflect the radiation, particularly in case one works in vacuum.

The regulation of temperature, just like the regulation of energy, is performed by means of shutter screens. To obtain a constant temperature, one controls the position of the screen by means of photoelectric cell pyrometers or thermoelectric piles. Programmed temperatures are obtained by analogous means.

A screen located, for example, as indicated in Figs. 11-6 and 11-10, makes it possible to decrease or increase the energy, and consequently to vary progressively the temperature of the mass treated.

C. DEVICES FOR MIXED HEATING OF SUBSTANCES BY HIGH-ENERGY LIGHT RADIATION AND BY PLASMA

High-energy light radiation furnished by different types of apparatus,

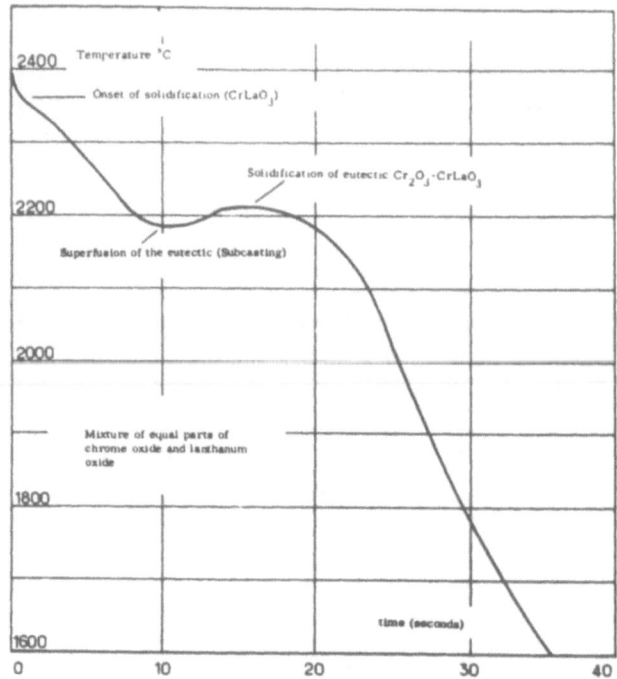

Fig. 11-13. Cooling curve of a mixture of equal parts (by weight) of chrome oxide and lanthanum oxide fused in a centrifugal furnace.

such as solar furnace and imaging furnaces, basically permits the heating to high temperature of condensed substances (solids or liquids) capable of absorbing the incident radiation either directly or by successive reflections. This type of apparatus is particularly well adapted for operations of fusion without contamination. It is much more difficult to heat gases or vapors by this means, radiation absorption being very low in this case. Heat exchange can be performed at hot surfaces placed in the focal zone, but the efficiencies obtained are generally low if the gas is heated to a high temperature near that of the surface.

In contrast, plasma tubes and torches heat the gas directly to very high temperatures, higher than those of the boiling points of the most refractory substances, but it is frequently relatively more difficult to use this type of apparatus with good efficiency for heating condensed substances.

So that the advantages of both methods can be combined, a number of heating devices incorporating the simultaneous use of plasmas and high-energy light radiation have been developed.

These mixed heating processes allow condensed product to be heated to a higher temperature than could be achieved by one of these methods alone. In fact, the temperature of hot gases in a plasma (15,000°C, for example) is generally much higher than the maximum theoretical temperature which can actually be obtained by means of imaging furnaces with an electrical source. Likewise, it should be noted that solar furnaces do not permit one to obtain a temperature in excess of that of the surface of the sun (about 5000°C),

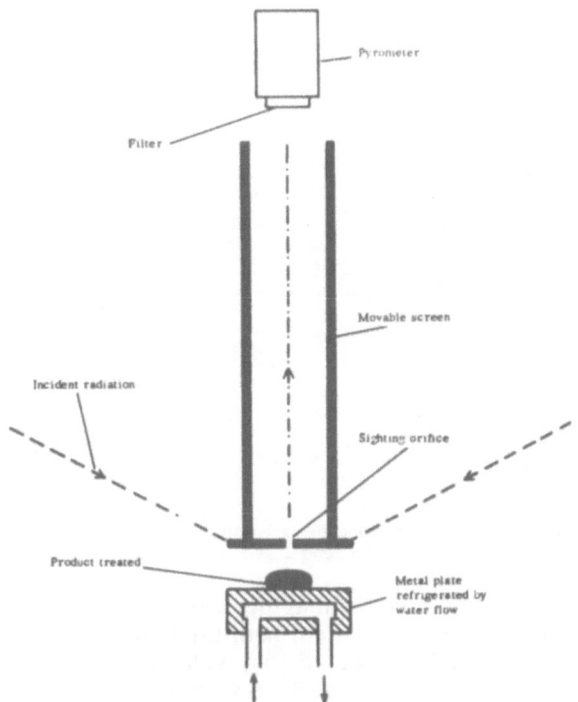

Fig. 11-14. Device for measuring the temperature of a fused product on a refrigerated plate, consisting of a photoelectric cell pyrometer and a cylindrical screen capable of shielding the radiation from the imaging furnace.

since at this temperature, the losses by radiation balance the energy received.*

In order to obtain a picture of the possibilities offered by such mixed installations, it is appropriate to consider not only the temperature of the plasmas, but also the energy which can in this way be transported to the product which is under treatment and is located, for example, in the focal plane. An imaging furnace or a classical solar furnace yields energy illumination of the order of 1 to 2 kW/cm², while plasma tubes (arc plasma) make it possible to supply energies corresponding to several tens of kW/cm².

It should be noted that "soft" plasmas—obtained with plasma torches heated by high frequency—yield lower energies, which frequently do not exceed several kilowatts or even several hundreds of watts/cm². Nevertheless, these latter plasmas are of particular interest, because they allow operation in the absence of electrodes—that is to say, under the best conditions of cleanliness—and use of gases such as oxygen or air. In addition, the speed of the gas current is considerably lower than with plasma tubes, which makes it possible to avoid abrasion of coherent substances and the violent elimination of pulverized products which are not agglomerated.

*By way of comparison, we note that the relatively cool flame of an ordinary torch limits considerably the maximum temperature which can be reached at the focus of an imaging furnace.

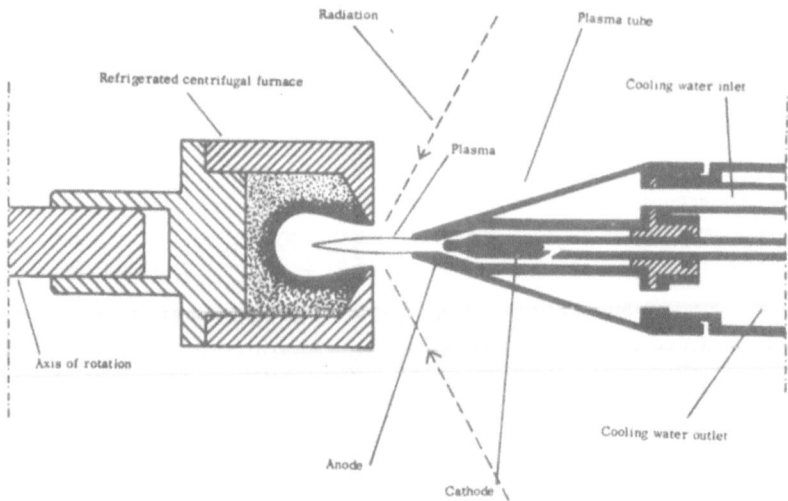

Fig. 11-15. Association of an imaging furnace with a plasma tube having an "elongated tip" for the treatment of a centrifuged product (cavity formation).

The temperature reached in imaging furnaces depends upon the emission factor of the substances treated, both for the wavelengths corresponding to the incident radiation and to that relative to the emitted radiation. It is for this reason that highly reflecting metals with unaltered surfaces or very white substances in a powdered state are difficult to heat to a high temperature by the action of short-wavelength radiation, similar to the radiation emitted by an arc or by the sun.

While plasmas transmit their energy to all substances in approximately the same manner, products having low emission factors and losing little energy by radiation are easier to heat to high temperatures by this method than substances which radiate strongly when hot. Thus, metals possessing an unaltered surface are frequently easier to melt by this method than are most other products.

The foregoing discussion shows under what conditions it is possible to combine the action of plasmas with that of high-energy light radiation to obtain higher temperatures. A high temperature is not the only advantage to be gained by these mixed processes. The following cases should also be cited.

First, by means of a plasma, one can preheat the white powdered products (thorium oxide, magnesium oxide) which poorly absorb the radiation in imaging furnaces. After being fritted at a high temperature or melted, these products become very absorbent, and their melting can be continued with the aid of only a radiating source.

Combination of the two means of heating makes it possible to obtain a number of chemical reactions between a gaseous phase and a condensed phase (solid or liquid). Various operations are realizable by this principle, and one can cite, for example, the reduction of different products by the action of strongly ionized hydrogen, the preparation of various nitrides by means of a nitrogen plasma, and the formation of carbides by means of hydrocarbides such as acetylene.

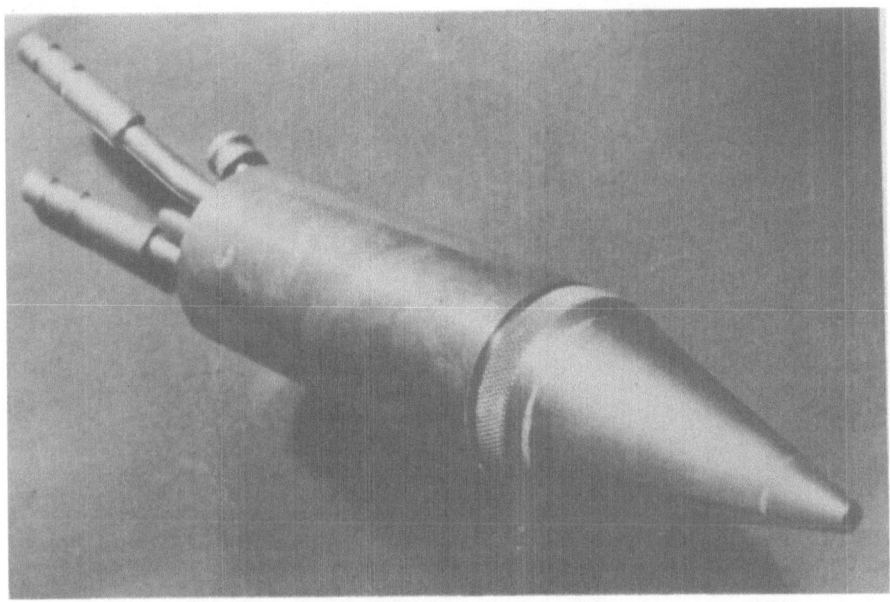

Fig. 11-16. Photograph of a plasma tube with "elongated tip" specially designed for mixed treatments with imaging furnaces (length of tube, 20 cm).

When the product receives sufficient energy in the condensed phase, it volatilizes into the vapor phase, and reacts with the high-temperature gas emanating from the plasma; this reaction may lead to a number of new products, particularly after quenching.

In the various cases described, the tube or torch used must not obscure an excessive portion of the incident radiation and must not make a shadow on the product treated.

Different installations can be used. The following types of apparatus are presented by way of example: To begin with, one can use a plasma tube provided with an elongated tip (Figs. 11-15 and 11-16). Note that the conical shape of the refrigerated volume surrounding the anode, is such that the solid angle thus realized is small compared with that of the cone formed by radiation. Moreover, the extremity of the tube is slightly removed from the focal plane so as to give better access to the latter. In such an installation, the treated product may be placed in the focal plane itself, but it is also possible to use a centrifugal furnace ([24]) possessing an orifice through which the plasma jet penetrates. In the latter case, the substance placed in the interior forms a cavity whose walls are brought to a high temperature.

The centrifugal furnaces may also have two orifices (Fig. 11-17), one for the radiation from the imaging furnace, and the other for the hot gases of the plasma. Under these conditions, the tube does not necessarily have an elongated shape.

Instead of a plasma tube operating by means of an arc, it is frequently advantageous to use a plasma heated by high-frequency induction in an insulating tube (of silicon, for example). Figure 11-18 presents a schematic view of the principle involved in this type of apparatus.

Now we shall examine how one can effect fusion of refractory products

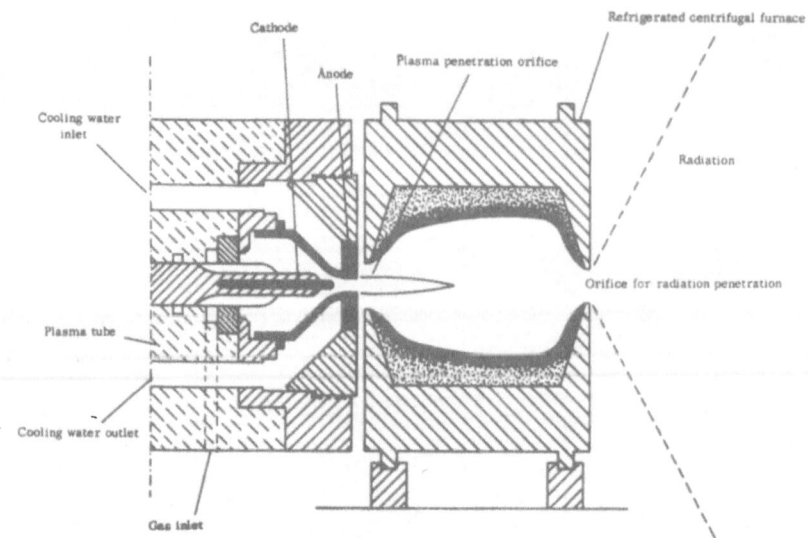

Fig. 11-17. Association of an imaging furnace with a plasma tube having a "flat front" for the treatment of a centrifuged product (cavity formation).

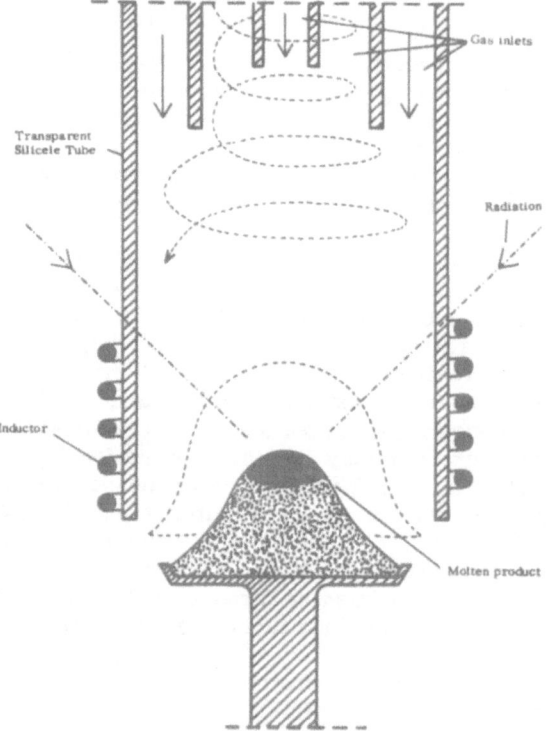

Fig. 11-18. Association of an imaging furnace with a "high-frequency" plasma torch for the fusion of a powdered product.

by means of the above mixed devices, especially the one with a plasma torch. Toward this end, it is interesting to compare the behavior that each of the two heating methods displays when used alone.

Imaging furnaces make it easy to melt nonagglomerated products. This is not the case if one operates with plasma torches, since the velocity of gases frequently causes considerable erosion of the products being treated when the latter have not been subjected to preliminary compression or fritting.

The expansion of gases in a cavity of sufficient dimensions, produced by centrifugation of a pulverized product (Figs. 11-15 and 11-17), considerably limits the erosion effects. Alternatively, one can obtain more or less spherical (uneroded) globules by rapidly passing the powder through the plasma.

It is easy to capture most of the radiation emitted by imaging furnaces by producing a cavity in the substance being treated (centrifugation). The efficiency obtained during direct treatment of a surface depends upon the absorption factor of the substances treated. This factor plays only a limited role in the case of heating by plasma, where the efficiency depends essentially on the importance of heat exchange between the product to be treated and the "flame." In order to increase the heat exchange surface and the contact time, it is always advantageous to project the plasma jet into a cavity formed by the substance, and one observes that the efficiency of the operation is maximum for relatively large cavity dimensions.

The use of a centrifugal furnace heated simultaneously by radiation and by plasma presents considerable advantages. The losses by radiation, during a fusion at a given temperature, depend only upon the dimensions of the aperture (Fig. 11-15) and are thus the same whether one uses a single or both means of heating. The losses by conduction through the walls formed by the centrifugated powder increase with the dimensions of the cavity. Under these conditions, the efficiency of fusion passes through a maximum at the instant the cavity, without being too large, possesses adequate dimensions to permit good exchange between the treated product and the plasma.

It is also of interest to combine the action of imaging furnaces with that of high-frequency heating. The method is applicable not only to electrically conducting substances, but also to all insulators whose conductivity becomes considerable above a certain temperature. The imaging furnace can also serve to initiate a fusion which is then carried out by input of electrical energy.

The above considerations provide a brief survey of certain possibilities of using different types of imaging furnaces. In particular, they display the promising future which appears to be offered by combining the techniques of radiation heating with other methods of producing high temperatures such as the use of plasma tubes.

REFERENCES

1. Glaser, P. E., "The Image Furnace: An Approach to High Temperature," Colloquium on Solar Energy, C.N.R.S., Montlouis (1958).
2. Butler, C. P., "Image Furnace Research," Proceedings of an International Symposium on High-Temperature Technology, Asilomar, November 1959 (McGraw-Hill, New York, 1960).
3. Parisot, J., "Obtaining High Temperatures by Concentrating the Energy Emitted by Intensive Arcs," Colloquium on the Chemistry of High Temperatures, C.N.R.S., Paris (1954).

4. Perrin, J., "Review of Imaging Furnaces," International Study Meeting on the Transmission of Heat, Paris (1961).
5. Foex, M., "Using High-Energy Light Radiation Produced by Imaging and Solar Furnaces," Meeting on Heat Transmission, organized by the French Institute of Combustibles and Energy, Paris (June 1961).
6. Trombe, F., "The Installations at Montlouis and the 1000-kW Solar Furnace of Odeillo-Font-Romeu," Colloquium on Solar Energy, C.N.R.S., Montlouis (1958).
7. Trombe, F., Foex, M., and La Blanchetais, C. Henry, "Conditions of Solar Energy Reception at the Focus of a Parabolic Mirror," Colloquium on Solar Energy, C.N.R.S., Montlouis (1958).
8. Trombe, F., and Foex, M., "Present Practical Applications of Solar Furnaces and their Economic Possibilities of Development," United Nations Conference on New Sources of Energy, Rome, Italy (August 1961).
9. Trombe, F., and Foex, M., Bull. soc. chim. France, 1315 (1954).
10. Glaser, P. E., "Industrial Applications—The Challenge to Solar Furnace Research," United Nations Conference on New Sources of Energy, Rome, Italy (August 1961).
11. Lazlo, T. S., "New Techniques and Possibilities from the Solar Furnace, "United Nations Conference on New Sources of Energy, Rome, Italy (August 1961).
12. Foex, M., "Some Considerations Concerning Present Applications of Solar Furnaces," Meeting on Very High Temperatures, organized by the Polytechnic Faculty of Mons (December 7 and 8, 1961).
13. Morning, D. O., "Arc Image Furnace," ARDC Contract AF 18 (603), 123, University of California (1959).
14. National Aeronautics and Space Administration, Solar Energy in Space, Washington (1962).
15. Gillette, R. B., Solar Energy, 24 (1960).
16. Hibbard, R. R., Solar Energy, 5:129 (1961).
17. Francia, G., "A New Collector for Solar Radiation: Theory and Experimental Verifications," United Nations Conference on New Sources of Energy, Rome, Italy (August 1961).
18. Arthur D. Little, Inc., Notice on Imaging Furnace (1959).
19. Glaser, P. E., and Blau, H. H., "On the Measurement of High-Temperature-Radiation Properties of Solids," Colloquium on Solar Energy, C.N.R.S., Montlouis (1958).
20. Blau, H. H., "Measurements of Flux, Emittance, and Related Properties," Proceedings of an International Symposium on High-Temperature Technology, Asilomar, November 1959 (McGraw-Hill, New York, 1960).
21. Foex, M., "Measurement of Temperatures and Thermal Analysis of Substances Treated with Solar Furnaces," (Communication presented to the IUPAC Meeting in Montreal, August 1961); Bull. soc. chim. France (January 1962).
22. Conn, W. M., and Braught, G., J. Opt. Soc. Am. 44:45 (1954).
23. Diamond, J. J., and Schneider, S. J., J. Am. Ceram. Soc. 43:1 (1960).
24. Foex, M., "Device for the Fusion of Refractory Powered Products by Plasma Tubes," French Patent 880,744 (1961).
25. Collongues, R., Bull. soc. franc. céram. 37 (1960).
26. Brenden, B. B., Newkirk, H. W., Jr., and Woodcock, S. H., Solar Energy 2:8 (1958).

Chapter 12

Thermal Image Description and Measurement

E. S. Cotton*

Pioneering Research Division
QM Research and Engineering Center, US Army
Natick, Massachusetts

A. INTRODUCTION

The optimum use of a thermal imaging source requires a knowledge of the irradiance distribution within the region of maximum beam convergence. In the case of a pulsed source, this requirement is even more severe, since there is no opportunity to reposition the sample during exposure in order to find the ideal geometry. Thermal damage studies nearly always employ sources which have rapid shutter systems producing pulses of variable length, and one of the major problems is that of characterizing and measuring the "image volume" within which experiments are to be carried out. It is desirable to know in advance the irradiance distribution which will impinge upon the sample surface, and also to know how this will be altered if damage is sustained and a part of the sample is vaporized.

At the Army laboratory in Natick, we operate several types of thermal imaging sources for which image description and measurement are necessary. These include a 5-kW tungsten filament projection lamp with an ellipsoidal mirror, a DC carbon arc with both single and double ellipsoidal mirrors, and a large, segmented-mirror solar furnace. All of these sources are equipped with blade-type shutter systems for pulsing the beam, and "rectangular" pulses as short as 0.1 sec can be produced. The solar furnace and the carbon arc are also equipped with vane-type shutters for superposing a time-varying pulse during the rectangular exposure. The use of short pulses imposes a mode of operation somewhat different from that employed in steady-state, high-temperature experiments. The materials under investigation do not generally achieve high temperatures, and the expected effect is usually an irreversible change caused by a time-dependent physical or chemical mechanism. As a result, the characteristics of the image and its reproducibility at each point in space are of prime importance in repeated exposures of samples.

B. IMAGE MEASUREMENT

The classification of an instrument used in measuring radiant thermal energy is usually made in terms of its time constant for attaining thermal equilibrium. As defined by Broido and Willoughby ([1]), a radiometer is an instrument which has a short time constant and gives readings which are

*Present address: M. I. T. Lincoln Laboratory, Lexington, Massachusetts.

proportional to the radiant flux per unit area falling on the detecting surface, or the irradiance. A radiation calorimeter, on the other hand, is an instrument which has a very long time constant compared with the length of the exposure to the beam. Thus, its readings are proportional to the integrated irradiance, or radiant exposure.

1. Radiometer

Measurement of the distribution of irradiance as a function of position within the beam is best performed with a radiometer which has a sensitive area small with respect to the image dimensions. Such an instrument was devised specifically for use in beams of high irradiance by Gardon ([2]) several years ago, and commercial models of the instrument have been made available more recently. As in many other laboratories, we have used these radiometers to measure the irradiance in the thermal beams of our sources. The results of these measurements on a plane perpendicular to the beam at the region of maximum convergence are shown in Fig. 12-1 for the solar furnace and for the carbon arc when imaged with a single ellipsoidal mirror. If one makes many such measurements on various planes, and also along the beam axis, a set of irradiance contours may be drawn, as shown in Fig. 12-2 for a horizontal plane centered in the solar furnace beam. These contours were drawn from many individual radiometer measurements, and also with the aid of image photographs to be discussed later; thus, they represent a composite measurement from which the beam may deviate slightly at a given time. The mapping indicated in Fig. 12-2 does not apply to all planes containing the system axis, since the optical concentrating surface of the solar imaging source is square in outline.

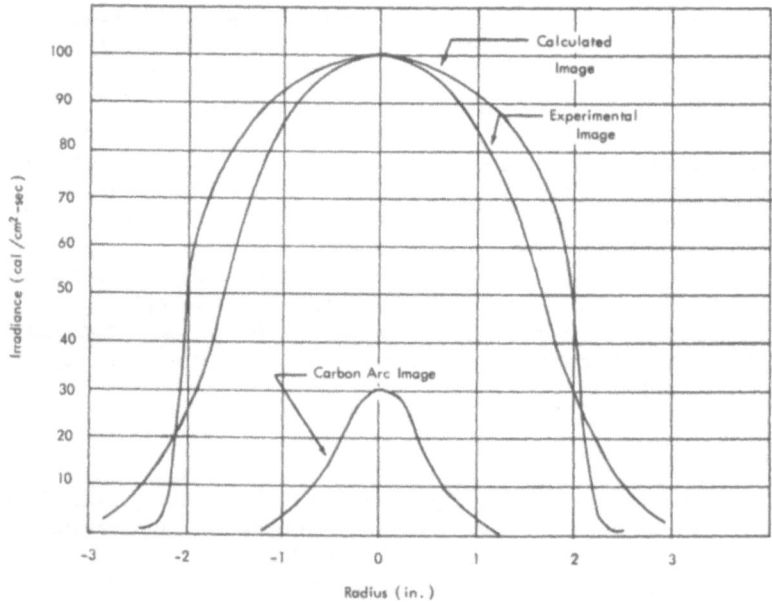

Fig. 12-1. Irradiance profiles at best image position.

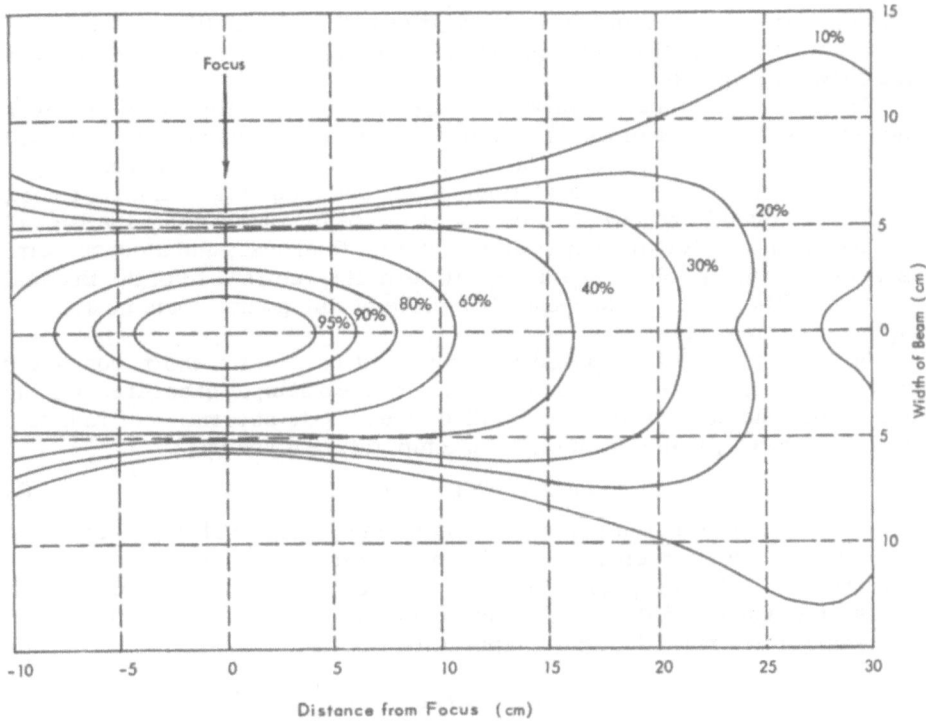

Fig. 12-2. Solar furnace irradiance contours in horizontal plane.

2. Calorimeter

The data shown in Fig. 12-1 and 12-2 are quite useful in determining the ideal position for particular sample assemblies. The criteria of maximum irradiance desired and degree of uniformity needed can be determined from such diagrams, and suitable sample holders may be devised to achieve the necessary exposure conditions. However, the actual irradiance levels experienced by the sample will be determined by the experimental procedure and also by source variations not predictable in advance. For routine exposures to a pulsed beam, the integrating type of instrument has been found to be more useful, for three reasons: (1) It can often be exposed to the same pulse as is the sample, thus eliminating absolute errors in timing of the shutters. (2) It integrates the beam not only in time but over a fixed area of the sample, giving the total radiant exposure delivered to that area. (3) It can be calibrated directly in absolute units which depend primarily on known physical properties of materials.

The radiation calorimeter used ([3-5]) in our laboratories utilizes a copper disc as a heat sink which absorbs the incident radiation. The geometry of this disc is arranged so that it interrupts a known area of the beam in a plane perpendicular to the axis of the system. The mass of the heat sink is quite large compared with the mass of the thermocouple which detects the temperature rise of the disc, thus ensuring quite a long time constant. Similar

devices have been used in other laboratories [1,5], although they differ somewhat in form and conditions of use. Three different types of our calorimeters
are shown in Fig. 12-3.

The discs are tapered on their edges at an angle slightly greater than
the convergence angle of the source beam, so that the incident radiation will
be absorbed only on the front face. Blackening of the discs is accomplished
by electrolytically-deposited platinum black and a coating of camphor soot,
which is frequently renewed. A constantan wire is soldered into the back of
the disc at the center to a depth of one-half its thickness, and a copper wire
is soldered into a hole near the edge to form the return lead of the thermal
junction. The discs are suspended on three needle points in the center of a
rigid brass ring [3].

In use, the disc is exposed to the radiant beam for a short, controlled
period of time, usually the same as that used for sample exposure. During
this pulse, the temperature of the disc rises due to absorption of radiation
by the front face, with the increase in temperature given by

$$dw/dt = aHA/mc - kw \tag{1}$$

where w is the rise in temperature, t is the time, a is the absorptance of the
surface, H is the irradiance, m is the mass of the disc, A is the area of the
front surface, c is the specific heat of copper, and k is a cooling constant. If
we assume that k and H are independent of time, and that c is independent of
the temperature rise, then the temperature rise is

$$w = (aHA/mck)(1 - e^{-kt}) \tag{2}$$

at any time during the heating pulse.

For discs of a practical size, the cooling constant, k, is not negligible, so
that cooling takes place during the heating of the disc as well as after the
heating pulse has been terminated. If the length of the rectangular heating
pulse is t_0, then the cooling curve has a slope

$$\left(\frac{dw}{dt}\right)_c = -\frac{aHA}{mc}(1 - e^{-kt_0}) e^{-k(t-t_0)} \tag{3}$$

For values of k normally encountered, this cooling curve can be extrapolated
back to a time $t_0/2$, giving at that point a w_0 value which is within 0.2% of the
value which would have been attained by the disc if there had been no cooling
at all.

In the use of practical calorimeters, we must also correct for the variation of specific heat with temperature and convert the observed thermocouple
emf to temperature rise. When these variations have been accounted for,
the radiant exposure measured by a copper calorimeter disc with the cold
junction between 20 and 30°C is given by

$$Q = Ht_0 = \frac{2.237m}{aA} E(1 - 0.02031E + 0.0007886E^2) \tag{4}$$

where E is the thermocouple emf. From equation (4) tables can be compiled

Fig. 12-3. Calorimeter assemblies: (1) carbon-arc mounting; (2) and (3) solar furnace mounting.

for discs of various sizes to convert directly from observed emf's to radiant exposure values.

A study ([5]) of the behavior of these calorimeters indicated that the effects of natural convection in the air surrounding the discs may be responsible for a 2-3% decrease in the absolute accuracy of such instruments, even when the other parameters are precisely known. However, the time variation in source intensity is often greater than this, and the calorimeters can normally be constructed so that they agree only to about 3%. Thus, the construction of a convection-free instrument has not yet been necessary for routine measurements. This disc calorimeter can also be used to measure the radiant exposure delivered by a variable pulse, where H varies in a known manner as a function of time. Such a measurement is quite valuable in simulation studies; of course, the extrapolation procedure must be altered in order to make it applicable in the case of variable H ([5]).

3. Radiation-Pressure Detector

The types of instruments just discussed, and also the various absolute instruments used to calibrate image sources, all use the heating effect of the incident radiation to activate a thermoelectric junction, whose output voltage in some way indicates the absorbed energy. We have recently constructed an experimental model of a radiation detector which utilizes instead the mechanical pressure exerted by electromagnetic radiation. The momentum transferred by a pulse of the incident radiation to a reflecting surface is measured in terms of the deflection of a pendulum, thus making possible a mechanical calibration of the beam irradiance.

The principle of this instrument is like that of the simple ballistic pendulum used to demonstrate the conservation of momentum. The beam is chopped or pulsed to produce a short burst of radiant energy which impinges upon the front surface of a disc forming part of a pendulum, as shown in Fig. 12-4. The system composed of the pendulum and the radiation pulse must conserve linear momentum; therefore, the pendulum acquires the momentum of the beam plus the momentum of the reflected radiation. The re-

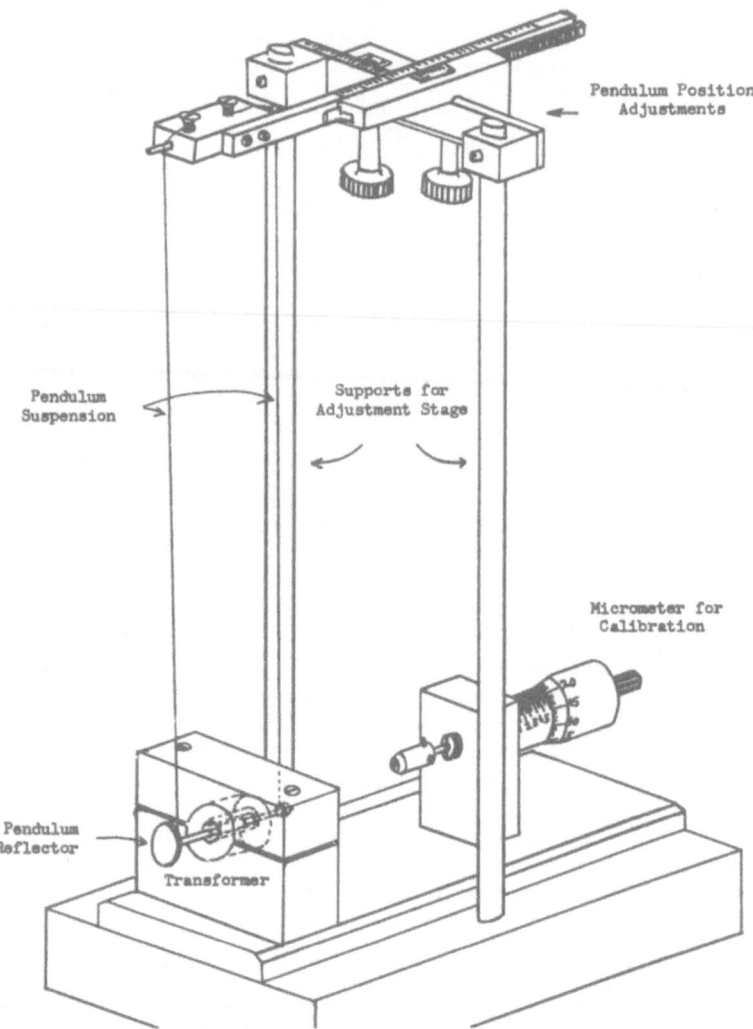

Fig. 12-4. Ballistic pendulum radiation detector.

flection of the pendulum is measured by means of the motion of its shaft, which carries the magnetic core of a linear, variable, differential transformer—a phase-sensitive detector of linear motion. The pendulum is suspended freely, and thus oscillates ballistically after the collision with the radiation pulse. From the amplitudes of the damped oscillations can be inferred the maximum deflection, which is calibrated mechanically with a precision micrometer.

The question of the classification of such an instrument can be raised, since it does not readily fit the descriptions usually laid down for heat sinks. However, the time-constant definition can still be used in a limited sense. If the time constant (period of oscillation) were short compared with the exposure time, the instrument would function as a radiometer, since the deflection would be proportional to the radiation pressure and thus to the flux density in the beam. However, heating effects would soon destroy the device under high irradiance conditions, so that such use is impractical. If the period is made long compared with the exposure pulse, the instrument can properly be called a radiation calorimeter, since its maximum deflection is proportional to the time integral of the force, or the momentum of the pulse.

If the pendulum suspension is made long, so that the above condition on the period of oscillation is fulfilled, then the oscillation can be assumed to be that of a damped simple harmonic oscillator. Under these conditions it can be shown that the radiation pressure is given by

$$p = \frac{a_1}{t_0 A}\left[w_0 m e^{\alpha_1}\left(\frac{a^2}{w_0^2} + 1\right)^{1/2}\right] \tag{5}$$

where a_1 is the amplitude of the first oscillation, t_0 is the pulse length, A is the area of the surface where the radiation impinges, m is the mass of the pendulum, w_0 is the natural frequency, t_1 is the time of the first maximum, and a is the damping constant. The mechanical pressure calculated in (5) can be converted into irradiance by the relation

$$H = \frac{pc}{K(1 + r)} \tag{6}$$

where H is the irradiance, c is the velocity of light, r is the reflectance of the surface upon which the radiation is incident, and K is a geometrical factor which varies from $\frac{1}{3}$ for the radiation in an isothermal enclosure to 1 for a perfectly parallel beam.

Initial trials of this detector have been made using the pulses emitted by photographic flashbulbs. The radiant pulses can be detected easily, but convective effects in the surrounding air and the need for precise balancing of the pendulum inside the transformer disturb the oscillations. We are hopeful that these problems can soon be solved so that we may evaluate usefulness of such a method and calibrate the device. An evacuated enclosure would, of course, remove convective effects, and elaborate shock-mounting would aid in balancing, but these are extreme measures we would prefer to avoid. At present, this calorimeter is wholly experimental.

C. IMAGE PHOTOGRAPHY

Although the most reliable description of the image produced by a thermal source is given by irradiance contour mapping, it is often extremely valuable to obtain instantaneous "photographs" of the thermal image. These photographs can be used to record changes in the image shape and focal length of the optical system, such as occur with a large solar device which is subject to geometrical and climatic variations. The photographs may also be used to reveal and record image characteristics in regions of the convergent beam where measurements are impractical or unduly tedious, as well as to locate positions where further measurement may be needed.

1. Photographic Emulsion

Several methods of obtaining photographs have suggested themselves, and two of them have been adopted for routine use and recording. The most obvious method for producing photographs of an optical image is the use of a commercial emulsion, where the image would be recorded as a variation in density of silver grains in the plane of a thin, developed emulsion. Since most thermal images encountered in practice are fairly large in extent, and include a large amount of visible light, the direct use of films having emulsions sensitive to daylight and incandescent light is possible without additional optics.

One such emulsion is Kodak Autopositive P.B. Film, designed for copying documents, which has also been used for solar photography with astronomical telescopes. The emulsion has a low sensitivity and a very fine grain, so that it is capable of resolving much detail. It has two bands of spectral sensitivity, making use of the Herschel effect: one of these is in the green to infrared and the other is in the blue to ultraviolet. The first of these removes density from the developed image, while the second adds density, after all density has been removed by exposure to yellow light. As supplied, the film has already been exposed so that it has maximum density.

For thermal imaging, we have used this film in a yellow band, where it functions as a positive imaging material. A piece of Kodak yellow sheeting is placed between the source and the image position in such a way that it filters the entire beam. The film, which can be handled in normal room light, is then suspended at the desired image plane and exposed to a pulse of radia-

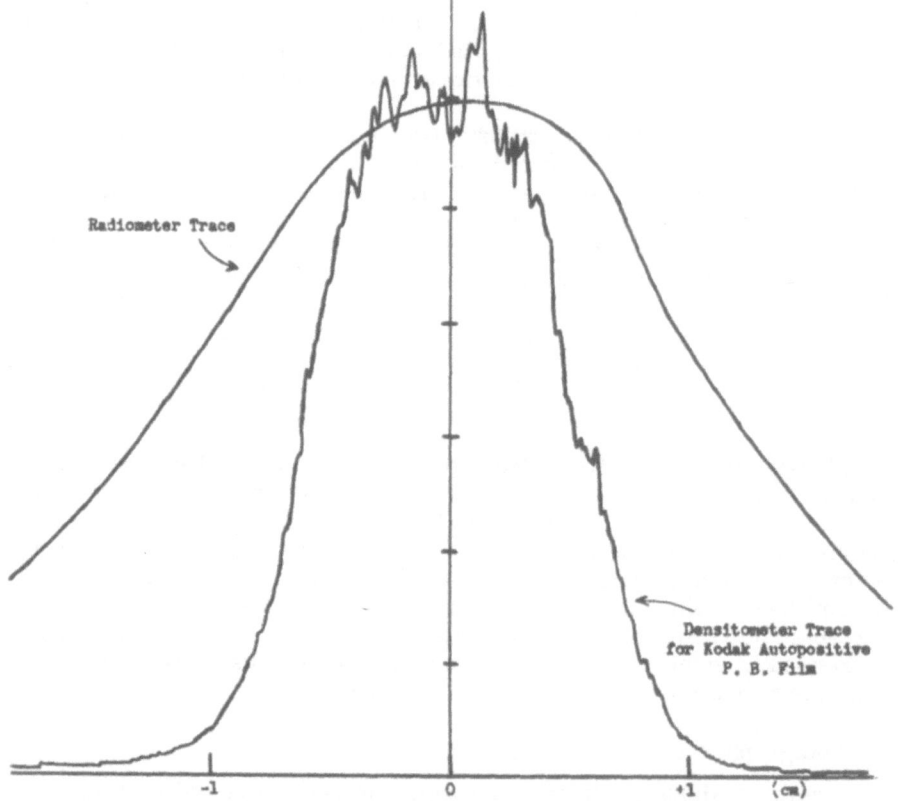

Radiometer Trace

Densitometer Trace
for Kodak Autopositive
P. B. Film

-1   0   +1   (cm)

Fig. 12-5. Densitometer and radiometer profiles for carbon-arc image.

tion. Developing, fixing, and washing are accomplished by routine photographic methods. The final image appears as a transparent region in an otherwise opaque medium. It can be used as a visual photograph of the image, or, with a densitometer, will give numerical data regarding the image shape. Figure 12-5 shows a comparison between the profile of the carbon-arc image obtained in this way and the radiometer profile.

The densitometer image gives the relative transmission of a beam of light through the emulsion, and is seen to have a much narrower profile than that measured by the radiometer. Much of this deviation is probably due to the low sensitivity of the film where the irradiance is low, a phenomenon which is characteristic of photographic emulsions. In addition, the sensitive band of the film makes it spectrally selective, whether it is used as a positive or as a negative emulsion. For mirrors of poor optical polish, such as those used for carbon arcs, the reflectance may be considerably lower for short wavelengths, due to the relative roughness of the surface. Thus, the image seen by the film (when used as a positive medium) at longer wavelengths than those which predominate in the arc spectrum may be considerably better focused than the actual thermal image. For these reasons, the images revealed by such simple photography are probably not reliable for quantitative thermal measurements, even though they may be very useful for alignment and as permanent records. Extensive research, using both spectral regions of sensitivity, could probably yield a technique which would be more quantitative.

2. Styrofoam

We were interested, however, in a form of rapid photography which would yield similar information without the need for development and would employ thermal effects to produce irreversible records. One such thermal effect is that of evaporation, which leaves no by-product and can yield three-dimensional "photographs" of the heated region. The material to be evaporated had to be one of low density, so that the response would be rapid and no other effects would interfere. We selected Styrofoam as a likely material, but found that the white variety was subject to charring and smoking, even though it could be evaporated quickly. Our attention was thus directed to Type 33 Styrofoam, which is a blue, fire-resistant variety used in the building industry. This material can be obtained in boards of various thicknesses up to 3 in.

Exposure of Type 33 Styrofoam to a pulsed beam of thermal radiation results in rapid evaporation of the material without flaming, leaving a clean depression whose depth is approximately proportional to the irradiance. The edge of this depression conforms quite well to a specific exposure contour for the pulse chosen. The range of exposures to which it is sensitive is from about 2 to 40 cal/cm^2, using a pulse of 1-sec duration. For higher irradiances, shorter pulses must be used. The depth of evaporation without melting at 20 cal/cm^2 is about 0.75 in. For higher exposures, droplets of melted polystyrene are formed in the intense portions of the photographs, although these may be tolerated in order to obtain data on the less intense outer portions of the image.

A set of Styrofoam blocks may be exposed very quickly and used to judge the performance of the imaging source. In addition, they may be filed as a

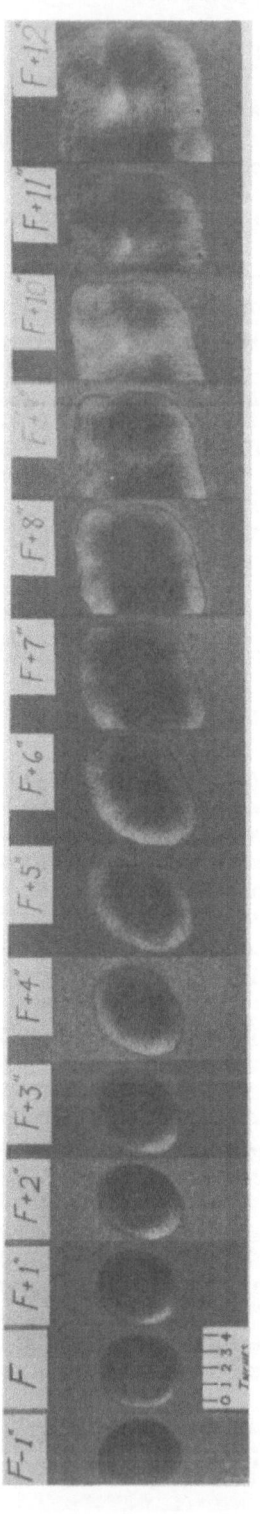

Fig. 12–6. Solar furnace beam along axis.

permanent record of the image contours at various positions in the beam. We have detected seasonal variations in the focal length and optical symmetry of our solar furnace concentrator in this way. A set of such photographs taken with 1-in. Type 33 Styrofoam is shown in Fig. 12-6, with separate planes perpendicular to the system axis indicated to show the variation in image shape over a range of 13 in. These photographs, taken in July 1962, reveal a typical astigmatism acquired in the summer months.

3. Heat-Sensitive Paper

It is also possible to photograph thermal images on a plane surface by using other irreversible reactions induced by heat. Commercial materials in the form of paints and waxes which melt or change color at specific temperatures are notable examples of such reactions. The materials can be painted on paper or cardboard sheets for exposure at selected positions in the beam. We have adopted an indicator which uses color changes occurring at low exposures, of the order of 2-10 cal/cm^2. This indicator is made available to US Army installations for a different purpose, that of detecting liquid vesicant agents in chemical warfare, but the paint which is used as the sensitive element responds equally well to heat. The indicator consists of an olive-green paint which contains a red dye; the dye is insoluble in the paint vehicle at room temperature, but when the temperature is raised beyond this point, the dye deposits as a glossy melt, and for levels beyond this, the paper itself begins to char. We use a manufactured indicator deposited on heavy paper for small images, and for larger images, we prepare our own indicator by painting on card stock.

The appearance of a specific degree of color change approximately follows the curve for achievement of a threshold temperature on the front surface of an insulated slab, as the pulse length is varied. This is probably true for most passive indicators which are coated on paper, and it means that the sensitivity threshold increases with pulse length. We do not use this paper for irradiance determinations, but have found it extremely valuable for checking on image position and mirror alignment. The automatic positioning system for our solar furnace does not incorporate automatic gain control to compensate for solar intensity, since on clear days this quantity changes very slowly. However, manual gain control is used only in steps, and thus the "image lag" of the servo system does change with time. We use the vesicant paper detector to show such image drift and also any drift caused by wind or miscellaneous factors. With a large image such as this, we can easily detect drift of the image center which exceeds 2'; this is equivalent to the accuracy of our general alignment and of the mirror segments, on the average.

D. IMAGE CHARACTERIZATION

The image contour most often presented in order to document the performance of a thermal source is the irradiance vs. displacement cross section on a plane which is perpendicular to the optical axis and is placed along the axis at the point of maximum irradiance or maximum beam convergence. In most cases, these latter conditions are coincident, although for some of the optical elements employed, they may be separated by a small distance. The resultant curve—or image "shape"—can be presumed to yield the most important parameters of the source for research applications, although the distribution in the whole volume of convergence is also quite important, as pointed out earlier.

1. Image Profile

Characterization of this image shape may be accomplished by comparing it to the radiance variation across the radiation source itself, or to an idealized uniform radiator which has the same dimensions as the source. This comparison requires that the focal length of the optical system be precisely known, so that the deviation of the image can be measured in terms of the ideal image diameter. We must usually rely upon the design for the particular optical system employed, or if it is sufficiently accurate, we may measure the focal length with a point source, and thus determine the ideal image diameter. If the source radiance can also be measured, the final ideal image profile can be drawn for relative comparison with the measured image. This comparison, however, is made in terms of the image formed by a perfect optical system and does not take into account the natural distortions which are imposed by any real optical elements, no matter how perfect their shapes. No absolute comparison can be made of these two image profiles, since the irradiance produced by a perfect optical system is still limited by the source geometry.

2. Performance Parameters

Evaluations of imaging performance by specific optical systems have been made, particularly for solar imaging devices. A parameter which evaluates the form of the optical system is the "concentration efficiency," which is the ratio of the flux received within the ideal image to the total flux reflected by the collecting optics [6]. This parameter requires that the ideal image diameter be known and that the flux collected be calculated or measured. For solar devices, the latter quantity is easily obtained, because of the great distance of the source, but for other optical systems, the variation of irradiance at the collecting surface must be determined. The concentration efficiency is valuable for choosing between different forms of optical systems for use with a given source.

Another parameter often used with solar imaging collectors in the "index of geometrical perfection," which is the ratio of the actual flux received within the area of the ideal image to the theoretical flux which would be collected by an ideal optical surface of the particular form used. For continuous surfaces of revolution, such as paraboloids, the theoretical flux can be calculated fairly simply [6], but for segmented, noncontinuous surfaces this task is more difficult, as has been shown for our solar furnace [7]. As with concentration efficiency, both the focal distance and the flux collected by each element of the collector must be known.

Most authors who have reported on the performance of thermal imaging sources in the literature have given detailed information on the actual irradiance distribution of the received image, usually at a predetermined focal plane. Such distributions may be compared in terms of their usefulness for producing a region of high flux density, without regard to the optical design of the imaging source or the value of the maximum irradiance in the image. A shape factor is desired which yields information enabling the reader to evaluate an image source for his own research. Intensity distributions of other kinds are often characterized by the half-width measured in terms of wavelength, magnetic field, or other parameter. Here, however, we are concerned with comparing distributions from various devices and with varying dimensions, so that a useful shape factor must be dimensionless. Furthermore, we need to know not only how rapidly the distribution falls off from its maximum, but also how badly the radiation is dispersed into regions where it cannot be used efficiently.

The following factors, therefore, might be considered for completely characterizing an image source, either separately or in combination:

> Maximum measured irradiance. . . . H_m
> Ideal image diameter. . . . . d
> Measured diameter at $0.9 H_m$ . . . . . . D_9
> Measured diameter at $0.5 H_m$ . . . . . . D_5
> Measured diameter at $0.1 H_m$ . . . . . . D_1

Measurement of H_m requires a calibrated instrument of some type, and the determination of d requires a measurement of the focal length of the imaging beam. The other three parameters may be determined from a relative measurement like that in Fig. 12-1. The quantity D_9 is an indicator of the uniformity of the image, since a sample which receives about 90% of H_m can be assumed to be nearly uniformly irradiated. D_5 corresponds to the half-width often used for designation of the amount of dispersion in an experimental measurement of line intensity. Its usefulness for images of extended sources is quite limited. D_1, on the other hand, is considered to be a very useful factor, since it designates an irradiance level beyond which heating effects are probably negligible for the sample under consideration. The choice of level is, of course, arbitrary, but our experience has shown that levels lower than $0.1 H_m$ are useless for irradiation purposes when the sample is exposed to the full image.

We may form shape factors from several combinations of d, D_9, D_5, and D_1. The easiest approach is to use simple ratios, such as D_9/D_5, D_9/D_1, d/D_1, d/D_5, and D_9/d. For a perfect image of a uniformly radiating source, all of these ratios would be 1.0. However, for actual sources, even if they are perfectly imaged, the ratios would not have this value. In Table 12-I, these ratios are shown for several idealized image forms, based on differing concepts of image shape. For instance, the radiance across the solar disc is nonuniform, as shown by Abetti [8], and if this variation were perfectly reproduced as an image, the ratios discussed above would have values differing from 1.0, as seen in Table 12-I.

TABLE 12-I
Trial Shape Factors

Image	d/D_5	D_9/D_1	d/D_1	d/D_5	D_9/d
Perfect image of uniform source	1.00	1.00	1.00	1.00	1.00
Perfect image of solar disc	0.59	0.58	1.00	1.03	0.58
Gaussian error function ($2\sigma = d$)	0.39	0.21	0.46	0.86	0.46
Gaussian error function ($4\sigma = d$)	0.39	0.21	0.92	1.72	0.23

Some authors have referred to the thermal image produced by a given source as "Gaussian" in shape, which presumably means that the image approximates the form of the error function

$$\phi(r) = (1/\sigma\sqrt{2\pi})\,e^{(-r^2/2\sigma^2)} \tag{7}$$

where r is the radial distance from the optical axis, and σ is the standard deviation of the normal distribution represented by equation (7), with the usual definition. However, such a function is meaningless unless it is related to the ideal image diameter. One could do this, for example, by assuming that the image diameter, d, is equal to 2σ, in which case the area under the curve within d would represent 68.3% of the total cross section. In Table 12-I are shown the values of the shape factors for this function, and it is seen that they are quite low. If we use $d = 4\sigma$ instead, those factors which involve d improve, of course.

An alternative method of comparing thermal images to a Gaussian function was given by Davis [9], who presented equation (7) in the form

$$H(r) = H_m e^{-a\pi r^2} \tag{8}$$

By fitting the irradiance value on the optical axis to the experimental curves in terms of r and H_m, one can assign a an approximate value which also becomes a shape parameter. However, this type of curve-fitting requires some arbitrary relation between the curves, and is equivalent to the above assumptions, except that the ideal image is not used. A more accurate method of curve-fitting was employed by Davies [10] as a means of calculating the loss of heat due to lateral flow in an irradiated sample, but this also was not related to the ideal image diameter. Such accurate image characterization, which requires a series of Bessel functions, becomes extremely complex for intercomparison of several sources.

In a physical sense, the error function does not apply to the entire image, but only to a determination of single points in the image, such as the edge. The transition from no illumination to full illumination within the image of a uniform source should follow a curve similar to the integral of equation (7), which is the familiar sigmoid curve of probability theory. The "edge gradient," or slope, is then indicative of the optical quality of the imaging system. The use of an edge gradient was suggested by Shack [11] and Cox [12] for use in optical design as an alternative approach to exclusive dependence on resolution of point and line images. Shack demonstrated that the derivative of the edge gradient has the same shape as the distribution of irradiance in a fine-line image. For precision optical systems, the edge gradient may be evaluated by ray-tracing or on the basis of diffraction theory and physical optics. However, these methods are seldom applicable to the systems employed in thermal imaging. The irradiance measurements made on thermal images are much less accurate than the detailed examination of an image with a knife-edge, and yet the general description in terms of the edge gradient seems to be very useful. However, it still remains for us to relate the ideal image diameter d, to the dispersion indicated by the edge gradient; this problem will be examined elsewhere.

TABLE 12-II

Published Data on Thermal Imaging Sources

Source	Maximum irradiance (H_m), cal/cm²-sec	Ideal image diameter (d), cm	D_1, cm	D_5, cm	D_9, cm	D_9/D_5	D_9/D_1	d/D_1	d/D_5	D_9/d
Avco Corp. 60-in. paraboloid solar furnace [13]	495	0.63	0.73	0.62	0.38	0.61	0.52	0.86	1.02	0.60
CIT lens-type solar furnace [14]										
1 central lens only	30	3.7*	4.7*	3.4*	1.8*	0.53	0.38	0.79	1.08	0.49
19 lenses	220	3.7*	6.7*	3.2*	0.9*	0.28	0.13	0.55	1.16	0.24
Curtis-Wright 60-in. paraboloid solar furnace [15]	600	0.63	0.85	0.53	0.28	0.53	0.33	0.74	1.19	0.44
Defense Research Board (Canada) solar furnace [16] (60-in. Cassegrain paraboloid)	30	2.3	3.0	2.3	1.3	0.56	0.43	0.77	1.00	0.56
MIT flat-segment solar furnace [17]	5.5	7.6	9.2	7.2	5.0	0.69	0.54	0.83	1.06	0.66
National Carbon Co. double-ellipsoid carbon-arc [18]	360	1.2	3.3	1.5	0.68	0.45	0.21	0.36	0.80	0.57
University of Rochester single-ellipsoid carbon-arc [19]	32	4.5	5.1	2.8	0.99	0.35	0.19	0.88	1.61	0.22
US Army spherical-segment solar furnace [7]	100	9.9	12.4	8.2	4.2	0.51	0.34	0.80	1.20	0.42
US Army single-ellipsoid solar furnace [4]	30	4.4	5.3	2.6	1.2	0.46	0.23	0.83	1.69	0.27
USNRDL double-paraboloid carbon-arc [19]	90	1.6	–	–	0.6	–	–	–	–	0.38
USNRDL condenser-relay carbon-arc [20]	28.5	5.3	–	–	2.5	–	–	–	–	0.47
USSR 10-m paraboloid solar furnace [21]	23	5.0	34.6	15.5	9.0	0.58	0.26	0.14	0.32	1.80
USSR 2-m paraboloid solar furnace [21]	864	0.74	1.18	0.50	0.25	0.50	0.21	0.63	1.48	0.34
USSR solar heat station model [21]	1.5	20.0	26.8	12.9	6.4	0.50	0.24	0.75	1.55	0.32
Sheffield University, Manchester College of Science and Technology carbon arc	300	1.6	2.32	1.42	0.57	0.40	0.25	0.69	1.13	0.36

*Relative units.

3. Shape Factors

Table 12-II lists a number of thermal imaging sources for which specific data have been published. We have shown the pertinent information for each of the reported images, where available, and the image shape factors. This table is not meant to be a compilation of all available data, but only serves to illustrate the application of image shape factors to several differing types of sources. The meanings of the various shape factors are indicated below.

D_9/D_5 . . . . This ratio yields information regarding the peaked quality of the measured image, whether it is caused by imperfections in the optics or the source (maximum value = 1.00).

D_9/d . . . . Same as D_9/D_5, but for the theoretical image.

D_9/D_1 . . . . The average edge gradient is indicated by this ratio, including the information given by D_9/D_5 (maximum value = 1.00).

d/D_1 . . . . This ratio indicates how much radiation has been lost from the ideal diameter, regardless of the region of loss. (Values exceeding 1.00 are not to be expected.)

d/D_5 . . . . The crossover point for the edge gradient is compared with the ideal diameter in this ratio. Good images should have a value near 1.0, but deviations will occur both above and below this ideal value.

It is difficult to select any single shape factor as being most sensitive in terms of the usefulness of the source. Likewise, it is not possible to select from Table 12-I any theoretical image which best describes a good thermal image. The MIT solar furnace, which does not employ focusing optics, should produce an image corresponding to a uniform source with an edge gradient. Similarly, a precise paraboloid solar furnace should produce an image which approximates the radiance of the solar disc. The Avco solar furnace is seen in Table 12-II to have values for D_9/D_5 and D_9/D_1 which are near those for the solar disc in Table 12-I. Both the MIT and Avco devices, however, deviate most from their ideal values in regard to the ratio d/D_1.

The low value of d/D_1 for the double ellipsoid carbon-arc source merits additional discussion. This system is unlike most of the others in that it employs two focusing surfaces to form the image, and it is thus expected that errors in curvature will have a cumulative effect in the image. The second optical surface must use as its source a dispersed thermal image which corresponds to those produced by the single ellipsoidal carbon-arc sources. It is interesting to note that this additional optical transfer function makes its principal change felt on the ratio d/D_1, leading to the conclusion that this ratio is the most sensitive to optical imperfections.

Without making a restrictive choice of parameters, it is clear that the specification of four factors is probably sufficient to describe the image of each source in general terms. These are: H_m —maximum irradiance, d —ideal image diameter, d/D_1 —radiation loss ratio, and D_9/D_5 —image peaking factor.

In summary, then, a measured thermal image cross section may be evaluated in three ways: first, it may be compared to the variation in radiance across a diameter of the source, if this is actually known; second, it may be compared to a model image having the same ideal diameter; third, it may be compared to other thermal images from similar optical systems. Numerical comparisons may be made in terms of the parameters previously described, emphasizing those which are most important to the user of the thermal radiation.

REFERENCES

1. Broido, A., and Willoughby, A.B., Measurement of Intense Beams of Thermal Radiation, Report No. USNRDL-TR-35, US Naval Radiological Defense Laboratory, San Francisco, California, 1 February 1955.
2. Gardon, R., Rev. Sci. Inst. 24:366 (1953).
3. Davies, J.M., Unpublished work (1953).
4. McQue, B., The QM Arc, QM Research and Development Command, Pioneering Research Division, Report No. T-6, 29 March 1956.
5. Cotton, E.S., and Levine, A., The QM Disc Calorimeter, Report No. T-28, Radiation Physics Laboratory, Pioneering Research Division, QM Research and Engineering Center, Natick, Mass., 29 August 1960.
6. Loh, E., Duwez, P., Hiester, N.K., and Tietz, T.E., Theoretical Considerations on Performance Characteristics of Solar Furnaces, Report No. 3, Technical Report No. I, Stanford Research Institute, 16 January 1956.
7. Cotton, E.S., Lynch, W., Zagieboylo, W., and Davies, J.M., "Image Quality and Use of the US Army Quartermaster Solar Furnace," Conference on New Sources of Energy, Rome, Italy, August 1961.
8. Abetti, G., The Sun (D. VanNostrand Co., New York, 1938).
9. Davis, T.P., The Carbon-Arc Image Furnace, Proc. Symposium on "High Temperature—A Tool for the Future" (Berkeley, California, June 25-27, 1956), pp. 10-15.
10. Davies, J.M., The Effect of Intense Thermal Radiation on Animal Skin; A Comparison of Calculated and Observed Burns, Report No. T-24, Radiation Physics Laboratory, Pioneering Research Division, QM Research and Engineering Center, Natick, Mass., 29 April 1959.
11. Shack, R.V., "A Proposed Approach to Image Evaluation," p. 275, National Bureau of Standards Circular 526: Optical Image Evaluation, US Dept. of Commerce, 1954.
12. Cox, A., "Image Evaluation by Edge Gradients," p. 267, National Bureau of Standards Circular 526: Optical Image Evaluation, US Dept. of Commerce, 1954.
13. Laszlo, T.S., Solar Energy 6:69 (1962).
14. Loh, E., Hiester, N.K., and Tietz, T.E., Solar Energy 1(No. 4):23 (1957).
15. Laszlo, T.S., Solar Energy 2(No. 3):78 (1957).
16. Wilson, L.G., and Drew, G., A Solar Furnace Using Cassegrain Optics, Defense Research Chemical Laboratories, Report No. 311, Ottawa (1959).
17. Gardon, R., Rev. Sci. Inst. 25:459 (1954).
18. Null, M.R., and Lozier, W.W., Rev. Sci. Inst. 29:163 (1958).
19. Davis, T.P., Krolak, L.J., Blakney, R.M., and Pearse, H.E., J. Opt. Soc. Am. 44:766 (1954).
20. Broido, T.R., The Production of Intense Beams of Thermal Radiation by Means of a High Current Carbon Arc and Relay-Condenser Optical System, US Naval Radiological Defense Laboratory, Report No. USNRDL-417, 24 November 1953.
21. Aparisi, R.R., Utilization of Solar Energy 1:151, Moscow (1957).

Chapter 13

Calibration of Sources for Imaging Furnaces

C. P. Butler
U. S. Naval Radiological Defense Laboratory
San Francisco, California

A. INTRODUCTION

A unique feature of the image furnace is that the heat flux available to an object at the focal plane is a constant regardless of the object's temperature. While this feature appears self-evident, it has not been widely exploited, in the solution of some heat flow problems, partly because of the small area of uniform irradiance, but also because the goal of most image furnace research has been to obtain the very highest possible temperatures, regardless of the area.

Before heat flow problems can be attacked with an image furnace, a knowledge of the absolute flux, its variation with time, and its areal distribution over the exposure plane must be gained. In some applications, the spectral distribution of the energy incident on the specimen must also be known, but this requirement will not be considered here, because we are concerned only with blackbody calibrations.

This paper is a description of techniques used in finding a focus at the plane of maximum area (with constant irradiance) for three types of image furnaces: the double parabolic, the single elliptical, and the relay-condenser. These are more familiarly known as the Navy 36-in. Searchlight, the Brenkert Projector, and the Mitchell Background Projector. These are conventional furnaces, and each one may be used within certain limits for simulating the infinite plane, unidirectional heat flow boundary condition.

B. IMAGE FURNACE OPTICS

The mirrors and lenses used in image furnaces are nowhere near as precise as those employed in conventional optical systems. Standards are less exacting partly because the grinding and polishing of large mirrors and lenses is so costly, but primarily because precision optics afford little gain in resolution, due to the fact that the distribution of energy in the plasma of an arc, and in fact in the sun itself ([1]), is not uniform in any single plane.

A combination of poor optics and an uncertain distribution of energy from the source means that classical ray-tracing methods are almost useless in predicting the behavior of a given image furnace. A great deal of work has been done to solve the problem of off-axis rays in parabolic and elliptical mirror systems, and while rigorous solutions have been offered, they should be used with caution.

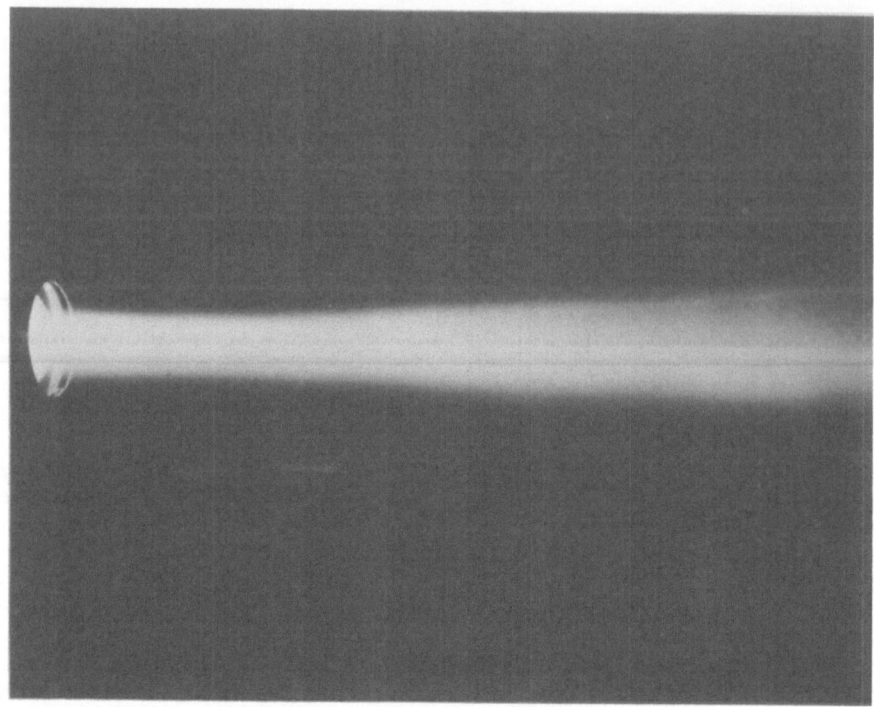

Fig. 13-1. Photograph of the caustic of the Breakert image furnace, operating with an irradiance of 4.0 W/cm^2 and an f-ratio of 7.

The Hartmann test was used at one time to align the lenses of the relay-condenser furnace and to find and map the contour of the irradiance at the exposure plane, but it proved inadequate ([2]). Subsequent mapping of the irradiance with a water-flow calorimeter showed that the plasma of the positive carbon could not be simulated by a plane surface. Similar attempts to find the best focus in the double parabolic mirror system proved fruitless.

The only way to assess the quality of the image in an image furnace is to operate the arc in a normal fashion, and then explore, with suitable instrumentation, the various exposure planes with full-beam irradiance.

It is common practice to refer to the focus, or the focal plane, of an image furnace, implying that this point coincides with a single plane normal to the optical axis. The focal point is the position at which the irradiance maximizes and is, of course, the location where the highest temperatures can be reached. The dimensions of the area of illumination at this point define the diameter of the spot, or the diameter of the melt of a refractory material. The actual area is often estimated from the size of the melt or from the color change in a heat-sensitive paint or paper, while such a method is quite adequate for preliminary work, it does not map the irradiance through the area so defined.

Actually, the converging rays of an image furnace do not focus in a single plane, but rather outline a roughly cylindrical volume of space bounded by the envelope of the caustic. The dimensions of this caustic depend on both the source and the optics, and can only be examined with a scanning device whose detecting area is smaller than the smallest dimension of the beam.

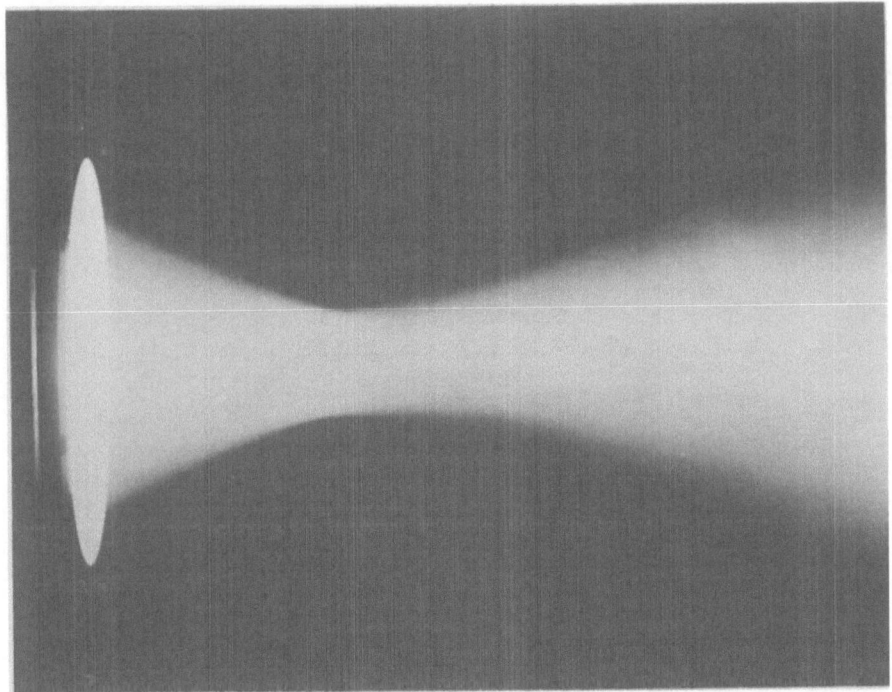

Fig. 13-2. Photograph of the caustic of the Mitchell image furnace operating with an irradiance of 125 W/cm² and an f-ratio of 0.8.

An interesting feature of the caustic is the conical hole appearing at both sides of the focus along the optical axis. This occurs in both refractive and reflective systems. At some distance from the arc, in the elliptical mirror furnace, only a great annular ring remains, with shadows crossing where the arc-supporting mechanism intervenes.

The outlines of the caustic for three different image furnaces are shown in Figs. 13-1, 13-2, and 13-3. We made these photographs by focusing a camera at right angles to the optical axis and spraying a fine powder from a flocking gun into the beam. As each particle crossed the beam, it left a trace on the emulsion, shown by the smooth outline of the envelope of the caustic. Pictures such as these are useful for qualitative analysis only and cannot be used for evaluation of the irradiance of the beam at any place.

C. IRRADIANCE MAPPING

The most satisfactory method of measuring the irradiance levels near the exposure plane is to use a water-cooled radiometer. This device is basically identical to the Gardon radiometer (³), except that the cold junction is maintained at a constant temperature, the sensitivities are lowered because the aperture is narrow, and silver rather than constantan is used for the receiving foil. A schematic view of one of these radiometers is shown in Fig. 13-4. A very useful radiometer for image furnace mapping has a 2.5-mm-diameter aperture and a sensitivity of 10 cal/cm²-sec-mV. The radiometer

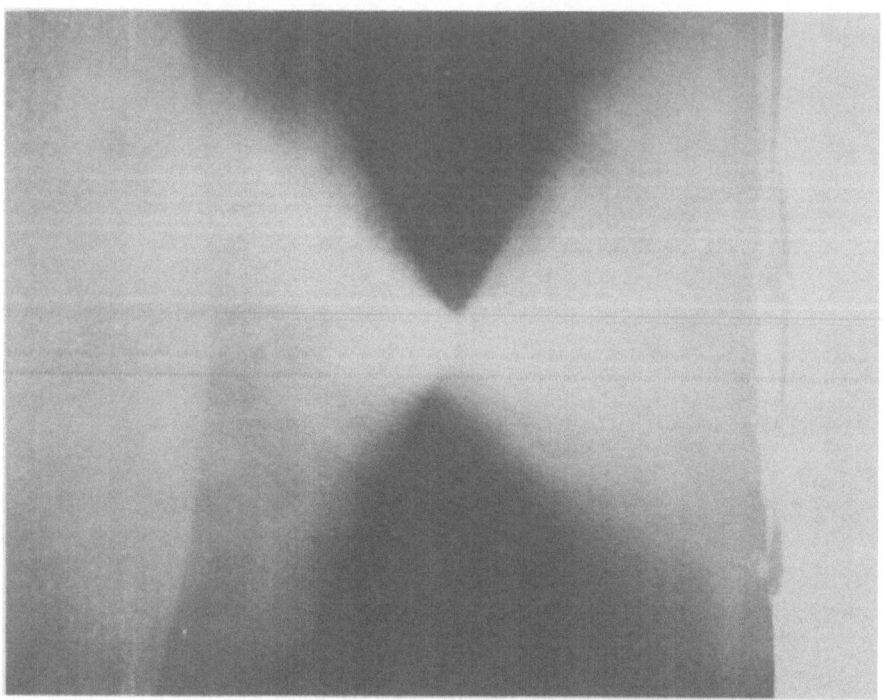

Fig. 13-3. Photograph of the caustic of the 36-in. Navy Searchlight image furnace, operating with an irradiation of 360 W/cm^2 and an f-ratio of 0.4.

output can be coupled directly to a recording potentiometer, and irradiances as high as 100 cal/cm^2-sec can be scanned with no danger to the foil.

Since the optical axis can only be approximated by laying out a central line normal to the mirror edge or lens rim, a three-dimensional motion must be used to map the envelope of the caustic. For this purpose the sample holder and bracket for the radiometer can be mounted on the table of a milling machine. Since the drive screws of a universal milling machine move the table in three perpendicular directions, the whole space within the caustic can be scanned.

In practice, some arbitrary definition of uniform irradiance must be established—for example, a 90% of the maximum irradiance contour. This means that an area of uniform irradiance will not vary in any place by more than 10% with no description of the locations of the high and low regions. Horizontal traverses are then made at some arbitrary location along the optical axis in the search for the location of the single plane of maximum irradiance or of maximum area. At this time, we look for the dip in the center of the caustic, because the greatest area of uniform irradiance occurs at just this position. Contour mapping of irradiance levels is a tedious job, because of time variations of the arc which are of the same order of magnitude as those under scrutiny for the shape of the exposure plane. For this reason, repeated traverses must be made, until the true outline of the caustic can be mapped.

Two maps of irradiance contours are shown in Fig. 13-5 for two lens

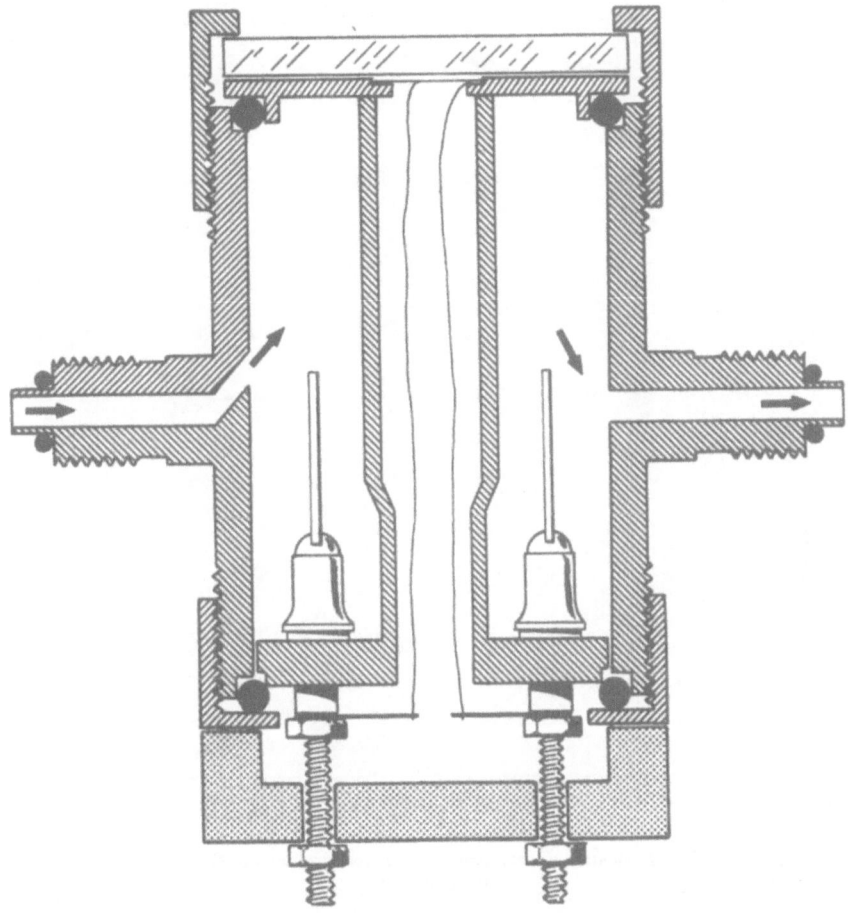

Fig. 13-4. Schematic drawing of water-flow radiometer.

positions of the relay-condenser system furnace. It will be noted that the
outline is far from smooth; the undulations are not due to arc variations, but
to a combination of the poor lens and the distribution of energy about the
plasma. It should be noted also that each of these contours is the outline
of the caustic viewed from a single plane perpendicular to the optic axis.
These maps do not reflect the dip, or the hole in the center mentioned earlier,
for by definition, an area of uniform irradiance includes everything within
10% of the maximum. A more detailed cross section of these irradiance contours would show a dip in each, not necessarily at the center.

It should be clear that if one were to map the exposure plane just described with a radiometer whose receiver diameter was, say, 1.2 cm, the
distribution would look quite different, and one might be justified in describing
the radial distribution as Gaussian. Mapping with this type of radiometer
has been done ([4]), but unless the resolution of the detector is very much
smaller than the minimum width of the beam, such a calibration has little
meaning other than to give a figure of the average flux density. If only the

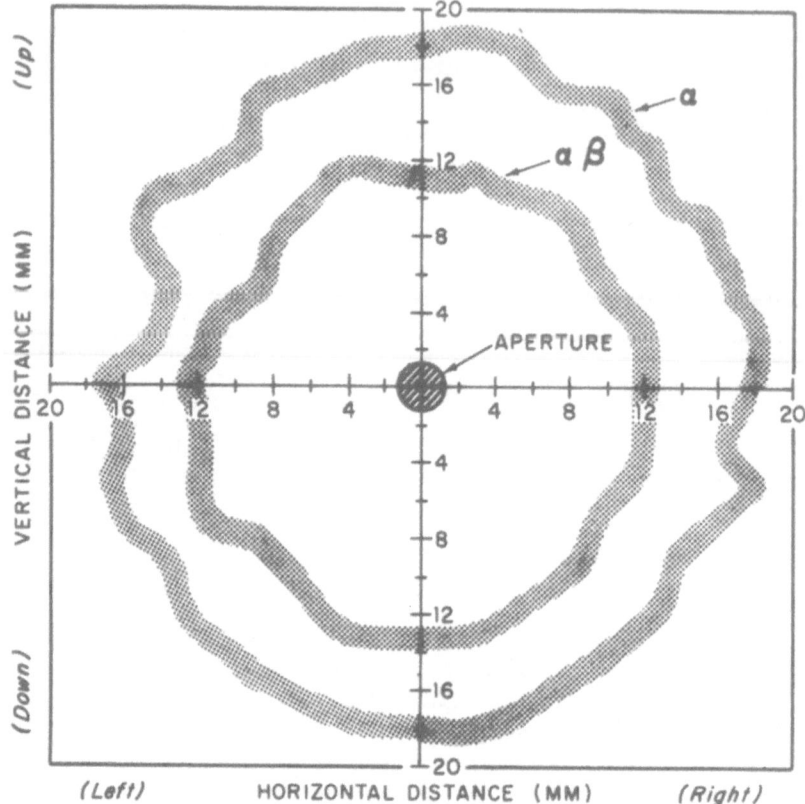

Fig. 13-5. Map of constant irradiance contour for two lens positions of the Mitchell image furnace.

average flux is required, then the ratio between the diameter of the detector and the diameter of the spot is not critical ([5]).

Anyone who has watched a rotating positive carbon closely will have noted progressive changes in the shape of the crater lip, changes which impart a wobbling motion to the plasma. In a nonrotating carbon, uneven burning of the crater lip throws the beam to one side quite rapidly. Such movements of the plasma are reflected in new spatial energy distributions at the exposure plane, so that the maximum irradiance area should be used with certain reservations. If the study to be made requires highly uniform irradiance, we recommend that not more than three quarters of the full measured diameter be used, because even the most meticulous mapping of a long series of carbons may not adequately describe the behavior of a new shipment or a new batch of carbons from another box.

The Mitchell Background Projector gives the most nearly uniform irradiance of any of the three types of furnaces. It is, in fact, inherently better than the other type, because the relay-condenser system is an optical system in which the actual plasma is always out of focus. The exposure plane in these furnaces is in focus for an imaginary plane inside the first set of lenses; hence, any small inhomogeneities in the crater of the carbon or of

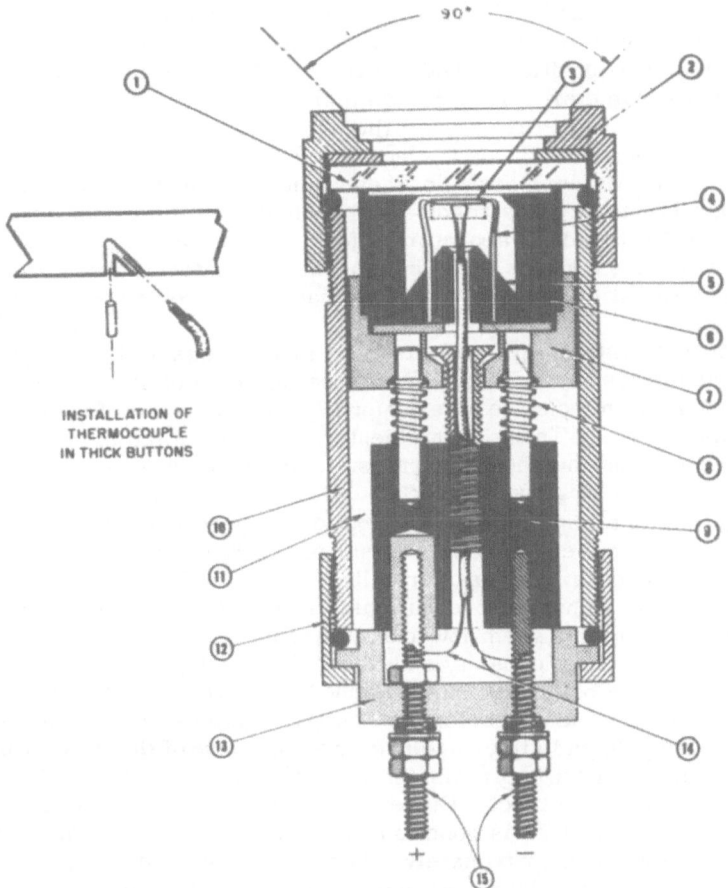

INSTALLATION OF
THERMOCOUPLE
IN THICK BUTTONS

Fig. 13-6. Schematic drawing of the NRDL field calorimeter.

the plasma are reflected in an overall change in brightness, rather than a change in any single area at the exposure plane.

The foregoing discussion of irradiance contour mapping deals solely with location of the exposure plane, and not with determination of the absolute flux density.   The reason for this is that radiometers have an uncertain time response, cannot be calibrated readily, and sometimes show anomalous changes in response.   As comparative instruments, they have no rival.

D. RADIATION STANDARDS FOR IMAGE FURNACES

For many years, the primary standard for measuring beam irradiance at the Naval Radiological Defense Laboratory has been the "Standard Water Flow Calorimeter" ([6]).   While this instrument is maintained at the NRDL, it is rarely used. What is actually used is a secondary standard calorimeter which was originally calibrated against the water flow instrument. The secondary standard is a portable instrument, readily adaptable to several collecting angles for different beam convergences.

Besides this secondary standard, we maintain a set of twelve field calorimeters, divided into four sets of three each. Each of these sets has a different set of sensitivities. These are occasionally compared against the secondary calorimeter. They are so maintained (i.e., by being sealed with lacquer) that unless they are opened, their constants should remain unchanged indefinitely. These standards are now about eight years old.

A schematic drawing of one of these field calorimeters is shown in Fig. 13-6. Our laboratory has made several hundred of these devices, and has furnished many to other laboratories for purposes of comparison and calibration of various sources. The bulk of this laboratory's data taken by this on thermal radiation from nuclear detonations was obtained with these instruments.

Source calibration with one of these instruments is limited to an area $3/8$ in. in diameter, or 0.7 cm^2. We chose this diameter in the early years of our work to match the area of uniform irradiance of the 36-in. Navy Searchlight source. These instruments are also limited to sources whose spectral emission matches the transmissivity of the quartz window with which all calorimeters are provided.

It appears that many laboratories build their own standard calorimeters, and design them to meet the requirements of a specific image furnace. Since the precision necessary for proper calibration of an image furnace is of the order of 5%, the elaborate precautions and the work entailed in building a water flow calorimeter seem now to be unnecessary. A simple button of pure metal—with known heat capacity and dimensions—suspended in the exposure plane and properly blackened is sufficient for the absolute calibrations.

Gold, silver, and occasionally copper buttons are used for these standards, but gold and silver are preferred because of the ease of blackening. It appears that the only satisfactory blackening technique is electrolytic application of platinum black. Our experience has been that results are reproducible, that the black is good to at least $10\,\mu$, and more important still, that it is stable to approximately 800°C. No other black approaches this figure without breaking down. All our standard calorimeters, as well as the secondary standards, are blackened with platinum.

E. ARC OPERATION

Some observations may be pertinent here as to the operation of carbon arcs and their stability during calibration. The actual operation of any arc is easily learned from detailed operating instructions which usually come with the machine. Such instructions, however, are written primarily for the man who operates the arc in a theater, where the requirements are considerably less severe than in the laboratory, where precise measurements are being made.

The sound of the arc, the shape of the tail flame, the degree of wobble of the plasma due to uneven crater rims, small fluctuations in the current and voltage all require the closest attention by an operator, even though they are automatically controlled. The vertical and horizontal adjustment of the electrodes with respect to each other should be watched with close attention. In the case of the horizontal positive, the length of the carbon remaining has a small effect on the calibration, because of the change in the obscuration as the carbon shortens. We have found that an experienced operator, i.e., one who understands what the basic problem is in obtaining a constant flux at the exposure plane, can make secondary corrections to the automatic feed mechanisms so that fluctuations in the irradiance can be kept to less than 5% over periods of 5 min and sometimes longer. This means that after a complete calibration, the energy level at the exposure plane for the next carbon will be within 5% of that predicted.

REFERENCES

1. Jose, Paul D., The Flux Through the Focal Spot of a Solar Furnace, HDRRM-TM-57-3, Holloman Air Force Base, 1957.
2. Broido, T. R., The Production of Intense Beams of Thermal Radiation by Means of a High Current Carbon Arc and Relay-Condenser Optical System, USNRDL-417, 1953.
3. Gardon, R., Rev. Sci. Instr. 24:366 (1953).
4. Mill, M. R., and Lozier, W. W., J. Soc. Motion Picture Television Engr., 68(No. 2):80 (1959).
5. Evans, P. E., and Wildsmith, G., Brit. J. Appl. Phys. 13:68 (1962).
6. Willoughby, A. B., Rev. Sci. Instr. 25(No. 7):667 (1954).

Crystal Growth

Chapter 14

Crystal Growth in the System FeO–MgO–Fe$_2$O$_3$*

F. A. Halden
Ceramics Research Section
Stanford Research Institute
Menlo Park, California

A. INTRODUCTION

Well-characterized single crystals of the double-oxide magnetic ferrites are of considerable interest to investigators concerned with the basic properties of these materials—particularly since they permit studies of the influence of possible cationic arrangements on magnetic characteristics. As a result, much effort has been devoted to obtaining reasonably large, pure crystals of these materials, primarily by the use of fusion techniques. Although frequent claims have been made for the growth of single crystals, unfortunately, except for a few compounds, careful examination reveals the presence of two phases, excessive ferrous iron contents, or both. The lack of phase-diagram information at the liquidus for the systems of interest reduces single-crystal growth attempts to an almost empirical approach.

With the development of arc image Verneuil growth techniques, it became possible to study ferrite compositions fused under a wide range of oxygen partial pressures while eliminating contamination normally introduced by hot furnace elements and melt containers. Magnesium ferrite was selected as a model system to illustrate the type of information that can be obtained using an arc image furnace. This material was chosen for several reasons: it is a major component in many ferrite compositions; it can be analyzed readily by chemical techniques; it has, to our knowledge, not been prepared as a single crystal using fusion techniques; and a considerable amount of phase equilibria data are available at low (<1300°C) temperatures.

Paladino ([1]) has summarized the available phase equilibria data for the MgO—FeO—Fe$_2$O$_3$ system up to 1300°C. Figure 14-1 indicates the results of his study, which shows that the composition MgFe$_2$O$_4$ falls outside of the single-phase spinel region, and that in the absence of ferrous iron the spinel region has no width. With the loss of oxygen, compositions near a stoichiometric mixture pass from the spinel-plus-hematite zone, through the single-phase spinel region, into a spinel-plus-magnesio-wustite area, as indicated by the light lines in Fig. 14-1. Paladino also indicated the influence of varying oxygen partial pressures on the extent of reduction (shown in the figure by the oxygen isobar). Woodhouse and White ([2]) studied this system to 1650°C in air, but were not concerned with the liquidus surface or with oxygen

*This work was supported by the U.S. Navy Office of Naval Research. Reproduction in whole or in part is permitted for any purpose of the United States Government.

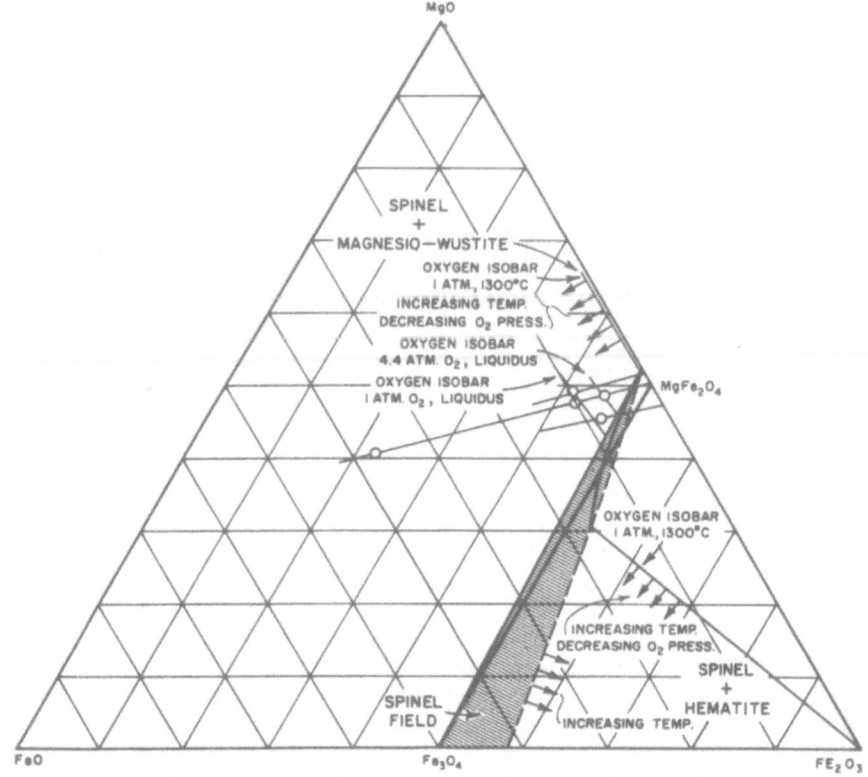

Fig. 14-1. Phase equilibria in the system FeO-MgO-Fe$_2$O$_3$.

partial pressures greater than that of air. The purpose of the present study is to extend these data to the liquidus and to oxygen pressures greater than 1 atm in order to indicate the conditions required for the growth of single crystals having low ferrous iron content.

B. EXPERIMENTAL PROCEDURE

The arc image crystal grower used in this study has been described in a previous publication ([3]). The furnace chamber has been modified, as shown in Fig. 14-2, to provide improved light transmission and to permit rotation of the boule during growth. By using a compression-type Teflon gasket and a thick-walled Pyrex chamber, pressures up to 50 psig can be maintained safely.

Ferrite powders were prepared by the oxalate precipitation technique using reagent-grade chemicals. Three powder compositions were employed to provide, respectively, approximate-stoichiometric MgFe$_2$O$_4$, slightly magnesium-rich, and slightly iron-rich materials. The purities and compositions of these powders are shown in Tables 14-I and 14-II. X-ray diffraction measurements on these powders revealed that, in agreement with Paladino, in all cases Fe$_2$O$_3$ was present in addition to the spinel ($a_0 = 8.38$ A).

Boules of the stoichiometric ferrite composition were grown by the normal Verneuil procedure under 1 atm of argon, 1 atm of oxygen, and 4.4 atm

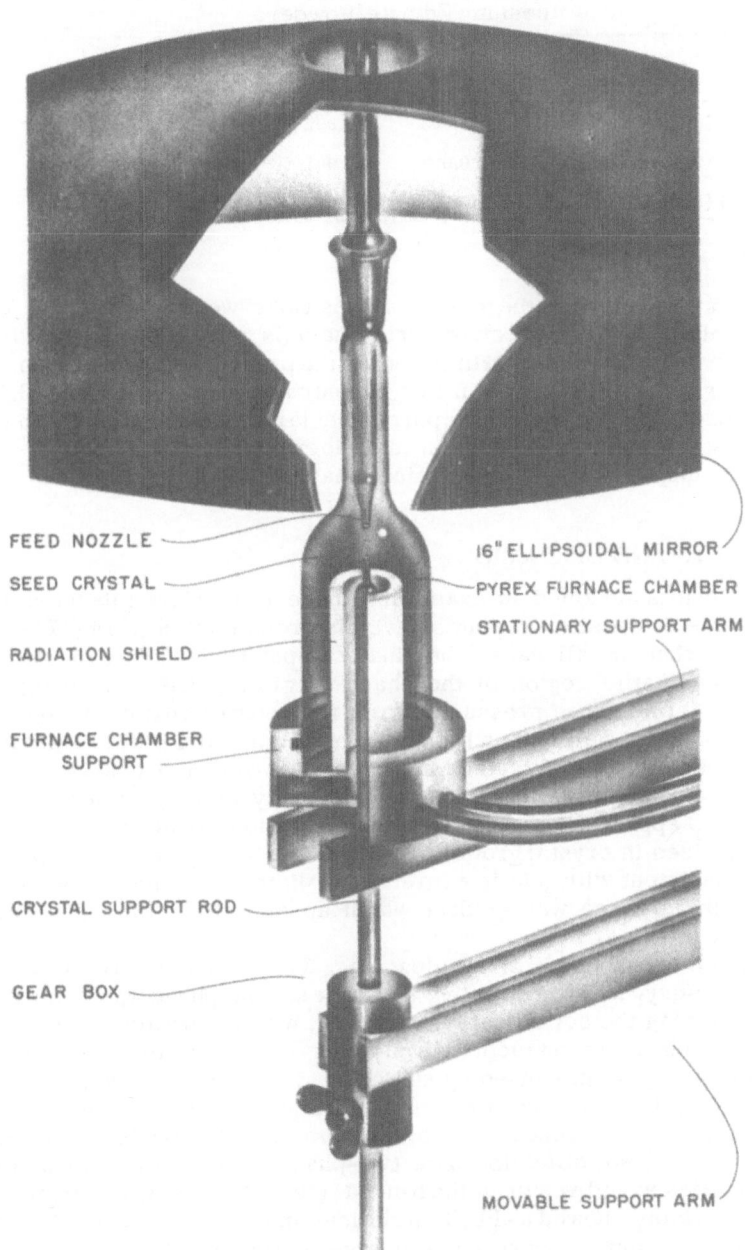

FEED NOZZLE

SEED CRYSTAL

RADIATION SHIELD

FURNACE CHAMBER
SUPPORT

CRYSTAL SUPPORT ROD

GEAR BOX

16" ELLIPSOIDAL MIRROR

PYREX FURNACE CHAMBER

STATIONARY SUPPORT ARM

MOVABLE SUPPORT ARM

Fig. 14-2. Arc image furnace chamber and lowering mechanism.

TABLE 14-I

Emission Spectrographic Analyses of Starting Magnesium Ferrite Powders

Powder No.	Impurities, wt. %			
	Al	Si	Ca	Cu
1 (magnesium-rich)	0.003	0.002	0.004	0.002
3 (stoichiometric)	0.003	0.007	0.007	0.001
4 (iron-rich)	0.003	0.005	0.004	0.002

of oxygen. Boules of the other compositions were grown under 1 atm of oxygen only. A mass spectrometric analysis of argon emerging from the furnace chamber during growth revealed the presence of 0.1% oxygen. Three typical ferrite boules grown in this manner are shown in Fig. 14-3. These boules were subsequently sliced perpendicular to the growth axis to provide specimens for chemical analysis, metallographic sections, and resistivity, hysteresis loop, and X-ray diffraction measurements.

C. RESULTS AND DISCUSSION

The results obtained in examining these boules are tabulated in Table 14-III. The compositional points have also been plotted in Fig. 14-1, which illustrates that in all cases the final compositions fall in the two-phase, spinel-plus-wustite region of the phase diagram. The three compositions grown at 1 atm oxygen pressure permit the construction of an approximate oxygen isobar at the liquidus temperature. A parallel isobar has also been drawn through the point obtained for 4.4 atm of oxygen pressure.

In all cases, X-ray diffraction revealed very faint MgO lines. The presence of MgO apparently results from equilibration during the relatively slow cooling utilized in crystal growth. The observed low spinel lattice parameter is also consistent with this interpretation. In the specimen grown in argon, a strong third phase was evident which appears to be the MgO–FeO solid solution ($a_0 = 4.25$ A).

Polished sections of the boules, Fig. 14-4, show the two phases quite clearly. In oxygen-grown boules, the dark second phase appears primarily in bands tracing the solidification interface, with a very fine precipitate occurring throughout the structure. The bands result from thermal fluctuations during crystal growth, caused by switching arc furnaces. Two arcs are used to provide continuous operation, and although no misalignment can be detected, it is almost impossible to duplicate the hot-zone shape in two machines. During solidification of a two-phase system, even minor thermal discontinuities would result in the banded structure observed. Oxygen-grown boules invariably showed α-Fe_2O_3 striations in a thin layer near the surface. These lines are quite apparent in the iron-rich composition in Fig. 14-4. In the argon-grown boules, the (MgFe)O phase is apparently continuous.

Resistivities of these boules were all quite low as a result of the ferrous iron concentrations. Coercive force measurements indicated a significant improvement resulting from the use of high oxygen pressure during growth.

TABLE 14-II

Chemical Analyses of Starting Magnesium Ferrite Powders

Powder No.	Weight percent total iron as Fe_2O_3	Weight percent MgO	Mole percent total iron as Fe_2O_3	Weight percent ferrous iron
1 (magnesium-rich)	79.00	21.14	48.5	<0.1
3 (stoichiometric)	79.51	20.10	49.9	<0.1
4 (iron-rich)	81.51	18.51	52.7	0.1
Theoretical $MgFe_2O_4$	79.85	20.16	50.0	0

No coercive force measurements were obtained on the argon-grown boules due to excessive internal cracking; however, a large increase with frequency would be expected as a result of the large ferrous iron content.

Figure 14-5 shows the effect of oxygen pressure on the ferrous iron content obtained at the liquidus. Paladino's values ([1]) obtained at lower temperatures are also presented for comparison. From Figs. 14-1 and 14-5, it is possible to estimate the conditions of oxygen pressure and starting composition which permit single crystal growth using a fusion technique. The values pertain only to the spinel-plus-wustite region of the phase diagram as the single-phase spinel region is approached, so that crystals can be grown under these conditions only near the phase boundary.

From these data, it appears that single crystals could be grown under 1 atm of oxygen pressure starting with a composition of 54.5 mol.% Fe_2O_3, 45.5 mol.% MgO. The resulting crystals would have an FeO concentration of approximately 10 mol.%. It would be necessary to maintain the boule above about 1200°C during growth and then cool it rapidly in order to prevent Fe_2O_3 precipitation near the surface. Control of growth conditions would be extremely critical, however, since one would be operating very close to the two-phase region.

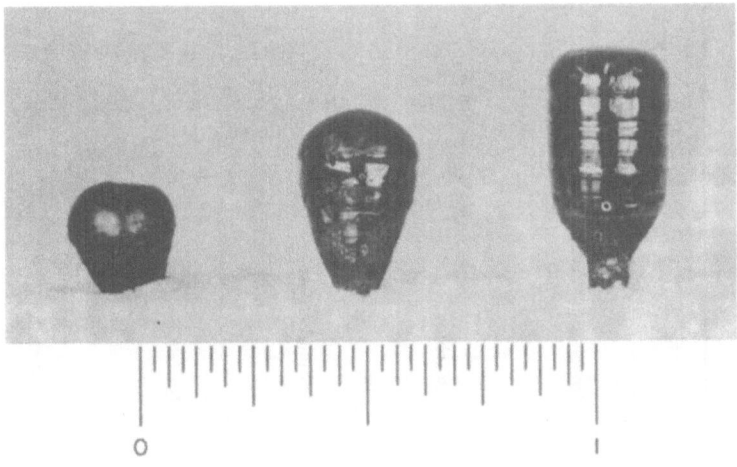

Fig. 14-3. Typical magnesium ferrite boules grown by arc image Verneuil technique.

TABLE 14-III
Summary of Data

Powder	Growth atmosphere	Phases	Boule composition, mol. %	Resistivity, ohm-cm	Coercive force, Oe
1 (magnesium-rich)	1 atm O_2	spinel (a_0 = 8.38 A) MgO (a_0 = 4.21 A)	MgO −48.9 FeO −9.7 Fe_2O_3 −41.4	0.44	1.0 kc −4.2 8.4 kc −12.0 50.0 kc −26.6
3 (stoichiometric)	1 atm A	spinel (a_0 = 8.38 A) MgO−FeO (a_0 = 4.25 A) MgO (a_0 = 4.21 A)	MgO −40.7 FeO −37.2 Fe_2O_3 −22.1	0.19	Boules cracked
3 (stoichiometric)	1 atm O_2	spinel (a_0 = 8.38 A) MgO (a_0 = 4.21 A)	MgO −47.4 FeO −10.3 Fe_2O_3 −42.2	1.1	1.0 kc −5.1 8.4 kc −5.3 50.0 kc −11.5
3 (stoichiometric)	4.4 atm O_2	spinel (a_0 = 8.38 A) MgO (a_0 = 4.21 A)	MgO −48.4 FeO −6.2 Fe_2O_3 −45.4	1.5	1.0 kc −3.3 8.4 kc −3.6 50.0 kc −7.2 90.0 kc −8.4
4 (iron-rich)	1 atm O_2	spinel (a_0 = 8.38 A) MgO (a_0 = 4.21 A)	MgO −45.4 FeO −8.0 Fe_2O_3 −46.5	0.38	1.0 kc −15.0 8.4 kc −18.2 50.0 kc −excessive

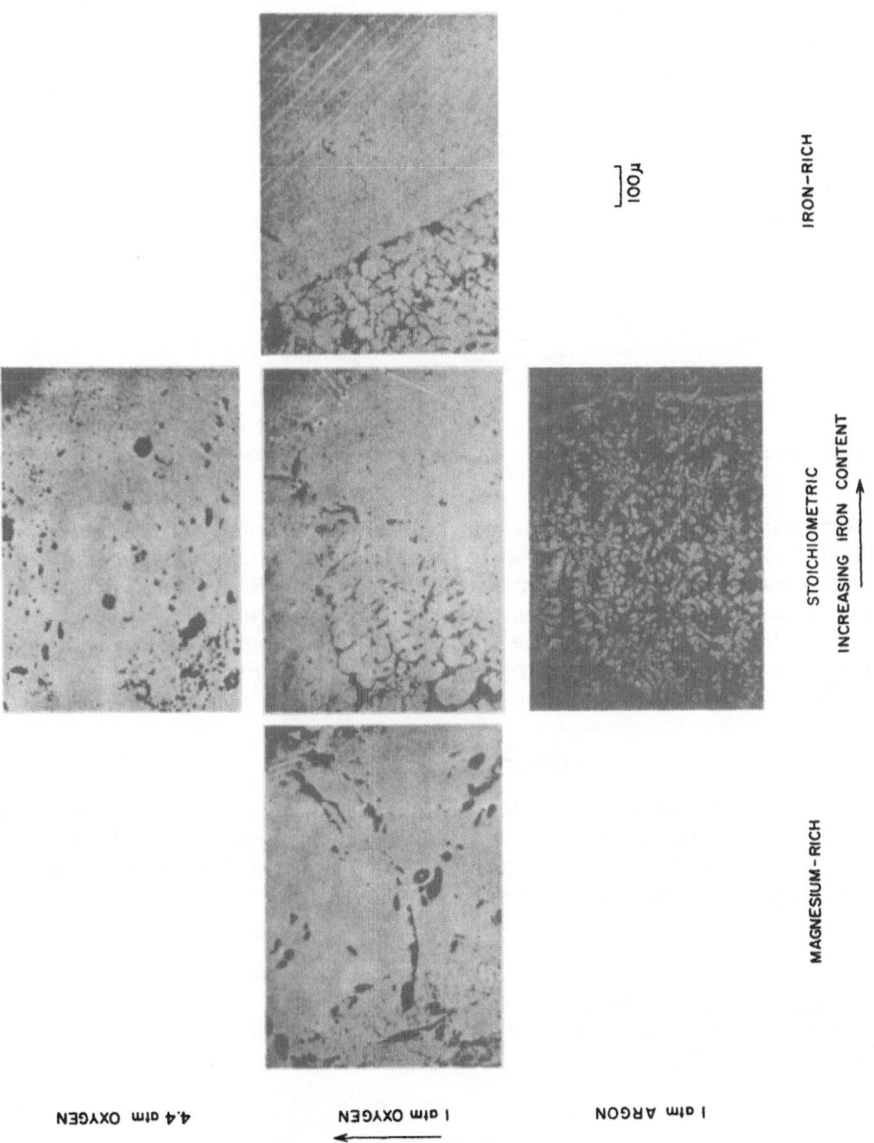

Fig. 14-4. Microstructure of magnesium ferrite boules.

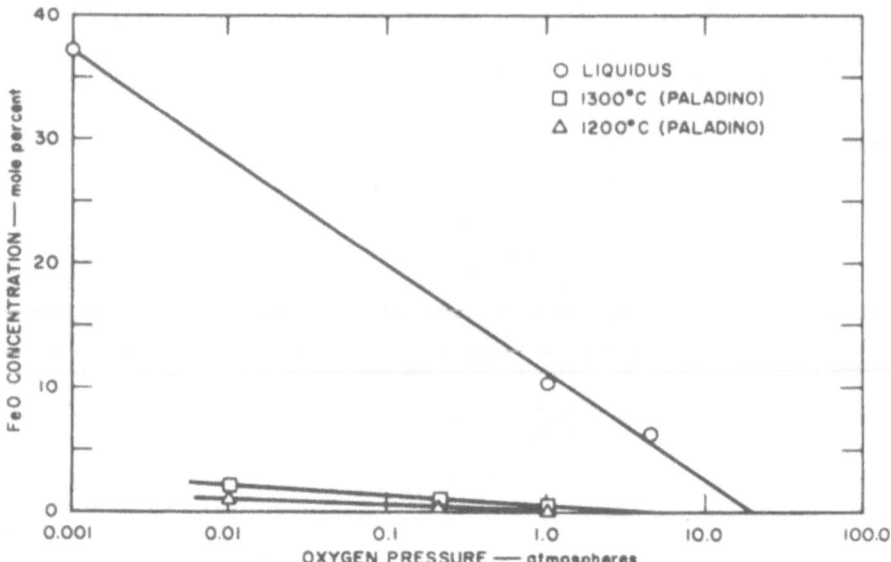

Fig. 14-5. Effect of oxygen pressure on ferrous iron content in spinel-plus-MgO [(MgFe)O] region of phase diagram.

With increasing oxygen pressure, the starting composition would shift until at 10 atm of oxygen pressure, single crystals having approximately 3 mol.% FeO should be grown from a stoichiometric $MgFe_2O_4$ starting powder. To obtain single crystals of the single-phase composition $(MgO)_{0.092}MgFe_2O_4$ having an FeO content less than 0.1 mol.%, a minimum of 20 atm of oxygen pressure would be required; if the logarithmic plot in Fig. 14-5 is not linear at the higher pressures, oxygen pressures up to 100 atm may be required.

ACKNOWLEDGMENTS

The author wishes to acknowledge the assistance of Mr. L. E. Sobon, who performed much of the crystal growth and specimen fabrication during this study. In addition, the contributions of Mr. W. J. Wiechec, Mrs. H. H. Johnson, and Dr. H. J. Eding are gratefully acknowledged.

REFERENCES

1. Paladino, A. E., Jr., J. Am. Ceram. Soc. 43:183 (1960).
2. Woodhouse, D., and White, J., Trans. Brit. Ceram. Soc. 54:333 (1955).
3. De La Rue, R. E., and Halden, F. A., Rev. Sci. Instr. 31:35 (1960).

Chapter 15

Carbon Arc Imaging Furnace and Its Application to Single-Crystal Growth

M. Kestigian,* G. J. Goldsmith, and M. Hopkins

Radio Corporation of America

Princeton, New Jersey

A. THE FURNACE

We shall describe the construction and operation of a carbon arc image furnace designed specifically for the growth of crystals, primarily by the powder-fusion technique in the temperature range between 2500 and 3500°C. A radiant energy system was chosen for this application because, while a number of other methods are available for the attainment of temperatures in this range under conditions suitable for crystal growth, there is none so versatile with regard to purity and control of the crystal environment.

The essential features of the system are shown in outline in Fig. 15-1. The primary heat source is a pair of high-intensity blown carbon arc motion picture projectors with ellipsoidal reflecting optics. The two sources are operated sequentially to provide continuous operation over arbitrary periods of time, since a single arc will burn a maximum of 20 min before the anode is consumed. The secondary focus of the primary reflector falls immediately outside the source housing, and the beam then diverges slightly to fill the water-cooled transfer mirror, which turns the beam through 90° toward the reimaging mirror. The image formed by the reimaging mirror at its primary focus has its large dimension in a horizontal plane; this is the more desirable configuration for powder fusion growth. The feed hopper is located above the reimaging mirror with its delivery tube passing through a hole to a point immediately above the focal plane. This hopper (Fig. 15-2) is an adaptation of the more conventional Verneuil hopper, in which provision is made for continuous adjustment of the feed rate and access to the interior during operation. As in the conventional system, the feed powder rests on a sieve which, in this system, is vibrated by a solenoid (doorbell) located on the outside wall and connected to the sieve through a phosphor bronze diaphragm, leaving the top lid free for access. Carrier gas is introduced beneath the sieve through an annular manifold. The carrier gas stream sweeps around the delivery cone carrying the feed powder with it. The feed rate is controlled by varying of the voltage applied to the solenoid, which in turn varies the amplitude of vibration of the sieve.

A photograph of the assembled furnace is shown in Fig. 15-3. The reimaging mirror, hopper, and crystal support are mounted on a separate as-

*Present address: Sperry Rand Research Center, Sudbury, Massachusetts.

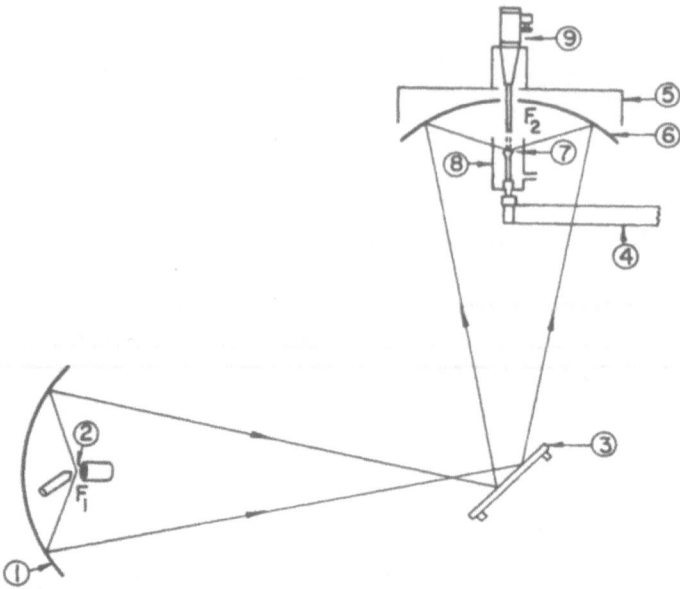

Fig. 15-1. Vertical arc image crystal growing furnace.

sembly attached rigidly to the ceiling girders. This type of mounting serves to isolate the growing crystal from vibrations produced by the operation of the arcs and transfer mirror. The hopper is mounted on a two-dimensional slide to provide accurate location of the feed flow relative to the reimaging mirror. The entire mirror–hopper assembly is mounted on a second two-dimensional slide for accurate positioning in the optical path. Further, there is a three-point tilt mechanism for the reimaging mirror. The crystal support pedestal is mounted in a chuck, which is attached to a rigid blade support. Positioning of the chuck in a horizontal plane is accomplished by means of a jeweler's lathe slide rest. Vertical motion is provided by a pair of lead screws, on which are mounted two preloaded ball bearing nuts. The lead screws are driven from a flexible coupling through a synchronizing bar. Two speed ranges are available, one at 15 in./min for coarse positioning and a second, slower, continuously variable rate from 0 to 4 in./hr for actual crystal growth. A pair of telescopes, 90° apart, with eyepiece crosshairs, is located in the plane of the image for precise positioning of the sample. Crystal growth is observed either through the telescopes or by means of a projected image produced by a sequence of prisms and lenses. The transfer mirror is constructed of a flat, $\frac{1}{8}$-in.-thick quartz disc, 6 in. in diameter, aluminized on its front surface. It is clamped to a water-cooled copper support, which has provisions for precise adjustment of the tilt angle. Rapid, 100 msec, rotation of the mirror from the first arc to the second (and vice versa) is accomplished by means of an inertial rotating system. It operates in the manner described below.

A magnetic clutch couples the mirror mount to a small gyro motor with a heavy flywheel. With the clutch disengaged, the motor is brought to a speed of about 10,000 rpm. Power is then removed from the motor and the clutch engaged, causing the mirror to rotate. Accurate positioning of the mirror

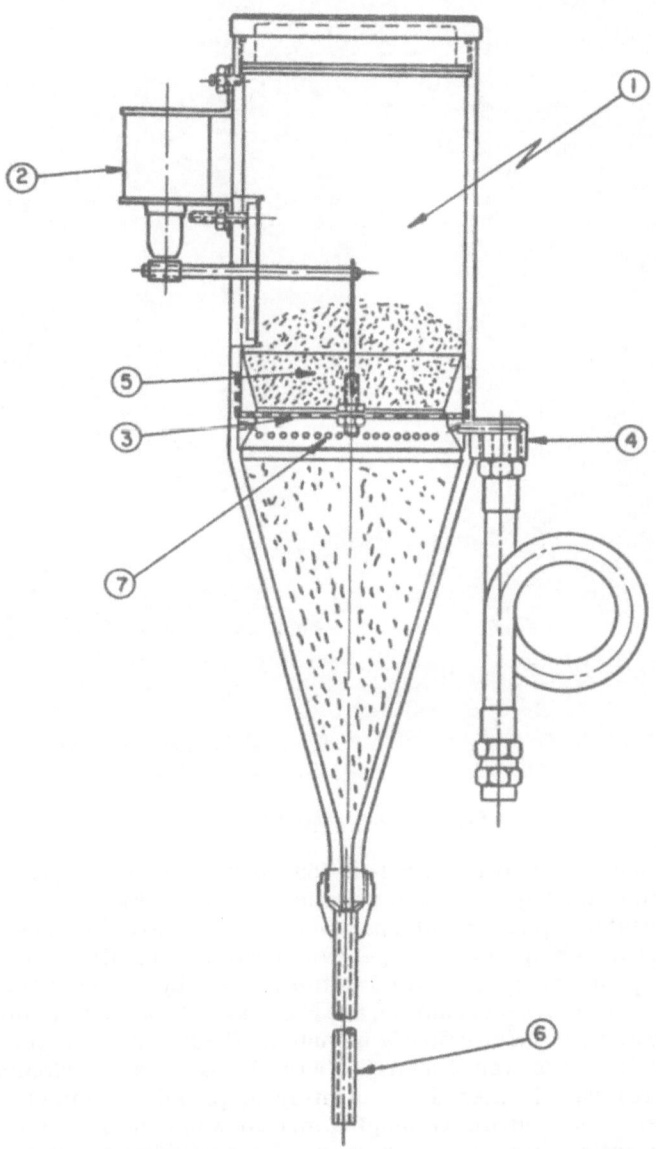

Fig. 15-2. Hopper assembly (1) feed hopper, (2) vibrator, (3) sieve, (4) gas inlet, (5) powder, (6) delivery tube, (7) holes allow flow of gas into delivery cone.

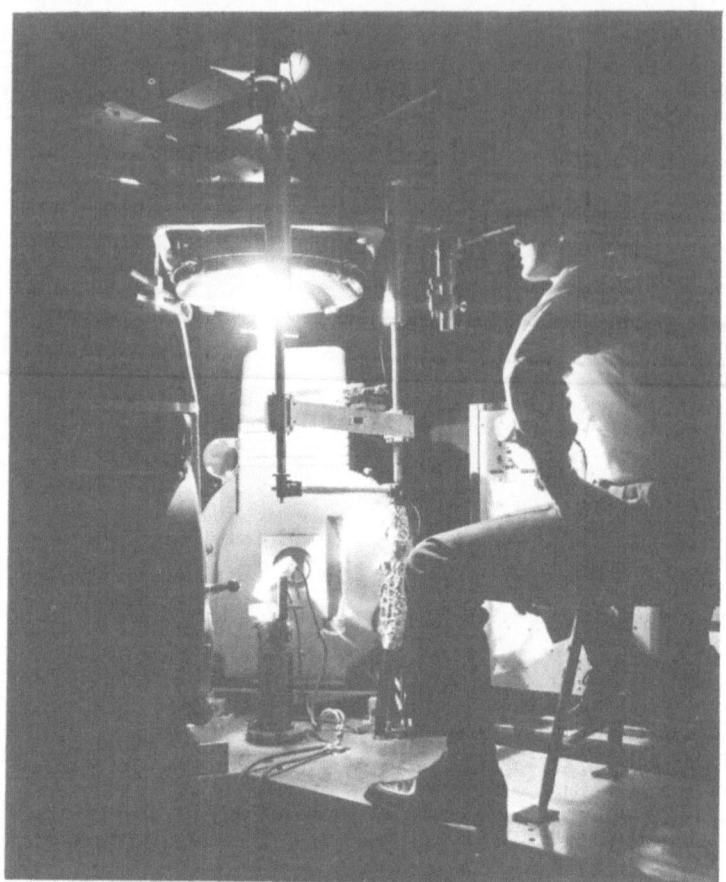

Fig. 15-3. Arc image furnace.

limits is obtained through the use of ball-loaded detents and magnetic clamps. The mirror rotation sequence is automatically programmed.

The successful application of a pair of independent arcs to crystal growth requires exact matching of the images with respect to position, size, and intensity. This problem is aggravated in this system by virtue of the fact that each element of the furnace, namely, arc No. 1, arc No. 2, and the mirror reimaging assembly, is independently mounted. Preliminary alignment of the optic axes of the arcs was performed with the aid of a cathetometer. The transfer mirror was located as accurately as possible symmetrically between the two arcs, and the reimaging mirror was centered above it. The next step was optical alignment, which was accomplished through the use of plastic light pipes mounted in place of the anode. The light pipes were illuminated by microscope lamps, and the entire optical system could be traced out. Superposition of the images produced by each of the arcs was determined with a "target" placed at the focal point of the reimaging mirror. Final adjustment was carried out by operation of the furnace and heating of a flat zirconia plate located at the image focal point. Satisfactory alignment was indicated when the size, shape, and location of the melted area in the zirconia plate were identical for both sources.

B. OPERATION OF THE FURNACE

The operating characteristics of the furnace were determined by means of a total irradiance calorimeter designed and constructed by Arthur D. Little, Inc. (Fig. 15-4). This calorimeter is of the flow type. Water is circulated through a blackened copper receiver at a known rate, and the differential temperature between the incoming and outgoing water is measured. A water-cooled guard ring surrounds the receiver. A block diagram of the heat flux measuring system is illustrated in Fig. 15-5. The system consists of the calorimeter, a thermostatted water reservoir, a flowmeter, a thermocouple amplifier, and a graphic recorder. The total heat flux and the time stability of the arcs is illustrated in Fig. 15-6. The maximum stable flux with an operating arc current of approximately 160 A was 230 cal/cm^2-sec., with a variation of less than ±2% over the life of a carbon. The two arcs are matched to within better than ±1%. This value was attained using first-surface aluminized mirrors throughout. Upon substitution of second-surface silver reflectors, the flux is reduced to 190 cal/cm^2-sec. These mirrors, however, were found to have substantially longer life than the former. Some preliminary observations have been made with first-surface multidielectric coated mirrors, which are claimed to be more durable than the front-surface aluminum. Heat fluxes from these appear to be about 20% higher than those obtained with first-surface aluminum.

Heat flux profiles of the image were measured by manipulation of the calorimeter in both a horizontal and vertical plane. In the horizontal plane, as shown in Fig. 15-7, there is approximately a Gaussian distribution of thermal energy. The effect of a vertical movement out of the focal point upon the horizontal profile is also shown. A larger image is obtained at a lower heat flux value. In the vertical plane (Fig. 15-8) the thermal gradient is considerably steeper.

Because suitable thermometers are not available to us for the temperature range attainable, initial experiments were conducted on melting mate-

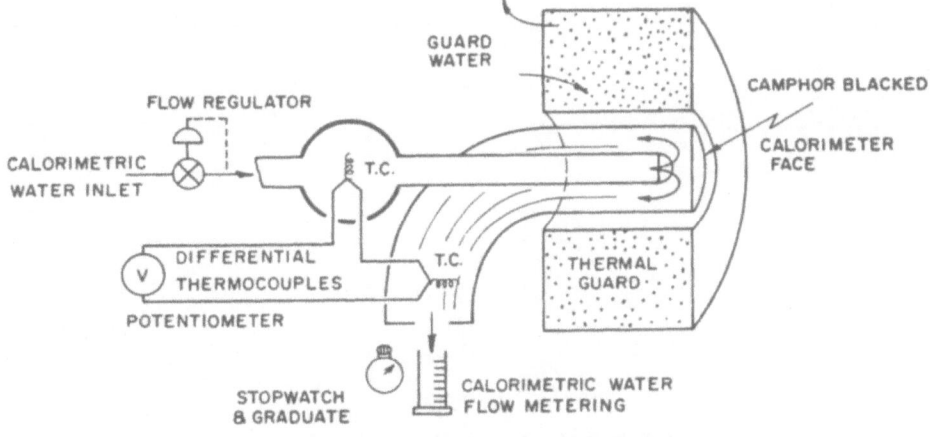

Fig. 15-4. Total irradiance calorimeter.

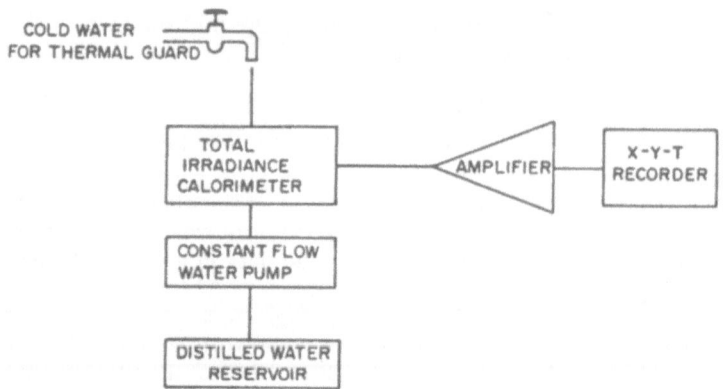

Fig. 15-5. Block diagram for irradiance measurement.

rials with "known" melting points. A second question which was of interest to us was that of the suitability of a radiant system such as this for the melting of highly reflective (white) compounds. Pressed pellets $\frac{3}{8}$ in. thick by $\frac{3}{8}$ in. in diameter of the following materials were inserted into the image, melted, cooled, and remelted: TiO_2, Al_2O_3, MgO, ZrO_2, and ThO_2. We can conclude that temperatures in the range of 3300°C can be reached in colorless compounds.

Initial crystal growth studies were conducted on titanium dioxide. While the extremely high temperature attainable is not required for this material, which melts at approximately 1840°C, it is of interest to determine whether nearly stoichiometric rutile can be grown from TiO_2 powder if a sufficiently oxidizing atmosphere is provided. Conventionally, rutile crystals grown by flame fusion are oxygen deficient and must be reoxidized. This system was also selected because we are very familiar with its growth characteristics. The pedestal on which the rutile crystals were grown was mounted inside a 3-in.-diameter pyrex cylinder, 8 in. in length. Standard rutile feed powder, obtained from the Linde Co., was supplied from the hopper, together with a

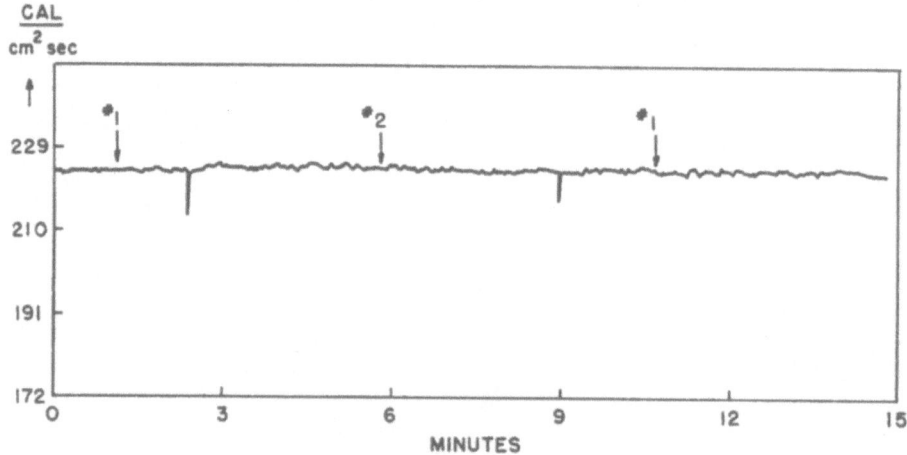

Fig. 15-6. Image furnace stability as a function of time.

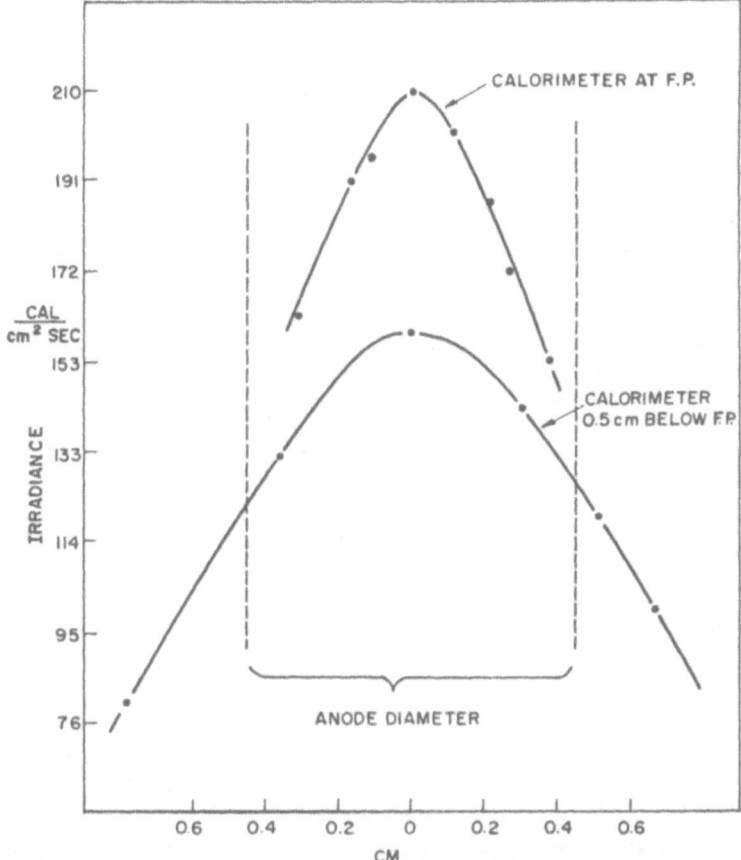

Fig. 15-7. Horizontal profile of focal point.

copious flow of oxygen. The crystals were grown at a rate of $\frac{1}{4}$ in./hr and formed colorless (stoichiometric) boules.

The most refractory material studied for crystal growth was zirconium dioxide stabilized with calcium oxide. Suitable feed material was prepared from zirconium lactate* or zirconium sulfate nonahydrate* as described below.

The desired molar composition of calcium oxide to zirconium dioxide was weighed out, with calcium carbonate and zirconium lactate as starting materials. These phases were ground together and fired in quartz vessels, first at low temperatures (400-500°C) to remove moisture and volatile components, and second, at high temperatures (1100°C) in air for twelve hours. The sample was cooled and remixed, and the calcination procedure was repeated. This treatment resulted in a feed powder which was satisfactory for single-crystal growth.

Single crystals of zirconium dioxide containing calcium oxide in the concentration range from 5 to 38 mol.% were grown. A typical crystal-growth experiment is as follows. The support rod is positioned just below the image

*Supplied by Titanium Alloy Manufacturing Co., Division of National Lead Co.

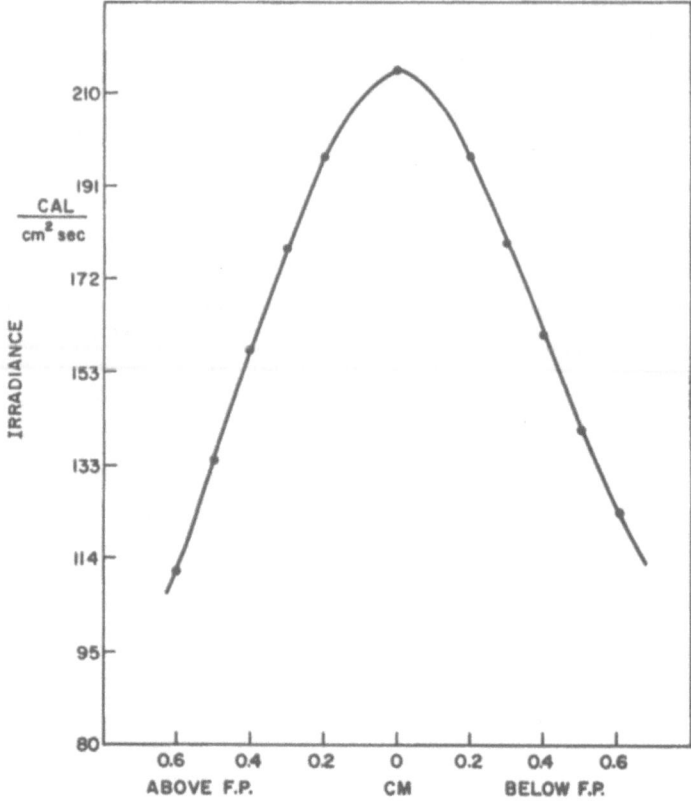

Fig. 15-8. Vertical profile of focal point.

focal point and the carrier gas turned on. The feed material is slowly introduced and the support rod raised toward the focal point until the falling
powder forms a sintered cone. The pedestal is further raised, until the tip
of the cone is melted.   The feed rate is slowly increased until the desired
diameter is reached.   The crystal is then lowered at a rate equal to the
growth rate.   Upon completion of the crystal growth, the boule is slowly
lowered out of the heat zone to prevent thermal shock. The largest single
crystal of stabilized zirconium dioxide that was grown was over 1 in. in
length by $3/16$ in. in diameter. The crystals are colorless and transparent,
and have face-centered cubic symmetry as determined by X rays. The crystals normally grew 17° off the (111) crystallographic axis.

The furnace has also been used for the diffusion study of chromium ions
in single-crystal magnesium oxide and for the preparation of several polycrystalline nitrides.

C. SUMMARY

Although this furnace has only been in operation for a short time and our
experience in its use is limited, we feel certain that it will become a valuable
tool in our high-temperature, solid state chemical investigations.

Chapter 16

Arc Imaging Furnace Crystal Growth

Robert P. Poplawsky*
Research Laboratories
General Motors Corporation
Warren, Michigan

A. INTRODUCTION

The advantages of replacing the chemical flame of the original Verneuil method for refractory oxide crystal growth with imaged energy from a high-intensity carbon arc have been adequately pointed out elsewhere, e.g., by De La Rue and Halden ([1]). It is also possible to make changes in the manner in which the polycrystalline material is fed to the growing boule. Arc image crystal growth with the powder feed of the Verneuil technique leads to several problems. In the process of dropping through the image point of the furnace, the powder is not likely to melt completely, unless it is very finely divided. If some of the incompletely melted powder particles reach the liquid–solid interface at the edge of the melt, unwanted crystallites may be seeded. This puts certain restrictions on the shape of the liquid surface, which may not always be readily satisfied. Furthermore, finely divided powder is difficult to feed in a well-controlled manner, particularly in a glass chamber filled with gas, which is being rapidly heated in such a manner as to give rise to a certain degree of turbulence. For these reasons, floating zone techniques have been applied; in these techniques, the polycrystalline material is fed to the melt in rod form. There is then, another, perhaps less often stated, advantage of the imaging method: it lends itself to the use of floating zone methods.

There are now several accounts of image furnace–floating zone crystal growing experiments in the literature. Poplawsky and Thomas ([2]) developed a technique which is convenient for use with a vertical image furnace and in which, in contrast to the usual case, the solid members above and below the molten zone are not of the same diameter. This will be called a "generalized" floating zone. Kooy and Couwenberg ([3]) have recently described a technique which is more like the conventional floating zone method, in which a molten zone may be passed through a uniform-diameter polycrystalline rod. They use a horizontal furnace and heat the side of the rotating rod. An advantage of this method is that it lends itself more directly to zone refining. In connection with crystal growing and crystal quality, however, one might expect less favorable radial temperature gradients near the molten zone, especially when working with poorly conducting refractory materials.

It is the purpose here to describe some crystal growing experiments and results which were obtained using the generalized floating zone arrangement on the Wayne State University Physics Department furnace. This furnace is

*Senior Research Physicist.

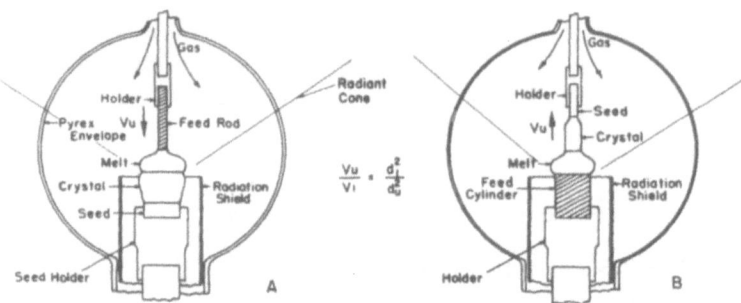

Fig. 16-1. Image furnace—floating zone arrangement. (a) Upward traveling zone; (b) downward traveling zone.

a vertical, double-paraboloid type. Very recent work on a rotating vane radiant power control, designed primarily to facilitate this type of experiment, will also be discussed. The early work was supported by the United States Air Force through a contractual agreement with the Air Force Office of Scientific Research, and recent work is being supported by the Research Laboratories Division of General Motors Corporation.

B. IMAGE FURNACE—FLOATING ZONE METHODS

Figure 16-1 is a schematic drawing of the generalized floating zone arrangement that Poplawsky and Thomas used with silicon. Drawing A depicts what is equivalent to an upward traveling zone, in which a long, relatively small-diameter rod is fed downward into a melt on a downward moving boule of substantially larger diameter. Drawing B shows the inverse of this, which is equivalent to a downward traveling zone, in which a smaller-diameter boule may be "pulled" from a melt on a substantially larger-diameter cylinder of polycrystalline material situated below it. In both cases, the molten region is supported primarily by the larger-diameter lower member, upon which most of the imaged energy is incident.

Since the volumes swept out per unit time by both upper and lower solid members must be equal for a boule of uniform diameter to be grown, the required relative speeds, v, of vertical travel are given by

$$\frac{v_{upper}}{v_{lower}} = \frac{d^2_{lower}}{d^2_{upper}} \tag{1}$$

where d represents diameter. It is desirable to rotate both upper and lower members of the floating zone arrangement to ensure that the radial temperature gradients will be axially symmetric.

A theoretical analysis of floating zones of this type has been given by Heywang[4]. His results give the length of a vertical molten zone which will be stable for given materials as a function of upper and lower member radii, r (Fig. 16-2). Heywang's relation for the total length, L, of a stable upward traveling, generalized molten zone is

$$L = \sqrt{2}\beta \left\{ \frac{1}{(1 + \beta^{1/2}/r_u)^{1/2}} + \frac{\sqrt{2}}{(1 + \beta^{1/2}/r_l)^{1/2}} \right\} \tag{2}$$

$$\beta = \frac{\sigma}{gD} \quad \text{and} \quad r_u \ll r_l$$

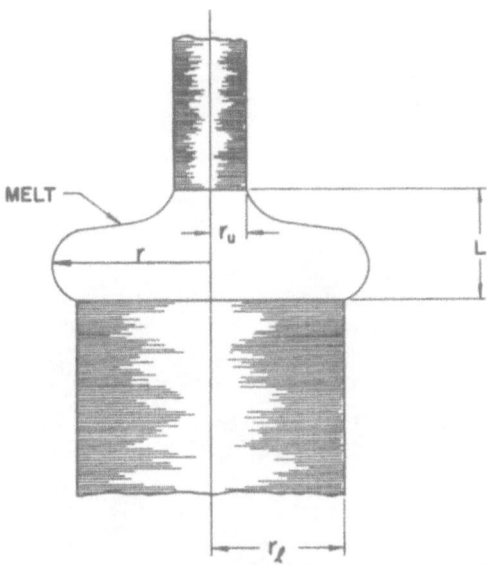

Fig. 16-2. Generalized floating zone showing pertinent
dimensions.

r_u and r_l are the radii of the upper and lower members, respectively, σ/D is
the surface tension to density ratio of the material, and g is the gravitational
acceleration. Equation (2) verifies what one might expect intuitively, that is,
that a large σ/D and a large lower-member diameter are desirable to ensure
stability of the zone. A large-diameter upper member is also desirable; however, this must be relatively small for practical reasons. Some numerical
estimates of molten zone lengths and volumes may be obtained if σ and D are
known.

C. RESULTS OF FLOATING ZONE EXPERIMENTS

Silicon has a favorable σ/D and was chosen for early crystal growing experiments using the methods outlined above. Figure 16-3 shows a silicon
boule as it appeared immediately after an upward traveling zone experiment.
The boule, rod remnant, and general mechanical arrangements are clearly
visible. Figure 16-4 shows examples of crystals obtained with both upward
and downward traveling zones. The crystals shown in part a of Fig. 16-4
were obtained by growing extensions onto 1.0-in.-diameter seed discs by
using an upward traveling zone. Part b shows crystals which were grown
onto rectangular seeds using a downward traveling zone. The uppermost of
these was grown without boule (upper member) rotation and is consequently
somewhat curved.

The lengths of all of these crystals were limited by the running time,
dictated by the lengths of the arc electrodes used. All silicon crystals were
grown in a flowing argon atmosphere. Their chemical and structural perfection compared favorably with those obtained by standard, induction-heated,
floating zone techniques.

Fig. 16-3. Silicon crystal immediately after an upward traveling zone
experiment.

 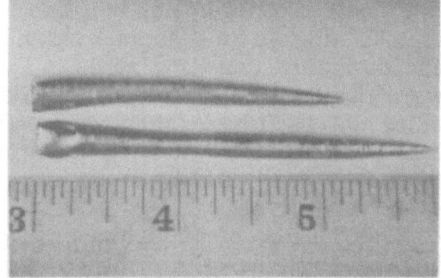

Fig. 16-4. Examples of silicon crystals grown by the image furnace—floating zone method: (a) Upward
traveling zone; (b) downward traveling zone.

Fig. 16-5. Manganese ferrite—downward traveling zone—experiment in progress.

The optical, thermal, and chemical properties of many of the transition metal oxides make arc image methods of obvious value in connection with their crystal growth. For this reason, interest in the properties of ferrites at the General Motors Laboratories led to the application of image furnace-floating zone methods to some of the ferri-spinels, e.g., magnetite, manganese ferrite, and nickel ferrite.

Figure 16-5 shows a downward zone, manganese ferrite, crystal growing run being conducted in air. The boule had been "necked in" and was in the growing out stage when the photograph was taken. The large object below the molten zone is a cylindrical radiation shield, which was made by evaporating gold onto a section of large-diameter pyrex tubing. Figure 16-6 shows examples of magnetite, manganese ferrite, and nickel ferrite boules which were pulled in this manner [5]. Many of these crystals, particularly those of manganese ferrite, grow with very well-developed faces, usually (111) planes. A section containing such a face will often be flat enough to furnish a ready-made specimen for etch pit studies.

As an illustration that image furnace—floating zone techniques are not limited to refractory materials, Fig. 16-7 shows two copper boules that were pulled from a melt which was established on a large, polycrystalline lower member.

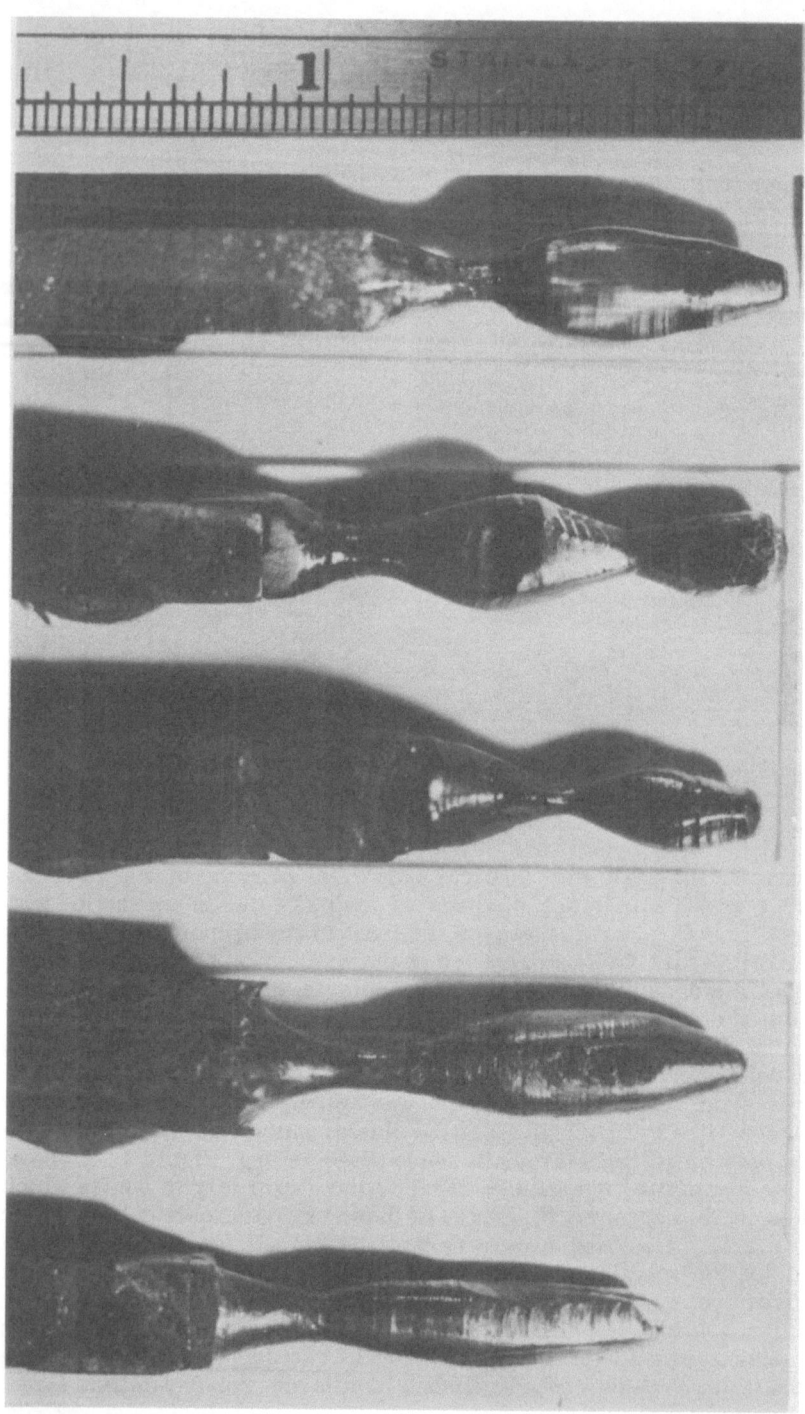

Fig. 16-6. Ferrite boules grown using a downward traveling zone.

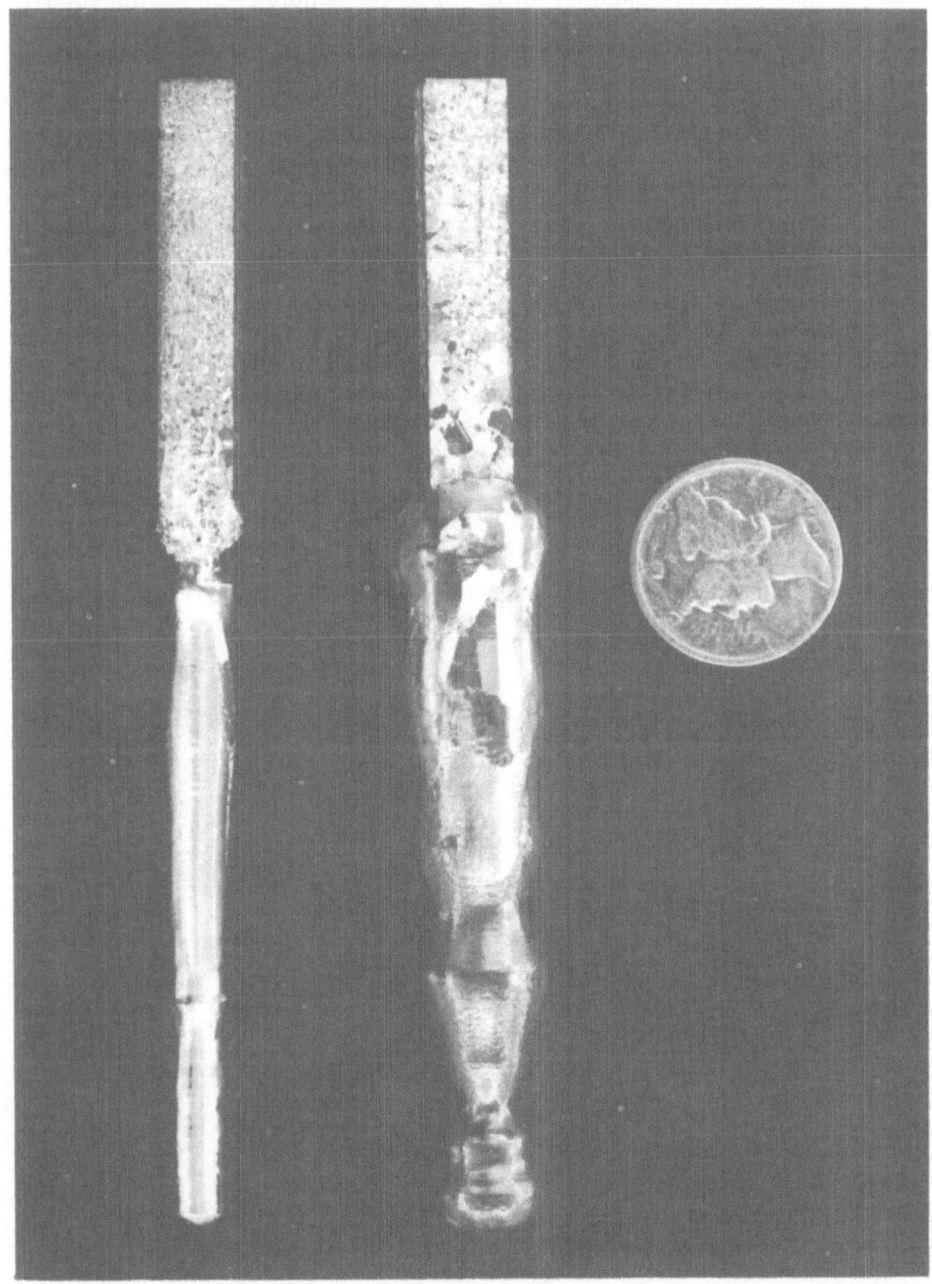

Fig. 16-7. Copper boules grown using a downward traveling zone.

During experiments with silicon and ferri-spinels, changes in the diameter of the growing member were made by simply changing the relative speeds (υ_u and υ_l) of upper and lower members of the floating zone arrangement. This had a twofold effect on melt temperature—that associated with the rate at which heat of fusion is given up, and that associated with the change in flux distribution on the surface of the melt as it changed its position with respect to the image plane. This is not a very satisfactory method for producing temperature changes to vary the boule diameter, because it is somewhat difficult to anticipate the magnitude of the effect on melt temperature. Unless the speed changes are very gradual, and a suitably long time is taken to neck in and grow out a boule, the floating zone is often lost. This is a very undesirable situation, because of the limited amount of growing time available with a furnace that has no provision for "continuous running."

The overall molten zone temperature and the small temperature variations which produced the changes in diameter apparent on the copper boules shown in Fig. 16-7 were obtained using an experimental, vane-type radiant power control. This device was constructed and evaluated by E. C. Shults as part of a master's degree program under the joint direction of Professor H. V. Bohm of Wayne State University's Physics Department and myself. A description and some of the characteristics of this power control will be given next.

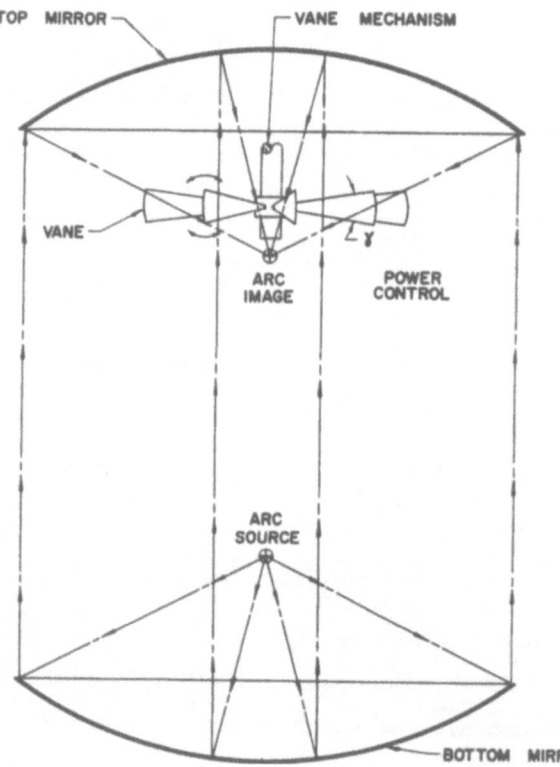

Fig. 16-8. Schematic drawing showing the vane-type radiant power control in the cone of radiation of a vertical, double-paraboloid-type image furnace.

D. VANE-TYPE RADIANT POWER CONTROL

The radiant power control properties which were considered desirable for crystal growing purposes were the following: the radiant power control, in its introduction and operation, should produce no significant changes in the furnace flux profile; it should be capable of setting the overall power level to accommodate materials of rather different melting points and at the same time of giving a fine control about this level; its use should not result in non-uniform heating of the uncooled paraboloids; and finally, its presence should not produce too large a fixed or "intrinsic" attenuation of the energy from the arc.

Although other type controls (e.g., mirror aperture controls) might well satisfy some of these requirements, it was decided to investigate the feasibility of a rotating-vane device placed in the converging cone of radiation of the double paraboloid configuration (see Fig. 16-8). A "rotating-vane" device in this case consisted of nine circular sectors which were arranged radially about the vertical optic axis of the furnace and of mechanism which was provided to rotate these sector-shaped vanes in synchronism about their respective radii, so as to change their angular position, while simultaneously rotating the whole vane assembly. The first rotation changes the solid angle subtended at the image point by the upper mirror, and thus produces the required change in radiant power. The latter rotation smooths or averages the effect over the entire flux-receiving area.

Figure 16-9 is a schematic drawing showing some of the details of design including the bevel-miter gear arrangement which was used to provide control of the vane angular position. The bevel gear rotation with respect to the overall device was accomplished by mounting a small motor onto the outer shell of the control unit and coupling to the hollow shaft carrying the bevel gear by means of a worm-gear arrangement.

It was necessary to place the control unit below the upper mirror on the Wayne State furnace, due to arc lamp obstructions above the lower mirror. In order not to obstruct the image point region, it was then necessary to place the unit high in the converging cone of radiation and to make the vanes rather long (16 in.).

E. EFFECT OF VANE-TYPE CONTROL ON FURNACE OPTICS

The change in temperature and in temperature distribution on a flux-receiving surface near a furnace image point will be closely related to the effect on total flux and flux profile of any radiant power control device which may be employed. These characteristics of the rotating vane control were therefore investigated. It will be noted that a design of the type described above allows complete freedom in choosing the fractional amount of radiant power which is to be controlled. Nine 40° vanes will control the entire amount remaining after the fixed loss due to unwanted shadows produced by the introduction of the control is absorbed. A full range of control is seldom necessary, however, and would most likely require water-cooling of the device. Vanes of 10° and 20° were found to be most useful for crystal-growing purposes. Finer control is obtained with the smaller vanes. In practice, the diameter of the lower member of the floating zone and its position with respect to the image plane can be chosen to accommodate the higher overall

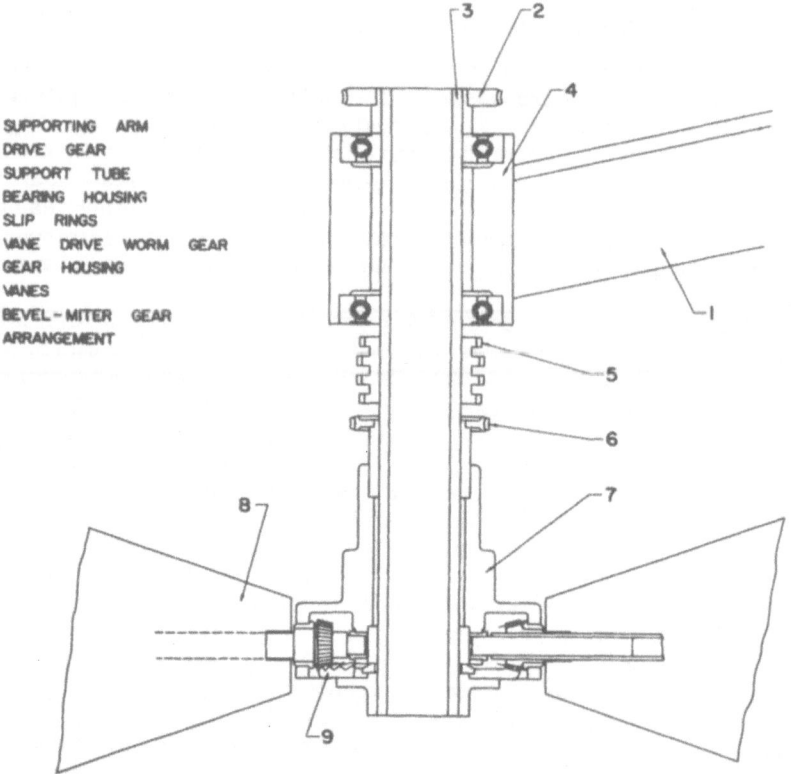

1. SUPPORTING ARM
2. DRIVE GEAR
3. SUPPORT TUBE
4. BEARING HOUSING
5. SLIP RINGS
6. VANE DRIVE WORM GEAR
7. GEAR HOUSING
8. VANES
9. BEVEL – MITER GEAR
   ARRANGEMENT

Fig. 16-9. Mechanical details of vane-type radiant power control.

power levels associated with small vanes, even with relatively low-melting-point materials.

A consideration of the change in solid angle subtended at the image point by the upper mirror, due to a change in angular position of the vanes on the control unit, indicates the manner in which the total flux, F, at the image plane should vary relative to its maximum, F_0. The expression obtained is

$$F/F_0 = 1 - \frac{9\,\gamma}{2\,\pi}\sin\theta \tag{3}$$

where γ is the vane angular width and θ is the angular position of the vanes with respect to the vertical. Figure 16-10 shows F/F_0 for several angular widths. Experimental points shown for the case $\gamma = 20°$ (full to half power) were obtained by using a large photovoltaic cell at the image plane in such a manner as to intercept all of the radiation from a disc-shaped incandescent lamp source of approximately uniform brightness. The discrepancy between the calculated and the experimental curves for the lower values of θ is probably due to the effect of off-axis points on the disc source. Because of its sinusoidal dependence on θ, the total flux does not deviate markedly from linearity for the lower values of θ. Furthermore F/F_0 is slowly varying for vane width angles (γ) of 20° or less.

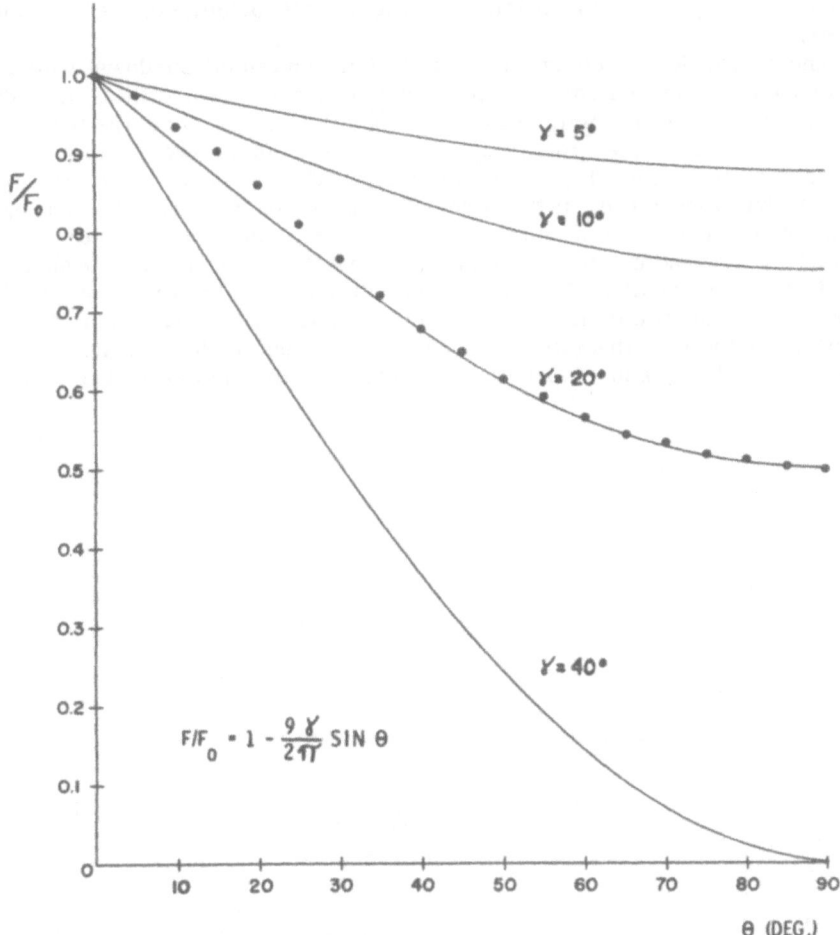

Fig. 16-10. Change in total flux, F, relative to its maximum, F_0, as a function of vane angular position θ.

In order to determine the effect of the power control on furnace flux profile, Mr. Shults probed the region near the image point with a very small photovoltaic cell. For practical reasons, the source in this case was a 6-V lamp with a filament small enough to be considered as an approximation to a point source. The entire vane assembly was rotated at 8 rpm. This gave a beam interruption rate of 72 per minute on the light-sensitive surface of the photocell. The peak voltage across a load resistance in the photocell circuit was taken as a measure of the peak irradiance at a particular cell position.

The profiles obtained on a plane 0.5 cm above the image plane, with the vanes set at several different angles, are shown in Fig. 16-11. A computed point source distribution for ideal optics is also given for purposes of comparison. It will be noted that all profiles have the same general shape. This characteristic somewhat simplifies the relationship between the total average flux reaching a receiving surface and the surface temperature, and should aid the furnace operator in anticipating the effects of adjustments in radiant power.

The power loss incurred due to various unwanted shadows, which are introduced by the control, with the vanes vertical, was a rather high 18% in the first version of the unit. Although this seems to preclude the use of this type of control in experiments where near-maximum power is required, an analysis of this loss shows that for the $\gamma = 20°$ case only, about half of this loss is "intrinsic," i. e., is due to the absorption or scattering of light by the vanes from off-axis points on the source. The remainder is due to shadows caused by protruding drive mechanisms and the support arm. It should be possible to practically eliminate these unwanted shadows by designs which make better use of the shadows already incorporated in a particular furnace. It will be noted also that the intrinsic loss decreases with a decrease in vane angular width, γ, and also with a decrease in the diameter of the source.

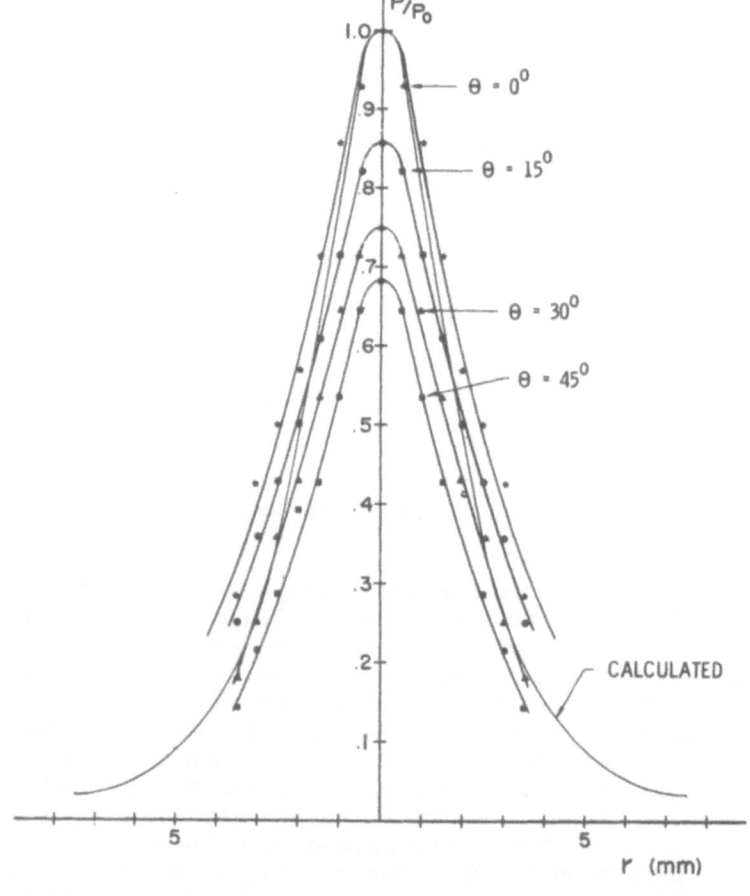

Fig. 16-11. Point source flux profiles for several angular positions of the vanes (θ). A calculated profile for an idealized system is included for comparison purposes.

F. SUMMARY

It has been shown that generalized floating zone techniques represent a sound approach to crucible-free crystal growth of a variety of materials in an image furnace. A radiant power control design which affects the image furnace optics in a manner compatible with these floating zone techniques has also been presented. Although this work has been carried out exclusively on a vertical, double-paraboloid-type furnace, many of the methods and results should apply equally well to other types of image furnaces.

REFERENCES

1. De La Rue, R. E., and Halden, F. A., Rev. Sci. Instr. 31:35 (1960).
2. Poplawsky, R., and Thomas, J. E., Jr., Rev. Sci. Instr. 31:1303 (1960).
3. Kooy, C., and Couwenberg, H. J. M., Philips Tech. Rev. 23:161 (1962).
4. Heywang, W., J. Naturforsch. 11a:238 (1956).
5. Poplawsky, R., J. Appl. Phys. 33:1616 (1962).

Experimental Procedures

Chapter 17

The Solubility of Water Vapor in Molten Alumina

J. J. Diamond and A. L. Dragoo

National Bureau of Standards
Washington, D. C.

The absorption of water by molten alumina was described by von Warten-berg in 1951 ([1]). He found that alumina melted in a tungsten arc in flowing hydrogen or in the inner part of an atomic hydrogen torch would spatter on solidifying. He also observed that alumina frothed when melted deep in an oxy-hydrogen flame, and that alumina melted in a ZrO_2-furnace formed foamy drops when water vapor was admitted into the furnace.

When we melted the end of an alumina rod in an oxy-hydrogen flame, we observed bubbles of gas escaping from the liquid, but, because of the flowing of the liquid in response to the force of the flame, could not decide whether this was due to the solid's melting, the liquid's solidifying, or the boiling of the liquid itself.

In order to observe the phenomenon, we decided to use an image furnace to melt the alumina because of the ease with which the molten alumina could be visually observed while it was melted and frozen in a variety of atmos-pheres, including water vapor. The furnace used was an ADL arc image furnace, with the arc operated at 150 A. The specimen rod was held in a spiral of platinum—rhodium wire and supported along the optical axis of the furnace in the middle of a 500-ml pyrex round-bottomed flask. The specimen was observed end-on through an optical pyrometer sighting along the optical axis of the furnace, and also at right angles to this axis. One atmosphere of water vapor was maintained by boiling distilled water, leading the vapor into the flask enclosing the specimen, and venting to air. One atmosphere of other gases was maintained by slowly flowing the tank gas through a desiccant into the sample flask and venting to air. The alumina specimen used was a rod of Morganite alumina, 0.25 in. in diameter. The purity of the material as supplied was about 99.7-99.9% Al_2O_3, the major impurities being Si, Fe, and Ga; melting it in a variety of atmospheres for about an hour resulted in preferential vaporization of the volatile impurities and an increase in purity to about 99.95-99.99% Al_2O_3.

When the end of the alumina rod was melted in water vapor, it formed a slightly pendant drop of liquid alumina. The surface of this drop was quiet and showed no sign of boiling. If the rod were then quickly drawn back from the focus, so as to partly freeze the molten alumina, the remaining liquid would boil, growing quite large bubbles in the process. After a few seconds, the liquid would stop boiling, any remaining bubble would be reabsorbed, and the diminished drop of molten alumina would again display a quiet surface.

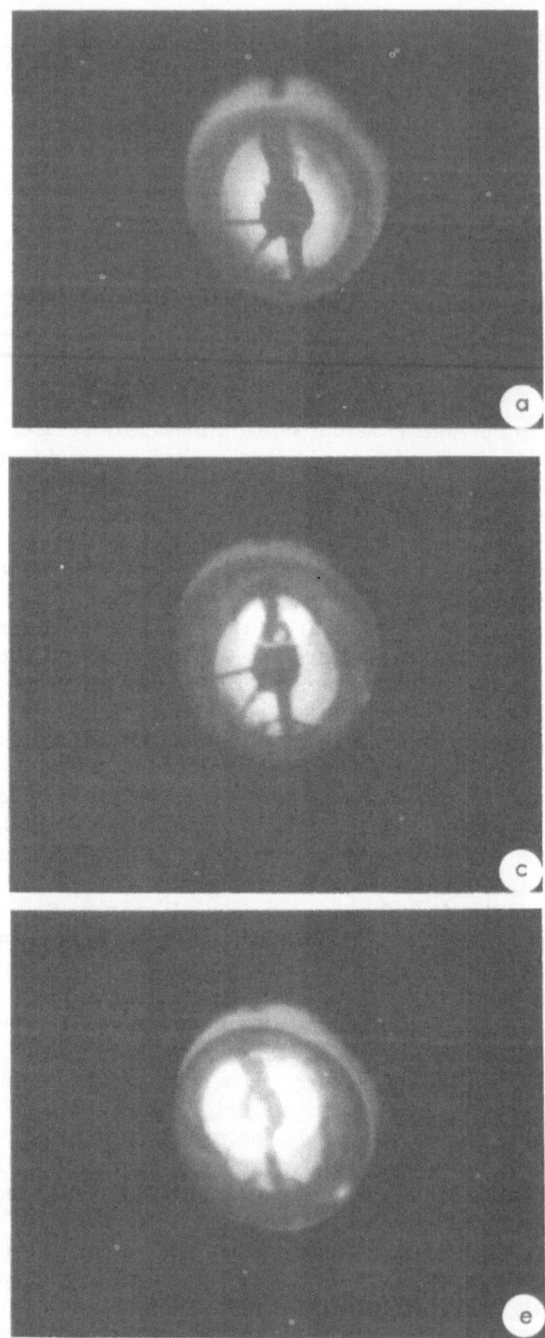

Fig. 17-1. A drop of liquid Al_2O_3 supported on the end of a rod of solid Al_2O_3: (a) end-on view: (b) side view; (c) end-on view showing small bubble of water vapor; (d) side view showing small bubble

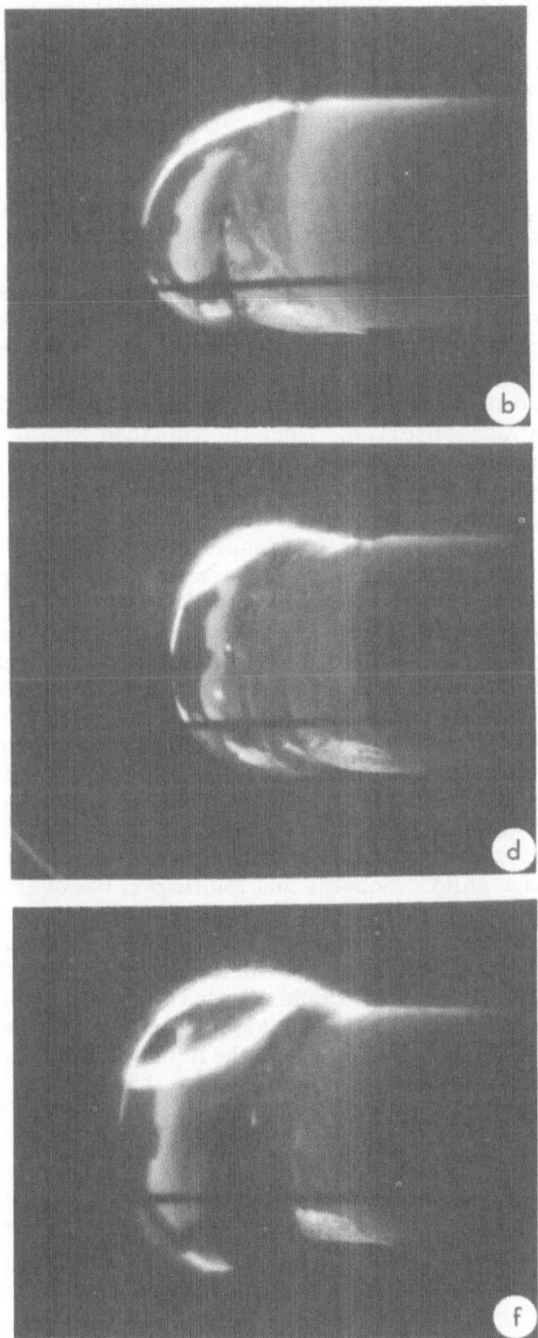

of water vapor; (e) end-on view showing large bubble of water vapor; (f) side view showing large bubble of water vapor.

The same behavior was observed in a static atmosphere of about 25 torr of water vapor free of any other gas, except that the boiling was less vigorous. In Fig. 17-1, part (c) shows an end-on and part (d) a profile view of early stages in the growth of such bubbles. Parts (e) and (f) show corresponding views of later stages in the growth of larger bubbles. Parts (a) and (b) show the quiet drops after boiling has stopped.

When the alumina was melted and then partly frozen in vacuum or in 1 atm of H_2, He, O_2, N_2, Ar, or air, no such boiling was observed. Partial freezing in the presence of H_2 sometimes resulted in behavior different from that in other gases, but nothing which could be called boiling.

When the alumina was melted in one of the above gases, and this was then replaced by 1 atm of water vapor, the amount of melt was seen to change. More of the rod melted and the drop increased in size when the replaced gas was H_2 or He, both of which have lower molecular weights and, therefore, higher thermal conductivities than H_2O. In this case no boiling of the drop of liquid alumina was observed. When the replaced gas was O_2, N_2, Ar, or air, some of the melt solidified and the drop decreased in size because of the greater thermal conductivity of the H_2O, and the remaining liquid then boiled. This indicates that the molten liquid picked up water rapidly enough during the period when the partial pressure of water vapor was building up to 1 atm so that the cooling and partial freezing of the specimen resulted in supersaturation.

When an atmosphere of water vapor around a molten alumina specimen was replaced by one of the other gases, the expected increase or decrease in amount of molten alumina was observed, but in no case was boiling observed.

It is concluded from these observations that water vapor is soluble in molten alumina in the 2300°C temperature region, and that it is more soluble in liquid than in solid alumina. Von Wartenberg [1] considered it unlikely that water vapor as such would be absorbed by melts above 2000°C and postulated the existence of a stable gaseous hydroxide which dissolves in the melt but not in the solid. Scholze and Mulfinger, however, found He to be soluble in Li_2O-SiO_2 melts at 1400°C [2], indicating that the possibility of the physical solution of water vapor in liquid alumina is not as unreasonable as von Wartenberg thought. We have no evidence as to whether the water vapor goes into physical solution in the liquid alumina or forms a soluble compound with it. However, we do have visual evidence that any alumina—water reaction product which might be dissolved in the liquid alumina is not so volatile as to noticeably increase the vaporization from the melt.

REFERENCES

1. von Wartenberg, H., Z. anorg. u. allgem. Chem. 264:226 (1951).
2. Scholze, H., and Mulfinger, H.-O., Angew. Chem. 74:75 (1962); Angew. Chem. Intern. Ed. Engl. 1:52 (1962).

Chapter 18

The Effect of Thermal Radiation on Textile Materials: Part I

Allan J. McQuade and Earl T. Waldron
Clothing and Organic Materials Division
U. S. Army Natick Laboratories
Natick, Massachusetts

A. HISTORICAL BACKGROUND

In considering the problem of protecting individuals from the effects of atomic weapons, the thermal properties of such weapons are a prime consideration, in their effect both the uncovered skin and the clothed parts of the body. In Glasstone's handbook ([1]), "The Effects of Nuclear Weapons," he has pictorially illustrated how the shape of clothing, its design or fabrication methods, and the properties inherent in the outer layer of clothing may all influence the severity of burn received by an individual exposed to the thermal radiation from an atomic weapon.

The possible significance of the thermal energy resulting from the atomic weapon compared with immediate nuclear radiation and blast effects can be seen in Fig. 18-1. The area affected by the detonation of a nominal-size weapon is depicted ([2]). In this diagram, a sizable area occurs near the center of the blast, shown as a solid area, wherein nuclear radiation hazards are such that protecting against thermal effects would not be of significant value.

A second zone occurs, shown as a cross-hatched area, wherein casualties result from ionizing radiation, thermal radiation, and blast effects. The third zone depicted is that area wherein the major effects are thermal and missile wounds. Obviously, a major reduction in casualties can be achieved if the individual can be protected in this third zone.

Providing protection against thermal radiation in this third zone becomes, in essence, a problem of rejecting, or negating, an instantaneous dose of thermal radiation. The characteristics of nuclear weapons have been previously described ([1]) but in general we are dealing with a massive fireball which emits at a blackbody temperature very nearly equal to that of the sun. The spectral distribution thus includes the range from approximately 0.3 to 3.0 μ, with a very small fraction of the total energy occurring in the near ultraviolet, a major fraction in the visible region, and a second major fraction in the infrared region.

The casualty-producing effects of this radiation have been studied in our own laboratories ([3]), by other Government agencies ([4]), and by investigators at the University of Rochester ([5]). The general consensus of these in-

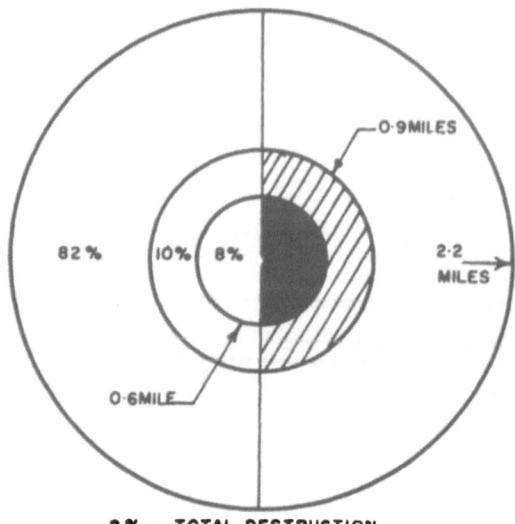

8 % — TOTAL DESTRUCTION

10%— SUBLETHAL IONIZING IRRADIATION SICKNESS,
BURNS, WOUNDS

82 % — THERMAL BURNS, WOUNDS

Fig. 18-1. Population at risk from an A-bomb.

vestigations is that approximately 3 to 5 cal/cm^2-sec imposed upon the bare skin will cause serious, and probably incapacitating, wounds, if the area of exposure is extensive. These investigators have similarly found that lightweight clothing in contact with the body provides very little increase in protection over that afforded by bare skin.

The first possible solution that one can envision, as a means of protecting against the 3 to 30 cal/cm^2-sec irradiance encountered in the third zone, is to provide a highly reflective garment. While this may be a suitable answer for the civilian, it poses many problems for the military. The combat soldier must have on him at all times whatever protective devices can be built into his ensemble. These devices, or systems, must be durable, permeable, and compatible with camouflage requirements. Obviously, an impermeable mirrorlike surface will not meet this requirement.

In addition to protecting the soldier against burns, it is also of military interest that the environmental protective qualities of the clothing be maintained after exposure to the weapon, so that the problem of resupply will be eliminated. This in itself poses a complex problem, since it is necessary to provide a garment that resists destruction as well as providing protection. Since the outer layer of a clothing ensemble will receive the unobstructed rays from a bomb detonation, it is apparent that this layer must serve as the prime deterrent to the passage of thermal energy and also resist the destructive effects of such radiation. Thus, a primary objective of our studies has been to find or develop materials, suitable for clothing, which will resist destruction and ignition.

A second objective has been to find materials and/or develop systems, that will permit us to isolate this outer layer of clothing from the man's body and thus restrict heat flow. The desirability of isolating the outer layer of the clothing ensemble by an air space was recognized in early studies by investigators at the University of Rochester. Mixter et al. [6, 7] have shown that significant increases in protectivity can be achieved by spacing the outer layer. These laboratory observations were subsequently confirmed in atomic field tests conducted at the Nevada Proving Grounds [8]. While this second objective is in itself of major significance, this presentation will be limited to a description of the materials research program aimed at reaching the first stated objective. As will be discussed, the factor of spacing the outer layer is considered in the techniques used in studying materials for the outer layer.

Early observations of the effectiveness of various ensembles showed that marked increases in protection could be achieved by application of fire retardant treatment to the cotton fabrics used in a spaced ensemble. However, it was subsequently recognized that the protection afforded by the fire-retardant-treated cotton would have a definite upper limit, and that in order to reach a new plateau other materials would be required. Consequently, a program was initiated in our laboratories whereby we would, in essence, "screen" the thermal capabilities of commercially available fibers and of some of the more exotic fibers under development in the various research laboratories of the country. The present discussion will be limited primarily to a description of some of the effects that we have observed with representative commercial fibers.

B. METHODS

1. Energy Source

In our work to determine the textile fibers best able to provide protection against thermal radiation, we have employed the carbon arc apparatus described by McQue [9] and by Davis et al. [10]. In brief, this is a 24-in. carbon arc searchlight which has been modified to give a converging beam of radiation. The arc is operated at approximately 180 A and 75 V, with a shutter mechanism mounted at the second focus of the 24-in. ellipsoidal mirror.

Samples to be irradiated are, in turn, placed behind the exposure shutters, and some measurement of "protectivity" or fabric effectiveness is taken.

2. Measurement of Fabric Effectiveness

Two techniques have been employed to measure the effectiveness of various clothing ensembles and materials. One involves the use of animals, wherein the clothing ensemble is interposed between the animal and the radiant energy source. The second utilizes a skin simulant, in lieu of the animal, and through temperature measuring devices imbedded therein, determines the surface temperature rise, or temperature profile in depth, and relates this to protectivity.

Both of these techniques require much greater quantities of experimental materials than are generally available in our program. Hence, a technique was developed that enables us to measure the heat flow from rather small samples. This technique has been previously described in detail ([2]) and is diagrammed in Fig. 18-2. In essence, a high-speed calorimeter disc is positioned 0.1 in. behind the sample, and heat flow resulting from a given irradiance imposed on the outer surface is measured. The relative effectiveness of various materials is judged by comparison of the heat flow to that observed with some standard material whose behavior with animals is known.

3. Paper Mat Techniques

In order that the various fibers might be compared, it was necessary that a single physical form be utilized. A technique for casting fibers into suitable form was developed in cooperation with the Textile Division of E. I. duPont de Nemours & Company, Inc. ([12]). The end result of this program was to cast each fiber, and fiber combination, into a paper mat weighing approximately 5 oz/yd^2 (equivalent to summer-weight clothing). Both mass and color were rigidly controlled, and to date, approximately 1000 entities have been produced for study.

4. Destruction Measurements

For evaluation of material damage, the circular discs of fiber mat materials used for measurement of energy transfer are employed. After the mats are subjected to the various irradiances, visual measurements are normally made of the degree of destruction sustained from a given exposure. Numerical values are assigned, ranging from 0 for no damage to 6 for complete destruction. However, in cases of particular interest, destruction is measured by weight-loss determinations.

5. Ignition Measurements

For measurement of the ignition characteristics of a material, a much larger sample holder is used than in the heat flow studies. In the heat flow

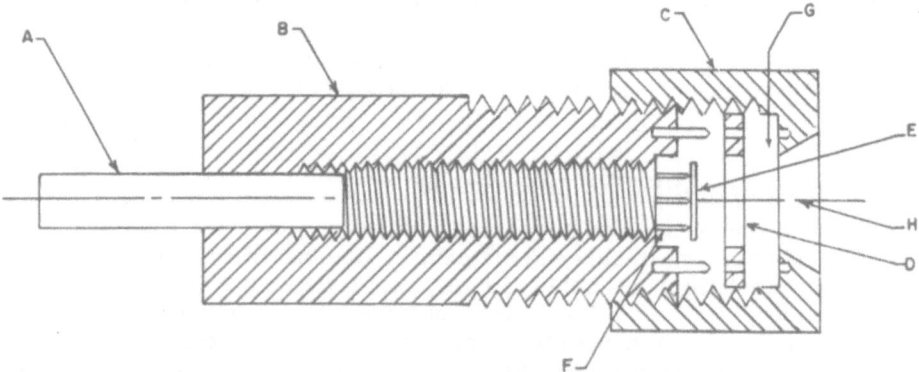

Fig. 18-2. Cross-sectional view of calorimeter assembly: A—threaded spindle, B—brass cylinder, C—threaded cap, D—brass spacer (0.1 in.), E—calorimeter button, F—supporting needles, G—sample location, H—aperture.

TABLE 18-I
Typical Fibers Prepared for Thermal Evaluation

Natural Fibers	Synthetic Fibers Commercially available
Cotton*	Cellulosics*
Wool*	Modified cellulosics*
Silk*	Polyamides (nylon)*
Asbestos*	Acrylics*
	Modacrylics*
	Polyester*
Inorganic	Vinyl chloride*
Glass	Nytrile*
Graphite	Limited availability
Quartz	
Potassium titanate	Fluorocarbon*
	Olefin*
	Pyrolyzed acrylic*

*Studied as paper mats, usually in blends with other fibers at 10% increments.

measurements, a $7/16$-in.-diameter aperture is used to ensure a fairly even distribution of energy across the target area since with a converging beam of radiation, the energy intensity drops off rapidly beyond this central area. While the $7/16$-in.-diameter aperture permits a more realistic definition of the energy distribution in heat flow measurements, its geometry does modify the ignition characteristics. Hence, when these characteristics are being defined, auxiliary measurements are made, in which comparatively large, unrestrained samples are exposed. We find that this technique enables us to measure the irradiance required to produce the sustained combustion typical of that observed when garments are exposed to the larger thermal sources.

C. OBSERVATIONS

1. Typical Fibers

More fibers are available today for characterization as to their thermal properties than ever before. Some of the more common fibers are listed in Table 18-I. Further, when one considers that many of these fiber types may represent three or four specific commercial fibers, each one slightly different from the others, this list more than doubles. Finally, if we are to consider many as they are used in two- or three-component blended fiber fabrics, the number of fibrous entities that can be considered for clothing fabric multiplies a hundredfold.

Nearly all of the materials in this list have been evaluated in fabric of fiber-mat form. Those which have been marked with a single asterisk represent fibers evaluated in the mat form. Others were in fabric form, and thus do not lend themselves to some of the comparisons shown later.

From work conducted to date with these fibers, individually and in blends, the general conclusions are that we are unable to predict from intrinsic fiber properties what to expect when these fibers are combined with

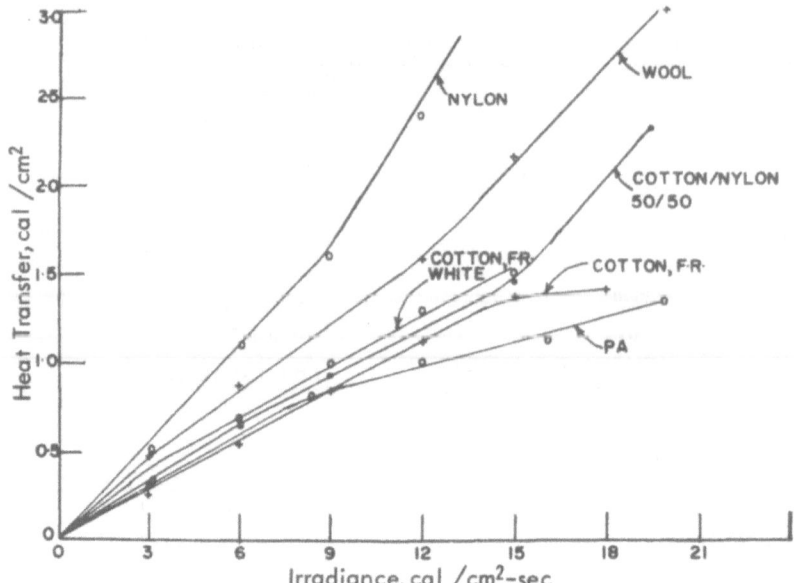

Fig. 18-3. Heat transferred from typical natural and synthetic fibers.

other fibers. Similarly, although fibers such as quartz and graphite are extremely stable, they lack the physical properties required to make lightweight garments. Hence, the basic fibers and combinations that we have concentrated on are those we expect will give the desirable textile fabric properties.

A typical spectrum of these fibers is shown in Fig. 18-3. Aside from showing that these materials differ with respect to energy transfer for irradiances of 1 sec, this drawing makes several other factors apparent.

a. Nylon. The data show that among typical conventional fibers, the highest heat flows are normally observed behind the pure synthetic fibers such as nylon. The response of these fibers differs basically from that observed with a fiber such as cotton. The upper limit of effectiveness for the synthetics is governed largely by the irradiance at which initial hole formation occurs. When this occurs, a physical collapse results, as the initial point of rupture expands throughout the entire exposed area. This collapse is frequently not accompanied by significant destruction of material. Rather, it is a function of the melting point and contraction of the heated material. In contrast, cotton, both untreated and fire-retardant-treated, undergoes successive degradation stages, which leave some residue intact at all but the combustion phase.

Obviously, when fabric collapse occurs with a synthetic such as nylon, large fractions of the incident energy are imposed directly upon the calorimeter button or—in a multilayered ensemble—directly upon the sublayers. The irradiance required for the destruction of the synthetics is low and varies with the color or absorptivity of the material. In general, we do not observe sustained combustion with the synthetics, but do observe flaring, which occurs during the actual thermal pulse.

b. Wool. The response of wool in this lightweight form, 5 oz/yd^2, somewhat parallels that of the synthetics such as nylon. However, as the irradiance increases, we observe that melting and fusing occur in depth, rather than engendering the complete collapse typical of the synthetics. The use of heavier and denser wool constructions will generally improve the performance of this fiber in relation to that of others.

c. Cotton. The performance of untreated cotton is not shown in Fig. 18-3, but one may find many reports as to its performance. Generally, it has heat transfer characteristics slightly better than those of fire-retardant cotton up to 8 to 10 cal/cm^2-sec irradiance. In this irradiance range, the untreated cotton will ignite and the resulting heat flow is high. Ignition will, of course, produce total destruction of the cotton, and use of darker colors causes flaming to occur at lower irradiances.

d. Fire-Retardant-Treated Cotton. The use of a flame retardant on cotton reduces heat transfer compared with that of either the synthetics or wool. Further, as shown in Fig. 18-3, its performance is scarcely influenced by color. The flame retardant eliminates the sustained flaming that occurs with untreated cotton, but this treatment increases the rate of fiber degradation. Hence, relatively low irradiances cause considerable damage to this material, but do produce an opaque charred layer that continues to protect the calorimeter button from the direct rays of the arc. The charred residue weighs approximately 30% of the material's original weight and is very friable.

e. Cotton/Nylon Blend. Blending of fibers to influence physical comfort and aesthetic properties of fabrics has long been generally practiced by the textile industry. Such blending can also affect thermal response, particularly the ignition and destruction characteristics, as is shown by the performance of the 50/50 blend of cotton with nylon. In this case, its heat transfer properties are similar to those of fire-retardant-treated cotton, while its resistance to ignition is greatly improved over that of untreated cotton.

The possibility of raising the ignition threshold and concomitantly obtaining greater resistance to destruction has caused us to concentrate upon fiber blends in our evaluation of materials. Specifically, as will be shown, it has caused us to be very much interested in the response of cotton/nylon blends, since they represent a means of substantially increasing the protective capability of our uniforms. At the same time, some of their important physical properties could be realized.

f. Pyrolyzed Acrylonitrile. The bottom curve in Fig. 18-3 gives the heat transfer data measured behind pyrolyzed acrylonitrile (PA) fibrous materials. Pyrolyzed acrylonitrile is representative of some of the less conventional fibers that are of interest. Here, the natural color of the fiber is black. It is extremely resistant to degradation and does not permit ignition or sustained flaming. However, gases from its decomposition may ignite, or flare, at high irradiances during the exposure. As shown by this figure, it has excellent resistance to heat transfer.

2. Blends of Synthetics

The positive responses that have been observed with some of the cotton/nylon blends are not characteristic of the majority of blends. In general,

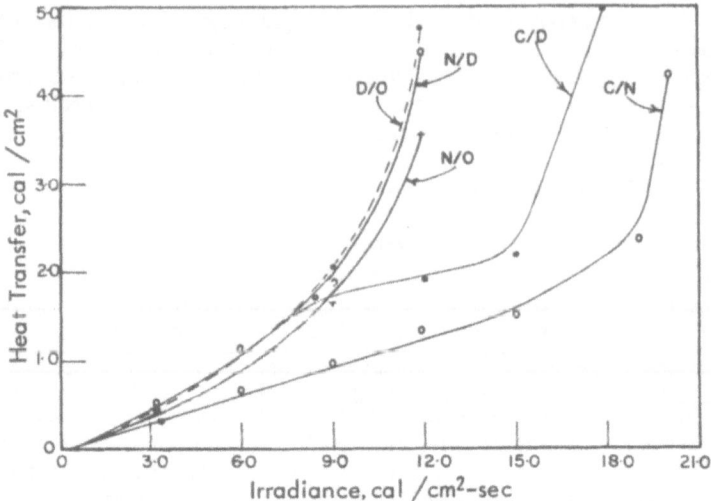

Fig. 18-4. Heat flow from typical 50/50 blends: D/O—dacron/orlon, N/O—
nylon/orlon, N/D—nylon/dacron, C/D—cotton/dacron, C/N—cotton/nylon.

we observe the greatest stability with blends of natural and synthetic fibers
and the least stability with blends of synthetics. Relationships for typical
50/50 blends are given in Fig. 18-4, where it is seen that the blends of
synthetics, such as polyester with acrylonitrile, polyamide (nylon) with
polyester, or polyamide with acrylonitrile, are essentially destroyed at
a comparatively low irradiance, with resultant high heat transfer.

It is similarly interesting to note the marked difference between the
responses of the cotton/polyester blend and the cotton/nylon blend. The
cotton/polyester, although more effective than the combination of synthetics,
is much less effective than the cotton/nylon, and is actually ignited at a lesser irradiance than that required for an equivalent all-cotton fabric.

3. Cotton/Nylon Blends

The consistently superior performance observed with some of the cot-
ton/nylon blends has led to rather extensive studies with these materials,
and we have observed that their resistance to degradation and ignition is
modified by the nylon content, the color of the nylon relative to that of the
cotton, and their physical relationship within the fabric. We have found that,
with all other things being equal, the lowest heat transfer values are observed with nylon contents of 35 to 40%. However, at this nylon content, the
ignition irradiance is a critical function of the relative colors of the two
fibers. This relationship is shown in Fig. 18-5, where ignition irradiance
for several cotton/nylon blends is plotted as a function of the ratio of the
reflectance of the nylon fiber to the reflectance of the cotton fiber.

The curves in Fig. 18-5 show that at a low nylon content, i.e., 20%, the
ignition response is dominated by the cotton fiber and that varying the
reflectance of the nylon fiber is of little consequence. At 40% nylon content,
a fabric could be designed wherein the nylon fiber would be slightly more
reflectant than the cotton fiber and would possess excellent resistance to

ignition. However, at 50% nylon content, we are permitted greater flexibility in adjusting the colors of the two fibers and still producing a fabric with high ignition resistance and acceptable heat transfer characteristics. This range of possible selections is extended at the 60% nylon content, but at this level of synthetic content, we encounter other problems, such as greater thermal destruction. Consequently, we have selected the 50/50 cotton/nylon blend as an optimum solution.

The physical relationship of the two fibers in the fabric similarly affects not response. This is seen from Fig. 18-6, where heat transfer as a function of irradiance is plotted for the two surfaces of a cotton/nylon sateen. This particular fabric was woven so that one surface was composed predominantly of cotton fibers and the other face predominantly of filament nylon fibers. As the curves show, with the cotton face exposed to the radiation, relatively high heat transfer and low ignition irradiance were observed. With the nylon face exposed to the radiation, a decrease in heat transfer and a substantial increase in resistance to ignition were observed.

4. Blends of Pyrolyzed Acrylonitrile

In addition to the cotton/nylon blend, the pyrolyzed acrylonitrile fiber has been the subject of considerable study. Here, the problem as dictated

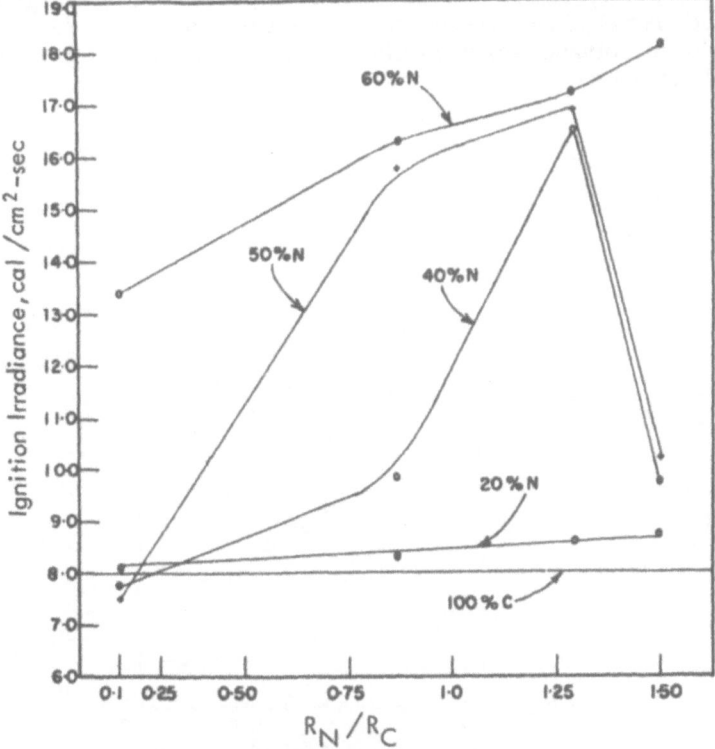

Fig. 18-5. Effect of relative color in the ignition characteristics of cotton/nylon blends: N—nylon, C—cotton.

by fiber physical properties was different. Both cotton and nylon have the necessary physical properties to allow us to attain strong, lightweight fabrics. Pyrolyzed acrylonitrile does not. While under optimum processing conditions strength approximately equal to wool has been obtained, the fiber lacks the extensibility characteristics of wool and will break under conditions of flex. Thus, with pyrolyzed acrylonitrile, the problem was one of finding other fibers for blending with it which would overcome these physical deficiencies, and at the same time permit it to retain its thermal protective properties.

Pyrolyzed acrylonitrile, as will be recalled from Fig. 18-3, showed the lowest heat transfer of the fibers considered. Furthermore, in ignition studies, there was no tendency for it to ignite at irradiances up to 25 cal/cm^2-sec. Lastly, it showed little evidence of destruction at this irradiance.

a. Blends of Nylon. The effect on energy transfer of blending with other fibers is illustrated in Fig. 18-7, which shows various blends of nylon with pyrolyzed acrylonitrile. From these data, it will be noted that as pyrolyzed acrylonitrile is added to nylon in increasing amounts, there is improvement in performance over that of 100% nylon, and it is indicated that with a 50/50 blend, performance equivalent to 100% pyrolyzed acrylonitrile might be obtained. With a greater amount of pyrolyzed acrylonitrile—for example, the 40/60 nylon/pyrolyzed acrylonitrile blend—improvement in energy transfer is realized, but physical strength is decreased. Not shown are the data for 20/80 nylon/pyrolyzed acrylonitrile which closely follow those for 100% pyrolyzed acrylonitrile.

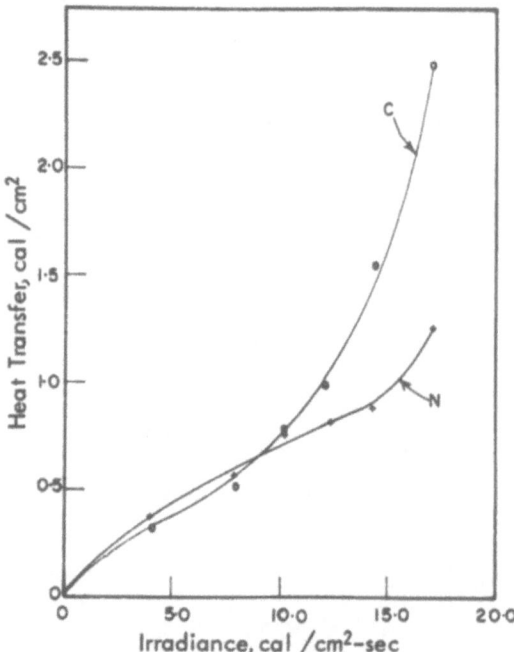

Fig. 18-6. Effect of fiber position in a cotton/nylon blend:
C—cotton facing radiation, N—nylon facing radiation.

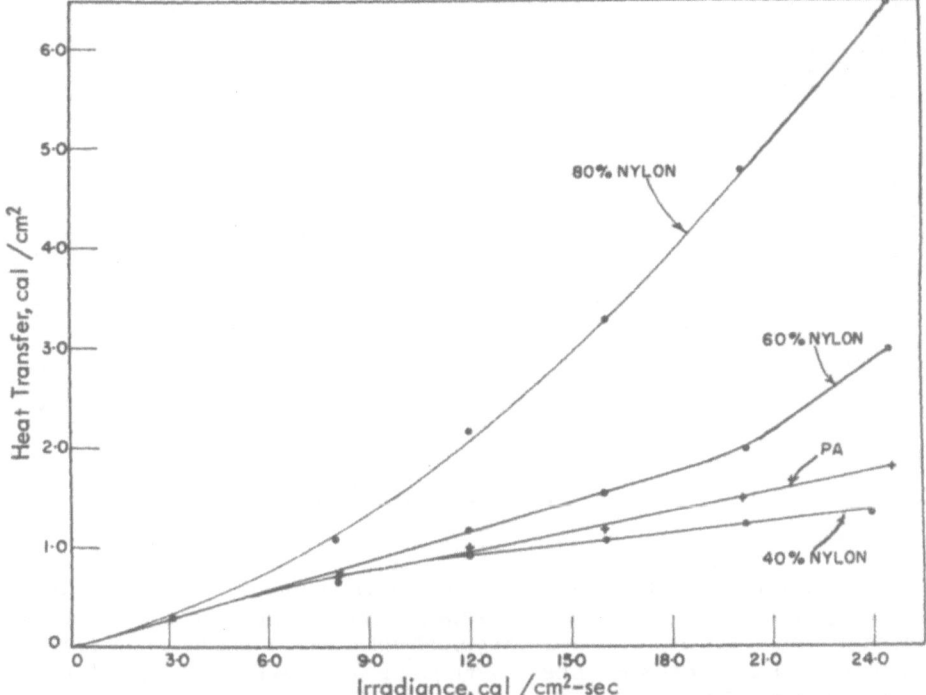

Fig. 18-7. Heat transfer from blends of nylon with pyrolyzed acrylonitrile.

b. Blends of Cotton and Wool. The commonly employed natural fibers, wool and cotton, were also examined in blends with pyrolyzed acrylonitrile, and the results in general were parallel to those for nylon. From the data of Fig. 18-8, it will be noted that, in 40% blends of cotton or wool fibers with pyrolyzed acrylonitrile, there is a reduction in heat transfer from that obtained with nylon. However, ignition studies showed that, at irradiances greater than 21 cal/cm^2-sec, sustained combustion was effected, and the two blends were destroyed.

It was further noted in this work that the color of the auxiliary fiber influences energy transfer from the pyrolyzed acrylonitrile blend. Figure 18-9 presents data obtained when either gray cotton or gray nylon was used with pyrolyzed acrylonitrile. Here, cotton is shown to be less effective than nylon. Further, as regards colored vs. uncolored auxiliary fiber, darkening of the fiber improved the performance of nylon, but effected higher energy transfer in the case of cotton. By the energy transfer criteria, the 60/40 blend of pyrolyzed acrylonitrile with gray colored nylon produced one of the more effective materials found to date.

Though inclusion of other fibers with pyrolyzed acrylonitrile produced this beneficial effect in heat transfer, it generally detracted from that fiber's good resistance to destruction. Combustion and flaming often occurred during exposure, and after irradiation, the material was extremely friable. Splitting and cracking of the fiber mass occurred, particularly when large areas of 4-in.-diameter were exposed to the solar furnace.

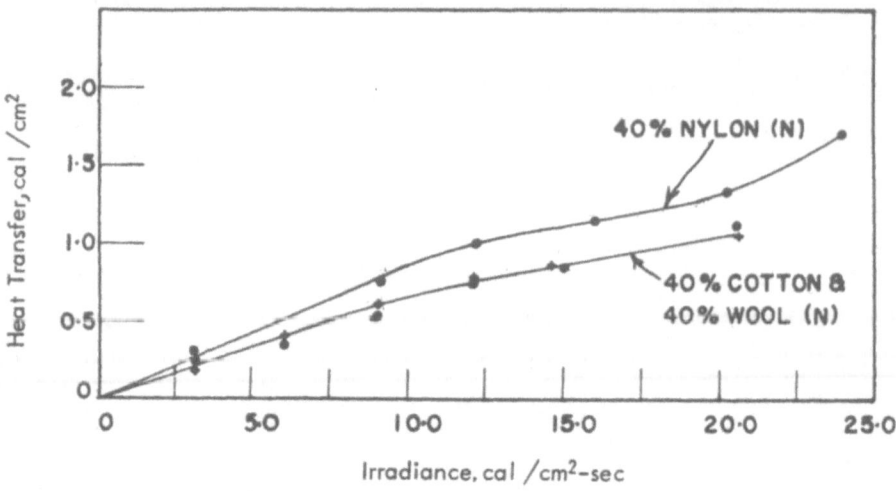

Fig. 18-8. Blends of pyrolyzed acrylonitrile with natural-colored fibers.

In summary, though pyrolyzed acrylonitrile blended with other fibers showed promise for inhibiting heat transfer, good performance was not obtained in those blends where its physical deficiencies might be overcome. Further, inclusion of other fibers induced greater decomposition, so that a multishot capability would not be obtained. Thus the fiber is no longer of major concern.

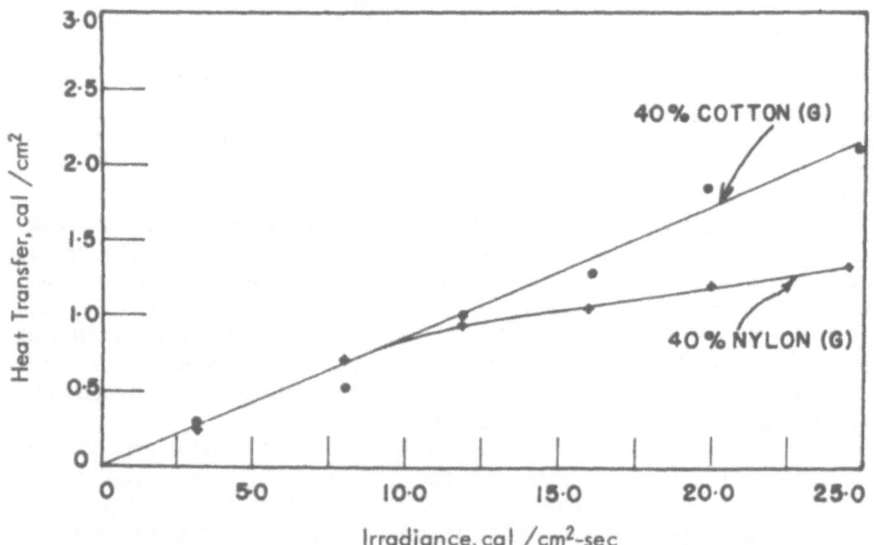

Fig. 18-9. Blends of pyrolyzed acrylonitrile with gray fibers.

5. Solar Furnace Studies

The work that has been described relates mainly to the laboratory analysis of the many fiber entities and characteristics which, of necessity, have had to be studied on a small scale. Actually, the atomic weapon is a large-scale device, and a much better evaluation of its capability can be obtained with our solar furnace.

With the solar furnace, we can irradiate areas of 4 in. diameter and thus include factors of fabric collapse, ignition, and area effects. These factors are not obtainable with the small-area carbon arc. As a consequence, any material that survives the combined paper mat—carbon arc analysis is woven into a fabric and re-examined on our solar furnace; animals are used as indicators of protection.

An illustration of the data obtained in this type of study is shown in Table 18-II. The data listed are measurements of the severity of burns sustained beneath multilayered systems exposed to the cited irradiances with our solar surface. In each case, the outer layer was separated from the T-shirt by a three-dimensional, open-weave spacer material. The two outer layers considered were a 9 oz/yd^2 cotton sateen, which had been fire-retardant-treated and a comparable cotton/nylon sateen whose design included many of the criteria found to be desirable in the paper mat studies. The burn severity code used is: 1 + M for a one plus mild burn, 1 + S for a one plus severe burn, 2 + I for a two plus intermediate burn, etc. The first

TABLE 18-II
Effectiveness of Spaced Ensembles which Utilize Either Fire-Retardant-Treated Cotton, a Cotton/Nylon Sateen, or a 50/50 Blend of Pyrolyzed Acrylonitrile with a Modified Polyamide

Sample	Total number of exposures	Burn severity							
		0	1+M	1+S	2+M	2+T	2+E	3+M	3+S
FR Sateen @ 10 cal/cm^2-sec	9	3		2	4				
Cotton/Nylon @ 10 cal/cm^2-sec	9	9							
FR Sateen @ 13 cal/cm^2-sec	12		1	1	6	2		1	
Cotton/Nylon @ 13 cal/cm^2-sec	12	9			2			1	
FR Sateen @ 15 cal/cm^2-sec	12				5	3		3	1
Cotton/Nylon @ 15 cal/cm^2-sec	12	9			3				
Cotton/Nylon @ 18 cal/cm^2-sec	11	7	1	1	2				
•PA/MP @ 16.4 cal/cm^2-sec	8	8							
•PA/MP @ 20.1 cal/cm^2-sec	10	10							
•PA/MP @ 23 cal/cm^2-sec	10	4		1	4		1		

*Pyrolyzed acrylonitrile with an experimental fiber.

column lists the particular outer layer fabrics considered and the irradiances to which the ensembles were exposed.

The data show that this particular fire retardant treatment on sateen offers protection in a limited number of cases at the 10-cal/cm²-sec level, while the untreated cotton/nylon sateen offered protection in all cases. Similarly, as the irradiance was increased, the number and severity of burns observed to occur underneath the fire-retardant-treated cotton increased, while the effectiveness of the cotton/nylon remained essentially constant throughout the irradiance range 13 to 18 cal/cm²-sec. When one considers that unprotected skin or skin in contact with lightweight clothing will be severely burned in the irradiance range 3 to 5 cal/cm²-sec, it is obvious that substantial protection can be offered by the careful engineering of fabrics and systems which utilize conventional materials.

It is similarly apparent that further increases in protection are feasible through utilization of some of the experimental fibers available today. This can be seen from the data listed for pyrolyzed acrylonitrile experimental fiber combination. While this particular combination lacks the physical strength required in a combat garment, it illustrates the potential in this type of approach.

REFERENCES

1. Glasstone, S., ed., The Effects of Nuclear Weapons, United States Atomic Energy Commission, US Govt. Printing Office, June 1957.
2. Pearse, H. E., and Kingsley, H. D., Thermal Burns from the Atomic Bomb, Rept. UR-254, The University of Rochester Atomic Energy Project, Rochester, N.Y., April 1953.
3. Cotton, E., Grey, M. G., and Penniman, F. G., A Comparison of Bare Skin Burns Induced by Thermal Radiation from the Carbon Arc and the Solar Furnace, Rept. T-25, Radiation Physics Laboratory, QM R & E Comd, Natick, Mass., August 1959.
4. Alpen, E. L., Butler, C. P., Martin, S. B., and Davis, A. K., Effects of Spectral Distribution of Radiant Energy on Cutaneous Burn Production in Man and the Rat, USNRDL-TR-46, NM 006-015, US Naval Radiological Defense Laboratory, April 1955.
5. Lyon, J. L., Davis, T. P., and Pearse, H. E., Studies of Flash Burns: The Relation of Thermal Energy Applied and Exposure Time to Burn Severity, Rept. UR-394, The University of Rochester Atomic Energy Project, Rochester, N.Y., April 1955.
6. Mixter, G., and Pearse, H., Studies on Flash Burns: The Protection Afforded by 2, 4, and 6 Layer Fabric Combinations, Rept. UR-261, The University of Rochester Atomic Energy Project, Rochester, N.Y., June 1953.
7. Mixter, G., Studies on Flash Burns: Further Reports on the Protective Qualities of Fabrics as Expressed by a Protective Index, Rept. UR-354, The University of Rochester Atomic Energy Project, Rochester, N.Y., October, 1954.
8. Babers, F. H., and McQuade, A. J., Thermal Protection of the Individual Soldier, Operation Blumbob Final Rept., Project 9.1.
9. McQue, B., The QM Arc, Special Projects Report T-6, QM R & E Ctr, March 1956.
10. Davis, T. P., Krolak, S. J., and Blaknew, R. M., Studies of Flash Burns: The Carbon Arc Source, Rept. UR-226, The University of Rochester Atomic Energy Project, Rochester, N.Y., November 1952.
11. Waldron, E. T., and Koza, W. J., Effectiveness of Textile Materials Against Intense Thermal Radiation, Part I—Experimental Technique, TFFL Rept. 176, QM & E Comd, Natick, Mass., December 1958.
12. McQuade, A. J., Waldron, E. T., and Farquhar, B. S., Annals of the New York Academy of Sciences 82:762 (1959).

Chapter 19

The Effect of Thermal Radiation on Textile
Materials: Part II

Joan B. Berkowitz-Mattuck
Arthur D. Little, Inc.
Cambridge, Massachusetts

A. INTRODUCTION

The preceding (The Effect of Thermal Radiation on Textile Materials) paper outlined the problem facing the Army in its attempt to provide protection against the thermal effects of atomic weapons. The use of the arc imaging furnace and the solar furnace to simulate thermal effects was described, and methods for measuring the effectiveness of potential clothing fabrics were summarized. A program aimed at the development of thermally stable fibers and fiber blends as a means of providing thermal protection was discussed.

B. STATEMENT OF THE PROBLEM AND EXPERIMENTAL PROCEDURE

We have investigated four alternative approaches to the problem of neutralizing 10 cal/cm^2 delivered in 0.5 to 2 sec to the outer layer of a clothing ensemble. These approaches were: (1) utilization of endothermic heats of fusion, vaporization, or sublimation; (2) utilization of endothermic heats of suitable chemical reactions; (3) expansion of the outer garment layer to form an insulating foam; and (4) instantaneous production of an insulating smoke by interaction between the thermal pulse and outer clothing layer. The base cloth was olive-green cotton poplin, 5 oz/yd^2, to which might be added a maximum weight of 1.5 oz/yd^2 of protective material.

A large number of fabric additives were screened in an ADL-Strong arc imaging furnace, fitted with a calorimeter assembly slightly modified from that shown in Fig. 18-2 (p. 232). For simulation of the thermal pulse from a nuclear explosion, a maximum flux of 25 cal/cm^2-sec must be delivered uniformly over as large an area as possible. The arc furnace delivers 350 cal/cm^2-sec over a 9-mm spot in the focal plane. So that the energy at the focal spot would be redistributed uniformly over a larger area, a mirrored box or kaleidoscope of $5/8$ in. square cross section ([1,2]) was interposed between the focal plane and the sample plane.

Approximately 60 fabric additives were selected for test, in anticipation that they would absorb thermal energy in one of the four prescribed ways. An additive was considered promising, and worth further investigation if the heat transferred through the treated 5-oz fabric during a 1-sec exposure to

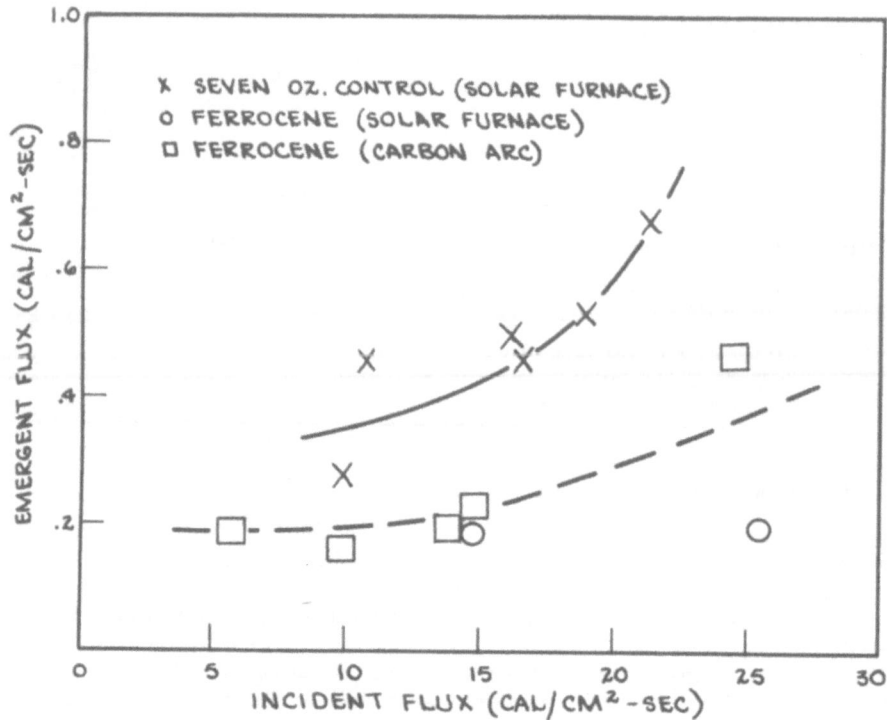

Fig. 19-1. Ferrocene-treated 5-oz. olive-green 107 poplin.

a flux of 10 cal/cm²-sec was less than that transferred through a 7-oz untreated cotton poplin under the same conditions. That is, additives were
sought which would be superior to heavier-weight cloth, since it is well
known that improved protection can be achieved with thicker clothing.

C. RESULTS

It was soon apparent that in nearly every exposure several of the phenomena under evaluation occurred simultaneously or sequentially during the
test. Thus, every sample tested smoked to some extent; a BCI alcohol-
soluble nylon that was expected to depolymerize with absorption of heat,
actually smoked, foamed, degraded, and possibly melted as well.

1. Utilization of Endothermic Heats of Physical Changes and Chemical Reactions

On the basis of the screening tests, it was concluded that the use of endothermic heats of fusion, vaporization, or sublimation is impractical. The
most attractive physical change of state, in that a large amount of energy is
required at accessible temperatures, is the sublimation of water, perhaps
from salt hydrates. The inorganic salts that were tried are listed in Table
19-I, along with their heats of decomposition to the anhydrous salt and water
vapor. It is to be noted that since the maximum permissible weight of salt
hydrate additive was limited to 30%, the decomposition of the salts could not

TABLE 19-I
The Attenuation of Thermal Radiation by Decomposition of Salt Hydrates

Compound	ΔH decomp., cal/g	Maximum heat absorption at 30% pickup, cal/cm^2	Incident flux, cal/cm^2-sec	Heat transfer, cal/cm^2	Heat transfer through control, cal/cm^2
$MgSO_4 \cdot 7H_2O$	596	3.0	13.0	0.33	0.39
$MgCl_2 \cdot 6H_2O$	469	2.4	11.0	0.38	0.37
$Na_2SO_4 \cdot 10H_2O$	386	2.0	11.0	0.51	0.37
$CaCl_2 \cdot 6H_2O$	384	2.0	11.1	0.37	0.38
$Na_2B_4O_7 \cdot 10H_2O$	368	1.9	11.0	0.57	0.37
$FeCl_3 \cdot 6H_2O$	332	1.7	13.5	0.26	0.40
$Cd(NO_3)_2 \cdot 4H_2O$	152	0.8	10.8	0.63	0.36

absorb any more heat than the values listed in column 3. Therefore, under the most favorable circumstances, a maximum of 3.0 cal/cm^2 of incident energy could be absorbed by the vaporization of water from a hydrate. In practice, as shown by the experimental results, the inorganic salt hydrates do not even come close to their theoretical potential. Column 4 lists the flux level at which each of the materials was tested. All of the salts were applied to 5-oz cotton poplin to a weight pickup of 1.5 oz/yd^2, and the exposure time in every experiment was 1 sec. The total amount of heat transferred to a blackened copper disc calorimeter behind the exposed sample is shown in column 5, and the amount of heat transferred through samples of 7-oz cotton poplin exposed under the same conditions is given in column 6. It is to be observed, first, that the heat transfer values in column 5 do not depend in an obvious way on the heats of decomposition given in column 2. Comparing columns 5 and 6, one finds that adding $MgSO_4 \cdot 7H_2O$, $MgCl_2 \cdot 6H_2O$, or $CaCl_2 \cdot 6H_2O$ to the 5-oz material is no more effective than increasing the weight of the cotton by 30%; $Na_2SO_4 \cdot 10H_2O$ and $Cd(NO_3)_2 \cdot 4H_2O$ actually appear to be detrimental, while only $FeCl_3 \cdot 6H_2O$ seems to hold any promise. In light of the other evidence, however, one can assume that it is probably not the heat absorbed in vaporizing the water that is responsible for the reduced heat transfer. It is, of course, possible that the dehydration of inorganic salts is kinetically unfavorable under the flash heating conditions of a nuclear explosion. However, it must be pointed out that if wet cloth were to be used instead of dry, the higher thermal conductivity of the wet cotton[3] would completely offset any of the beneficial effects of endothermic vaporization. State changes other than the vaporization of water involve energies that are much too small to be worth considering under the severe weight limitation imposed. Heat absorbed in an endothermic chemical reaction could theoretically be much larger than heat absorbed in a change of physical state. However, in the experiments done under the present program, no clear-cut case was found in which an endothermic chemical reaction could be held responsible for beneficial effects.

2. Utilization of Insulating Foams and Smokes

In principle, just as increasing the bulk of clothing in general—even without increasing the weight—is an effective means of providing thermal protection, so the formation of an insulating foam at the outer garment surface should provide excellent protection. In practice, a material was not found that would form a stable foam for the duration of the thermal pulse.

The formation of a smoke screen is a superb means of reducing skin burn. On the basis of a simple theoretical model [4], it can be shown that formation of a black smoke screen, 4 cm thick, in front of a 0.04-cm-thick cloth with an emissivity of 0.75, can reduce the amount of radiation transferred through the rear surface of the cloth by more than a factor of two. The theoretical conclusion is supported by field tests in which a dramatic attenuation of thermal radiation has been observed as a result of scattering by atmospheric haze, moisture, dust, and by artificial white chemical smokes. A dense smoke between the point of burst and the target can reduce incident thermal energy at the target to as little as one-tenth of that which would otherwise have been received. Materials screened in the present program that were found to provide dense smoke layers on exposure to a thermal pulse include ferrocene and related organometallics, the metal acetyl acetonates, camphor, and the anthraquinone derivatives. Results with ferrocene, which has also been reported as a promising organic coating for space vehicles [5], are plotted in Fig. 19-1. The major problem encountered with smokes is that the particles may ignite, depending upon the conditions of oxygen concentration and ambient temperature and the density of the smoke. The ignition limits of smokes is a subject of practical interest worthy of further investigation.

REFERENCES

1. Chen, M. M., Berkowitz-Mattuck, J. B., and Glaser, P. E., Appl. Opt. 2:265 (1963).
2. Glaser, P. E., Chen, M. M., and Berkowitz-Mattuck, J., Solar Energy 7:12 (1963).
3. Fuel Research Lab., MIT, Cambridge, Mass. Reports under Noni-1841(37), Project No. NR 051-237, UNR.
4. "Development of a Means of Neutralize Intense Thermal Radiation," A. D. Little, Inc., Final Report under Contract DA 19-129-QM-1087, Oct. 31, 1960.
5. Missiles and Rockets, Dec. 5, 1960, p. 18.

Chapter 20

Ablation and Analytical Measurements
Using an Arc Image Furnace

J. E. Brownsword, J. K. Phillips, and M. T. Conger
The Goodyear Tire & Rubber Company
Akron, Ohio

A. INTRODUCTION

The Goodyear Tire & Rubber Company has been using an ADL arc imaging furnace* for research purposes for the past two years. This paper discusses two applications for which we have found the arc imaging furnace to be well suited. These are (1) study of ablative insulation and (2) polymer identification. The work concerning ablative insulation will be presented first.

B. ABLATIVE INSULATION FOR SOLID PROPELLANT ROCKET MOTORS

While ablative insulation is used to insulate several missile components, the Goodyear arc imaging furnace was purchased specifically for use in a program for studying ablative insulator materials for insulating the motor cases of solid propellant rocket motors. It has been used to evaluate over 600 different insulator materials.

The structural component (case) of a solid propellant rocket motor is a thin-walled cylinder with hemispherical ends. The propulsion nozzles are attached to this part, and its walls transmit the propulsion thrust loads. This cylindrical container also serves as the fuel container and as the combustion chamber. It must, therefore, withstand the high temperature and high gas pressures that are developed by propellant combustion.

No single material available today has both the strength and temperature resistance required for motor cases. The solution to this dilemma is to make the motor case exterior from a material chosen for its high strength and to incorporate, on the inside, a lining of material chosen for its ability to protect the exterior component from the hot combustion gases (Fig. 20-1). For many reasons that cannot be discussed within the scope of this paper ablative, elastomeric materials are superior to all other materials for this motor case insulating liner.

An ablative insulator is an insulator that absorbs a large amount of heat through its sacrificial destruction. The phenomena associated with the sacrificial decomposition of a material are a function of the ablative environ-

---

*The arc imaging furnace used in this investigation is a general-purpose model, Serial No. 53109, designed by Arthur D. Little, Inc., Cambridge, Mass., and manufactured by Strong Electric Co., Toledo, Ohio.

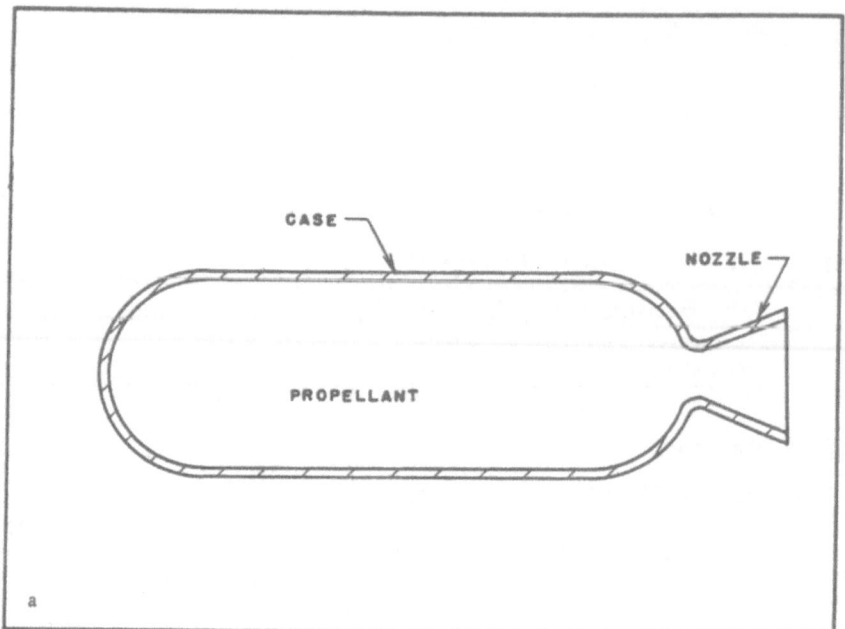

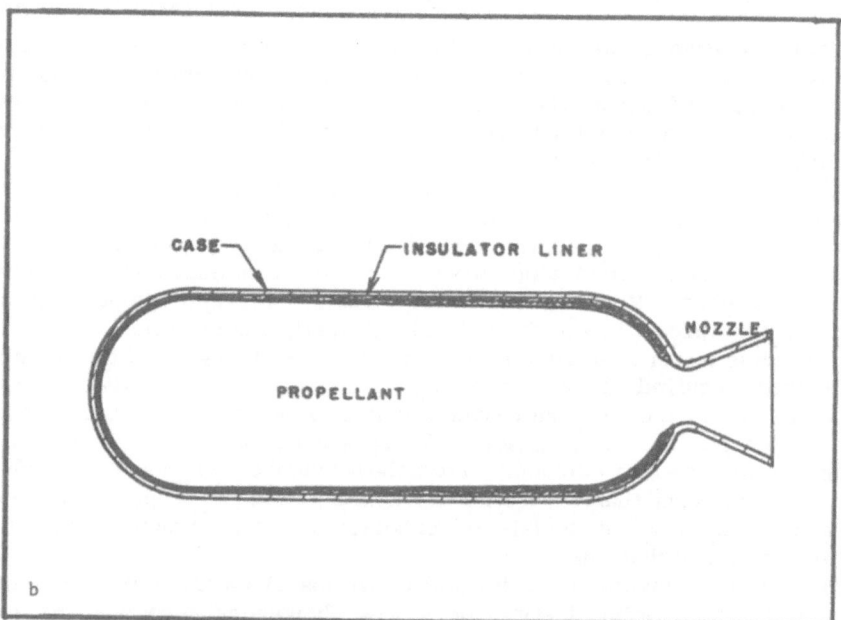

Fig. 20-1. Rocket motor: (a) without liner; (b) with liner.

ment. To simulate simultaneously all of the firing chamber environmental conditions using laboratory equipment is an impossible task. Acetylene torches, plasma jets, and arc imaging furnaces are being used, however, for the preliminary evaluation of ablative insulation. Goodyear's choice of an arc imaging furnace for this purpose was based on a number of factors. First of all, it was found that consumption rates, measured with the arc imaging furnace, correlated with those measured by rocket motor tests in the low-velocity areas. At the time we selected the arc imaging furnace, the only application for rubber base insulation was in the low-velocity areas. It was also expected that an arc imaging furnace would be ideal for studying ablation fundamentals, since the ablation products could be collected for analysis.

1. Test Procedures

It has been found that certain test procedures are of particular value in studying ablative insulation when an arc imaging furnace is being used.

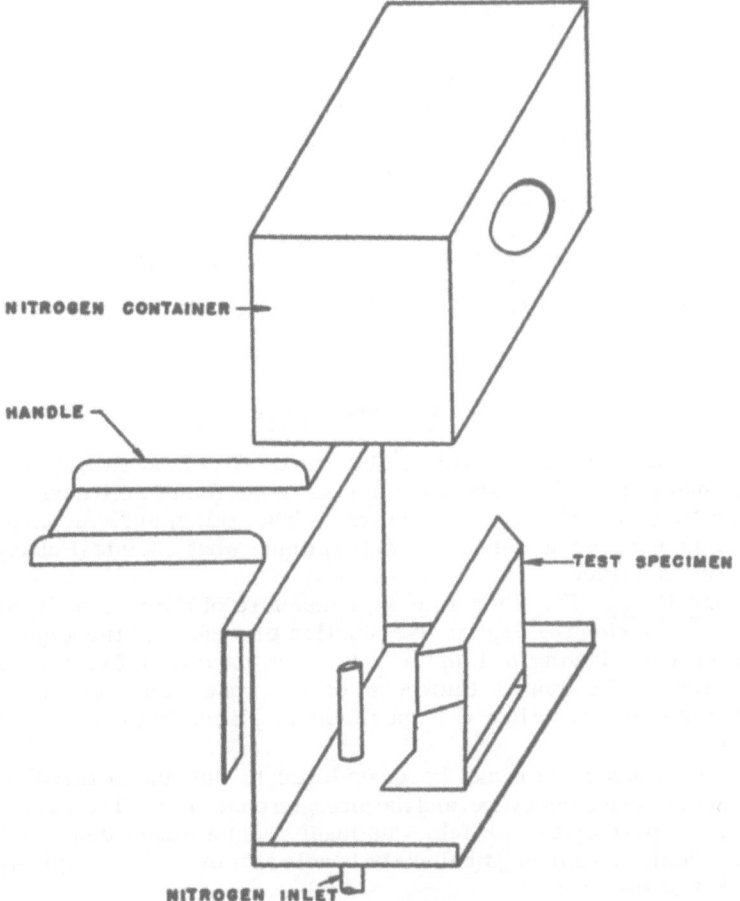

Fig. 20-2. Specimen holder.

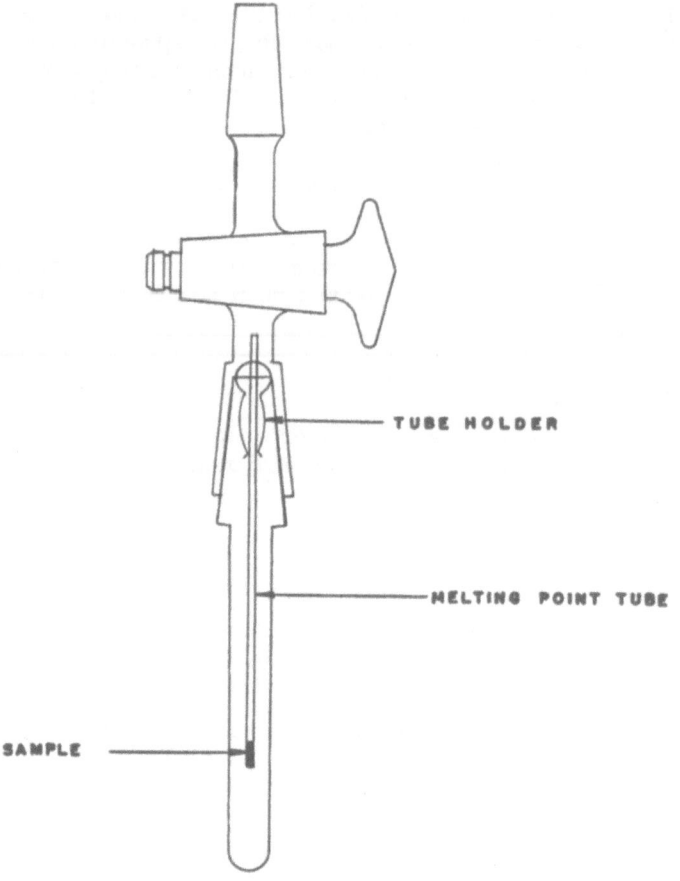

Fig. 20-3. Sampling tube.

Three of these procedures will be discussed. The first two, char rate and relative insulating value, are used to determine the effectiveness of materials for use as ablative insulators. The third, analysis of pyrolysis products, is used as a means for determining what chemical changes take place during ablation.

a. Char Rate. The char rate is a measure of the rate of consumption of a material during the destructive ablation process. Three-eighths-inch-thick specimens, having a 1 in. by 1 in. face, have been found to be a convenient size. The initial thickness of the specimen is measured to the nearest 0.001 in. and then the specimen is placed in the holder shown in Fig. 20-2.

The specimen holder can be considered to consist of three parts: the holder, the nitrogen inlet tube, and the nitrogen container. The holder, shown in the lower part of the sketch, was made from a single piece of 16 gauge stainless steel. Two prongs below the handle slip over the sample-positioning bracket of the furnace.

The nitrogen inlet is copper tubing. The nitrogen container is shown in a raised position. It is a metal box having a 1-in.-diameter hole in one side and an open bottom. When in position, the horizontal portion of the holder completes the enclosure.

After being exposed to the radiant heat for a specified time, the specimen is quickly transferred to a container partially filled with dry ice.. The carbon dioxide atmosphere prevents burning, and contact with the dry ice cools the sample to prevent further decomposition.

The surface of the test specimen is then scraped to remove the ash formed during ablation. The thickness is again measured and, by comparison of this figure with the initial thickness, the thickness of material pyrolyzed is determined. This thickness, expressed in mils, divided by the exposure time in seconds is the char rate.

b. Relative Insulating Value. The char rate test is, in reality, a measure of the thermal stability of a material. The relative insulating value test is a measure of the material's heat blocking ability. The procedure used is identical to that for determining char rate, except that thermocouples are used to measure temperature changes during the ablation. A thermocouple is inserted in a hole which is drilled parallel to the face of the test specimen; the hole extends to the center of the face of the sample. A recording poten-

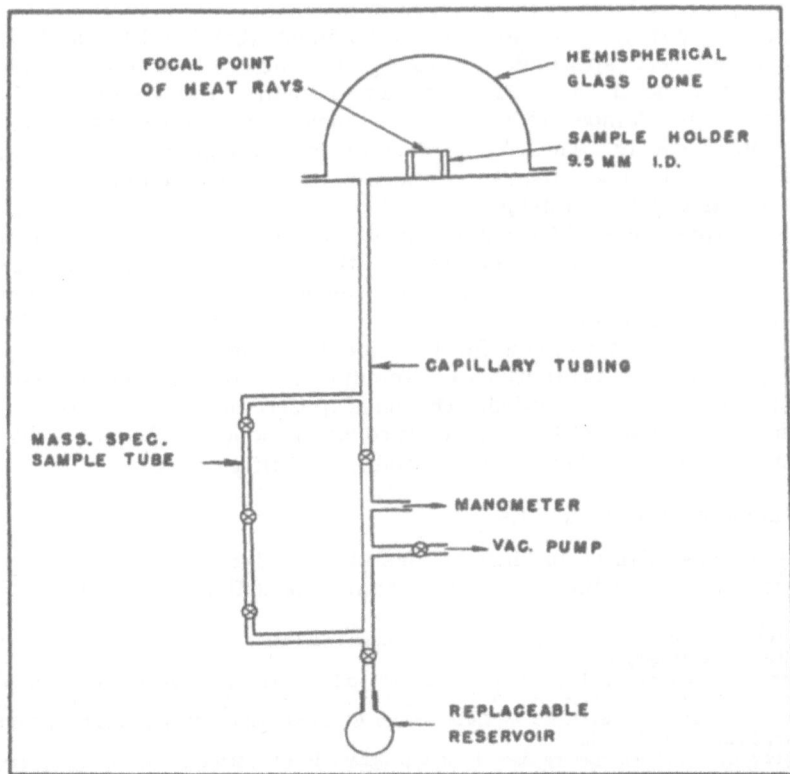

Fig. 20-4. Gas generator and collector.

tiometer[*] is used to record the temperature over the range from room temperature to 400°F. The time—temperature curve obtained is used to compare the insulating capability of one material with that of another.

c. Analysis of Pyrolysis Products. The equipment described in Fig. 20-3 is used to completely pyrolyze samples weighing 3 to 5 mg in vacuum. The 0.0625-in.-OD capillary melting point tube is open at both ends to permit free flow of the gaseous pyrolyzates away from the heated zone during the pyrolysis.

The tube holder is a metal clip made from 10-mil stainless steel sheet which is bent to fit the inside of the glass tube and which also serves to clamp the capillary tube, thus holding it in the outer glass tube.

A small sample of the material to be pyrolyzed (3 to 5 mg) is inserted into the end of the capillary tube and suspended by means of the metal clip in the center of the outer glass tube, which is then attached to an adapter. After being evacuated approximately 10^{-6} mm Hg, the sampling tube is precisely mounted in the arc imaging furnace in such a way that the small sample in the capillary tube is centered in the focal point. The carbon arc is then started and allowed to stabilize. Subsequently, the douser is opened, exposing the sample to the high-intensity radiant energy for 2 sec. A timer is used which automatically turns off the carbon arc after the 2-sec exposure period. After the pyrolysis, the sampling tube is mounted on a mass spectrometer[†] for analysis of the gaseous pyrolyzates.

For long pyrolysis periods, the equipment shown in Fig. 20-4 is used. The 9.5-mm test specimen, $\frac{1}{4}$ to $\frac{1}{2}$ in. long, is held in a graphite tube attached to a metal base. To form an enclosure, a flanged hemispherical pyrex dome is clamped to the base by means of a split clamping ring and a soft rubber gasket. The balance of the equipment consists of a mass spectrometer sample tube, reservoir, manometer, and vacuum pump, connected together with capillary tubing.

After adjustment of the equipment so that the face of the specimen is at the focal point of the reimaging mirror, the equipment is evacuated to 3 in. Hg. The vacuum pump valve is then closed and the equipment allowed to stand for 15 min to determine if there are any leaks. If not, the sample is pyrolyzed with a specific heat flux for a specific length of time. After pyrolysis, sufficient time is allowed for the equipment to reach room temperature, when the pressure is recorded. The mass spectrometer sample tube is then removed for analysis of the gaseous pyrolysis products. The ash is analyzed by means of both emission spectra[‡] and X-ray diffraction.[§]

2. Discussion of Test Results

Figure 20-5 illustrates how an elastomeric ablative insulator functions. When the insulating liner is exposed to high heat flux, its surface immedi-

---

[*]The recorder used in this program is a Speedomax recorder manufactured by Leeds & Northrup Co., Philadelphia, Pennsylvania.
[†]The mass spectrometer used in this study is an industrial model instrument equipped with an automatic peak digitizer and printer. The mass spectrometer is Model 21-103C with Mascot peak digitizer No. 34-201, manufactured by Consolidated Electrodynamics Corp., Pasadena, California, with Clary Printer No. 34-002.
[‡]The ultraviolet emission spectra used in this program is a Bausch & Lomb Littrow Spectrograph manufactured by Bausch & Lomb Optical Co., Rochester, New York.
[§]The X-ray machine used in this study is manufactured by General Electric X-Ray Corporation, Milwaukee, Wisconsin.

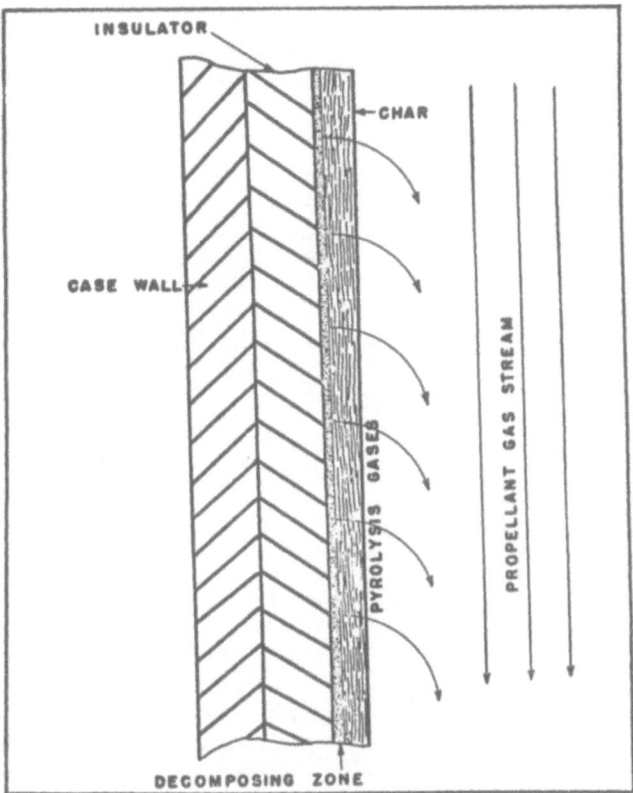

Fig. 20-5. Ablation of elastomer liner.

ately begins to decompose. The initial decomposition rate is high. A char or ash layer develops quickly, and the gaseous decomposition products pass through this layer. As these gaseous products pass through the char layer, they are further decomposed to lower molecular weight fractions, absorbing additional heat. As these low-molecular-weight decomposition products exit from the char layer, they cause a thickening of the gaseous boundary layer, and this adds to the effectiveness of the insulator. Stabilization occurs after a few seconds, and the ablation rate approaches a constant, which is much lower than the initial rate. The ablation rates that occur are a function of the specific ablation environment, of which heat flux is one important factor.

In Fig. 20-6, char rate data obtained with the arc imaging furnace are shown as a function of exposure time for two heat flux levels. As data show, the char rate decreases with both exposure time and heat flux; this result is in agreement with the theoretical discussion presented above. From such data, it is possible to select test conditions for a specific rocket motor application. For such a selection, a material is tested in the motor, and the char rate is found. With the same exposure time used, a heat flux that will result in the same char rate is selected for the arc imaging furnace test.

As indicated in Fig. 20-6 by the dotted line, which shows the effect of fore and aft out-of-focus measurements, focusing errors can be substantial and must be avoided.

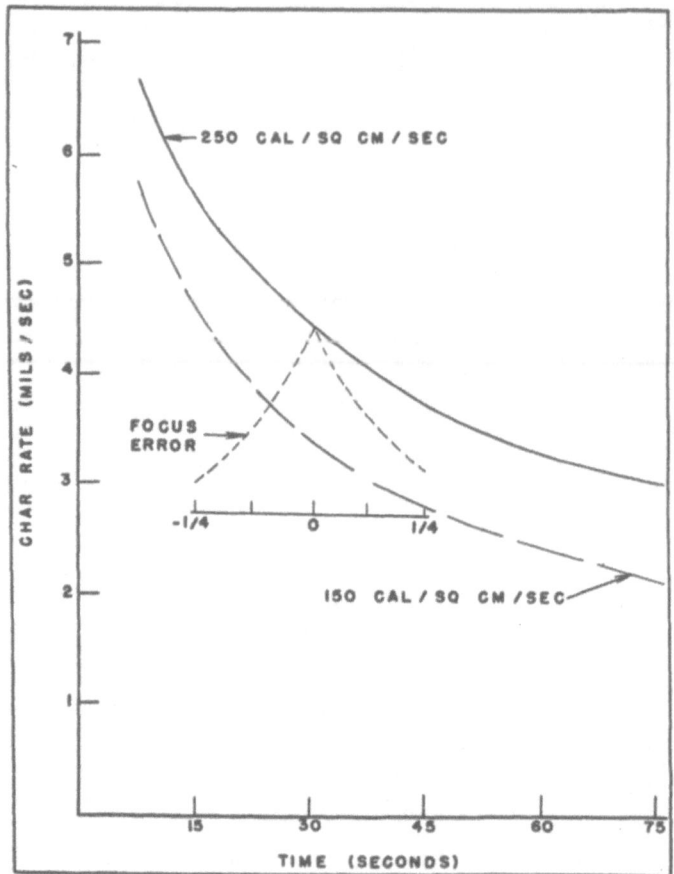

Fig. 20-6. Char rate vs. time.

Another source of error which must be avoided is the effect of the surrounding atmosphere. It has been found, for example, that certain ablative materials have char rates abnormally higher than those the motor test indicates when the propellant gases are reducing. A close correlation is obtained when the arc imaging furnace test is carried out in a nitrogen atmosphere rather than in air.

As an illustration of the degree of correlation that can be obtained, our experience concerning the selection of a material for the low-velocity area of a rocket motor is cited. In this instance, it was decided to use the arc imaging furnace char rate as the basis for selecting materials for expensive rocket motor tests. Forty-five materials were selected. Thirty-nine of these were expected, from the arc imaging furnace test, to have char rates lower than an established standard, and six were expected to have higher char rates. The arc imaging furnace data proved to be correct for 38 out of 39 of the low-char-rate specimens and for 6 out of 6 of the high-char-rate specimens.

It should be emphasized that the above correlation was between the static arc image furnace char rates and char rates measured on a low-velocity area of the rocket motor. There is no precise way of relating low-velocity char rate data with those that would be obtained at high velocity.

The purpose of an ablative insulator is to insulate the structural components of the case from the hot combustion gases. No evaluation of an insulator would be complete without some method of determining at least a relative insulation value. The blocking of heat transfer by an ablative insulator occurs through several mechanisms, the combination of which gives the overall insulating effect. To assign definite values to each of these mechanisms would be extremely difficult and expensive. However, the determination of a relative, overall effectiveness value is quick and inexpensive, if the arc imaging furnace relative insulating value test is used.

Figure 20-7 shows the time vs. temperature curves for two materials with identical char rates. As can be seen from these data, the overall insulating effectiveness of these materials is vastly different. It can also be noted that their relative effectiveness varies according to the temperature limit that is chosen. For a 200°F temperature limit and a $\frac{1}{4}$-in. thickness, the nitrile rubber compound will protect for 75 sec, and the phenolic-asbestos

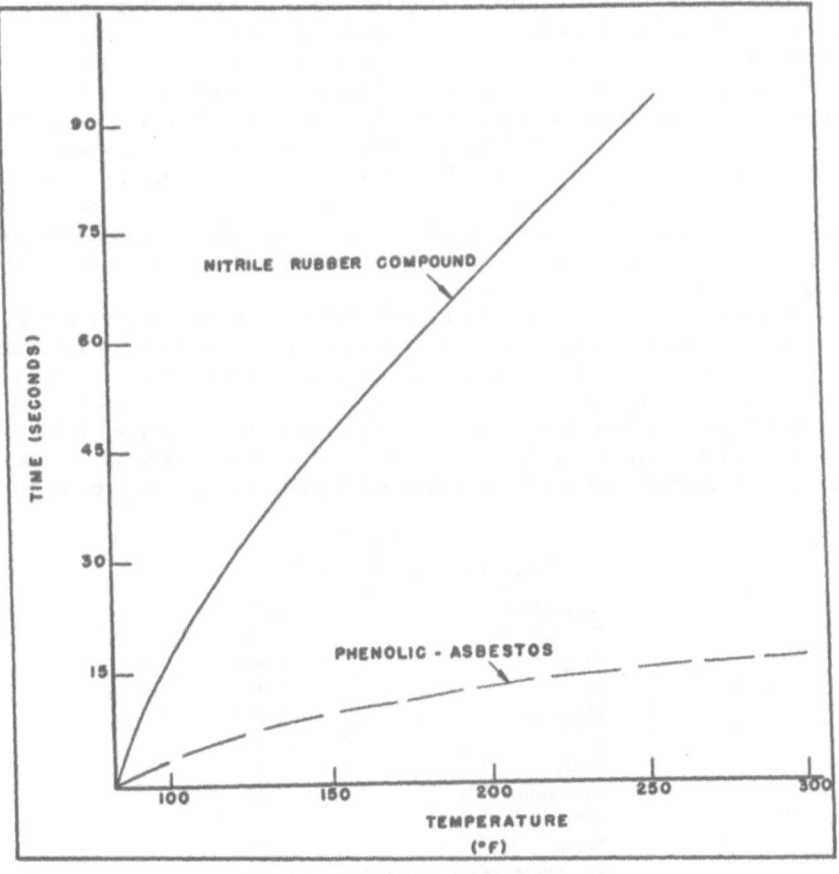

Fig. 20-7. Time vs. temperature $\frac{1}{4}$-in. sample.

TABLE 20-I
Pyrolysis Gas Composition

Gas	Exposure time, sec	
	2	50
Hydrogen	26.3	60.9
Methane	8.5	11.2
Acetylene	2.8	5.1
Hydrogen cyanide	3.4	4.3
Carbon monoxide	18.3	10.6
Propylene	6.6	1.3
Acrylonitrile	4.5	0.5
1,3-Butadiene	10.3	0.7
Miscellaneous gases	19.3	5.4

Note: Gas identifications shown are tentative; positive identifications are being made.

composition will protect for only 15 sec. Under these conditions, the relative insulating value is said to be 5.

A limited study of the composition of the gases given off during the pyrolysis of ablative insulators has been conducted. The clean, variable, and controllable heat generated by the arc imaging furnace makes this instrument a valuable tool in a study of this kind. The goal of this study is to examine some of the chemical phenomena associated with the ablation of elastomeric compositions and to collect data associated with these phenomena, in the belief that such information will permit the selection of more efficient insulation for future rocket applications. The study of the ablation gases, along with the char analysis and other fundamental data, such as differential thermal analysis, should give an insight into the chemical reactions taking place during ablation.

Table 20-I shows an approximate analysis of a gaseous pyrolyzate. The material pyrolyzed was an elastomeric composition containing nitrile rubber and silica as the principal ingredients. For a 2-sec exposure time, the specimen was small enough to be completely pyrolyzed without the formation of a thick ash layer. It is assumed that these products of decomposition are the same as those formed in the initial stage of exposure of a larger sample prior to the accumulation of a surface ash layer. For the 50-sec exposure

TABLE 20-II
Char Residue Analysis

Aluminum	0.45
Calcium	0.25
Copper	0.002
Iron	0.06
Magnesium	0.07
Silicon	60.00
Sodium	1.00
Titanium	0.20
Zinc	0.50
Boron	0.07
Remainder assumed to be mostly carbon	

time, a larger specimen was incompletely pyrolyzed. In this instance, a significant portion of the gases passed through a hot char layer. The additional decomposition which occurred in the char layer is indicated by the increased percentage of low-molecular-weight components in the gaseous pyrolyzate.

The ash layer formed during the pyrolysis of the nitrile rubber–silica composition was analyzed. The emission spectra data of Table 20-II show the presence of certain metals, all of which are present in insignificant amounts except silicon. Further analyses have been made by means of X-ray diffraction. This technique is used for qualitative analysis of crystalline components. Work along this line has indicated that recombination of some of the pyrolysis products occurs within the char. The formation of metallic carbides is noticeable. It was also found that the composition of the ash varies, depending upon the depth within the char.

Crystalline carbon, presumably in the form of graphite, has been found to be present in varying amounts in the char residue of ablative insulator materials. Those materials which appear to form carbon which is more crystalline in the residue, as opposed to amorphous carbon, also appear to have stronger char structure. Strong char structures are essential for ablative applications involving high velocity.

C. POLYMER IDENTIFICATION

The information on the arc imaging furnace in connection with polymer identification was taken from a paper "Identification of Rubber Polymers by Mass Spectrometry," by J. K. Phillips. This paper was presented at the 140th American Chemical Society Meeting, Rubber Division, Chicago, Illinois, September 7, 1961 and has been published in the Applied Spectroscopy magazine.*

The identity of an elastomer or elastomers in a rubber compound can be determined within a few minutes by the use of a new analytical procedure. It involves high-temperature pyrolysis for 2 sec, in vacuum, with an arc imaging furnace, followed by analysis of the gaseous pyrolyzate with a mass spectrometer. Numerical values, derived from the mass spectra of standards, permit semiquantitative analysis of some mixtures of elastomers in vulcanized products. Qualitative analyses have also been made on most of the common textile polymers and a small number of miscellaneous polymers. The method of pyrolysis has been discussed previously in this paper. It was the method used for the pyrolysis of 3 to 5 mg specimens in vacuum.

Although the entire mass spectrum is normally run on all pyrolyzates, the peak heights for only a few mass numbers are needed for elastomer identification. Fortunately, most of the elastomers studied to date split off large amounts of their respective monomers during the pyrolysis in the arc imaging furnace. Consequently, the peak heights of the mass numbers corresponding to the molecular weight of the monomers are very useful in the elastomer identification.

An empirical ratio, R1, used to identify most elastomers, is obtained by division of the sum of the peak heights at mass numbers 53 and 54 by the sum of the peak heights at mass numbers 68 and 104. This ratio emphasizes the
*Applied Spectroscopy 17:No. 1 (Jan. 1963).

TABLE 20-III
Empirical Ratios for Common Rubber
Polymers

Rubber type	R1 Values
SBR (styrene/butadiene)	20.68
Natural	1.44
PBD (polybutadiene)	38.61
NBR (acrylonitrile/butadiene)	18.20

peak heights at the mass numbers corresponding to the molecular weight of the respective monomers. Butadiene is 54, isoprene is 68, and styrene is 104. Mass 53 is an intense peak in the isoprene pattern, resulting from the splitting off of the methyl group from the isoprene molecule.

Empirical ratios, R1 values, for several common rubber types are shown in Table 20-III. These values were obtained from the gaseous pyrolyzates of compounded and cured rubber samples. Supplemental observations were also made to verify the identifications indicated by the R1 values. Natural rubber, for example, is present if the ratio of mass number 53 to mass number 54 is one or greater, and is accompanied by a high peak at mass 68.

Semiquantitative analyses can be made on simple elastomer mixtures. After it is determined which elastomers are present in the mixture, the relative amount of each can be read from prepared curves. Figure 20-8 shows typical curves prepared from data obtained on known mixtures of compounded vulcanizates.

The analytical procedure described has been used primarily to identify rubber polymers. However, limited tests show that essentially the same procedure can be used to identify many other types of polymers, including the acetates, chlorides, ethers, and textiles.

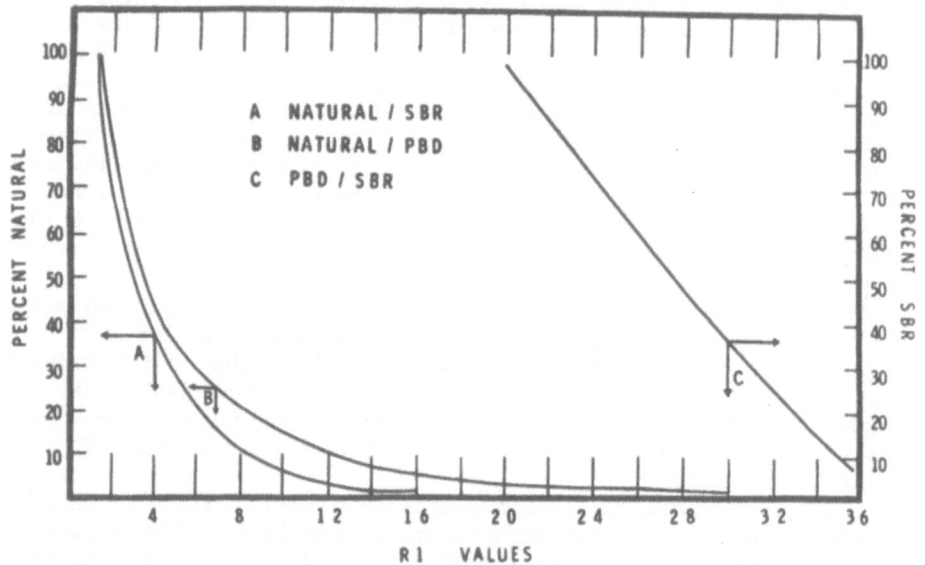

Fig. 20-8.

All materials to be pyrolyzed must be black or dark-colored to absorb the radiant energy from the arc imaging furnace. Because of this, the procedure described lends itself readily to the analysis of black loaded vulcanizates, eliminating the time-consuming carbon-black separations required for some analytical procedures.

It is postulated that the gases liberated from the polymer during pyrolysis move immediately into the cool portions of the evacuated tube away from the heated zone. The cooling of the gases greatly reduces additional pyrolytic fractionation, which occurs in conventional pyrolytic procedures.

Preliminary studies show that compounding ingredients used in vulcanizates have very little effect upon the analytical results, because most of the compounding ingredients either are nonvolatile under these test conditions or are converted to nonvolatile components during the high-temperature pyrolysis. Also, the small number of peaks used in the empirical ratios greatly minimizes the possibility of interference.

D. CONCLUSIONS

Our satisfaction with the arc imaging furnace as a laboratory tool is in no small measure influenced by the fact that through its use we have been able to save many thousands of dollars. Rocket motor tests cost from $1000 upward per shot, depending upon the size of motor used. By replacing some of the expensive rocket testing with the arc imaging furnace testing, we have already saved, according to estimates, in excess of $100,000.

We have not made a similar estimate concerning the polymer identification work. This would be difficult to do since, prior to the use of the arc imaging furnace method, polymer identification was so expensive that it was impractical to identify polymers where funds and time were limited. We now find that, with this procedure available, polymer identification is being used not only for polymer research projects, but also in connection with a wide range of problems, including factory quality control.

Chapter 21

Heat Capacities of Boron Nitride and Aluminum Oxide Using an Arc Imaging Furnace*

H. Prophet and D. R. Stull

Thermal Laboratory
Dow Chemical Co.
Midland, Michigan

A. INTRODUCTION

A novel technique for the determination of relative heat capacities in the temperature range above 1300°K has been developed for use with arc imaging furnaces ([1,2]). Essentially, cooling rates of different materials were compared in a standard container, whose heat loss rate was a function only of its absolute temperature. The furnace, a double-ellipsoidal-mirror-type arc imaging furnace, operated at a maximum power input of 21 kW and delivered approximately 240 cal/sec at the image plane. The $\frac{3}{8}$-in. anode crater, maintained at a brightness temperature of 3800°K, irradiated a sample area slightly greater than 1 cm^2. The essentials of the furnace are shown schematically in Fig. 21-1; a full description has been given elsewhere ([3]). The chief advantages of such a heat source were the ease of atmosphere control, the absence of hot furnace walls, and the rapid attainment of high temperatures. The major disadvantages were the small total power output, which limited sample size, and the high energy concentration, which created both thermal shock problems and large temperature gradients.

B. COOLING RATE METHOD

The method was based on the fact that at high temperatures a standard container of unvarying emissivity, E, radiated heat according to the relation $W = E\sigma A T^4$, where W is the heat flux per unit area, σ the Stefan–Boltzmann constant, A the area, and T its absolute temperature. At a given temperature, the rate of radiative heat loss was dependent only on the external dimensions and properties and the absolute temperature. By maintaining the container dimensions constant and obtaining a highly reproducible surface, we were able to make the heat loss rate at a given temperature a function only of that temperature. The conductive and convective losses were made functions of temperature, since we maintained their sink temperatures relatively constant. These losses were arranged to be minor compared with radiative transfer. (Conduction losses were roughly 10% of the total heat loss.) Under these conditions, the heat loss rate at any temperature was independent of the contents of the vessel. Measuring the rate of temperature decrease of the

*This work was supported by the Advanced Research Projects Agency under Air Force Contract No. AF 33(616)-6149.

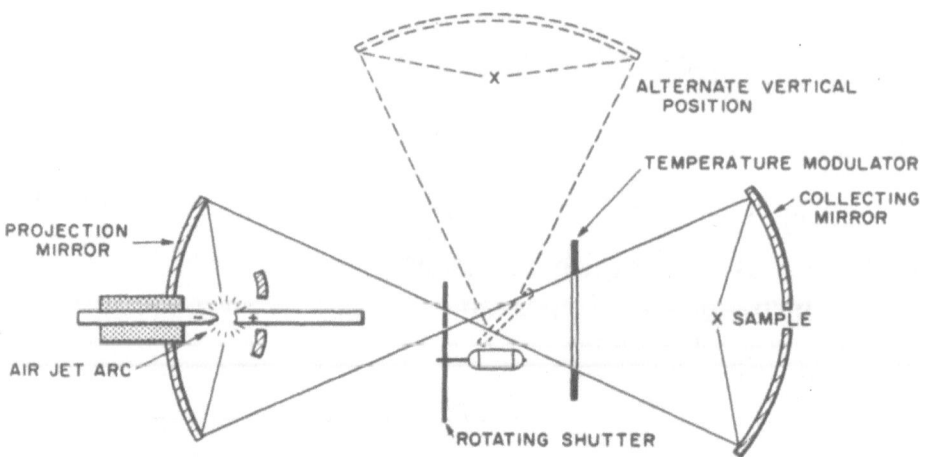

Fig. 21-1. Schematic diagram of the arc imaging furnace—normally operated in the alternate vertical position.

container and contents allowed the heat capacity to be determined ideally, as shown below:

$$H_T = (m_2 C_p + m' C_p') R'$$

for calibration with known C_p' and

$$H_T = (m_3 C_p + m'' C_p'')R''$$

for determination of C_p''. Thus

$$C_p'' = [C_p(m_2 R' - m_3 R'') + m' C_p' R']/m'' R''$$

where H_T is the rate of loss of heat at temperature T; m_2 and m_3 are the masses of the empty vessel in the two series; m' and m'' are the masses of the calibration material and sample; C_p, C_p', and C_p'' are the gram heat capacities of vessel, calibration, and sample materials; and R' and R'' are the rates of decrease of temperature at temperature T for calibration and sample series. However, in practice, the system did not behave ideally, deviating by 2-8%. In order to correct for this deviation, it was assumed that the vessel's temperature, as read, varied with its cooling rate. So that the correction could be determined, the cooling rate of the empty vessel was also determined, and the assumption was made that the temperatures in the calibration runs were correct. If T is the true temperature, S is the observed temperature, and $S = T + K$ for a small temperature interval, then at a temperature T_1,

$$\left(\frac{dT}{dt}\right)_{T = T_1} = R_{S = T_1} + K \left(\frac{dR}{dS}\right)_{S = T_1} + \frac{K^2}{2} \left(\frac{d^2 R}{dS^2}\right)_{S = T_1}$$

where $R = dS/dt$ for the empty vessel.

If it is so arranged that the vessel and calibration material are the same, then the ratio of the cooling rates of the empty and calibration series is given by

$$\frac{(dT/dt)_{T = T_1}}{R'} = \frac{m_2 + m'}{m_1}$$

where m_1 and m_2 are the masses of the empty vessel in the empty and calibration series, and m' is the mass of the calibration material.

The value of K for the empty vessel was found through solution of the quadratic equation at each temperature of interest. The values (dR/dS) and (d^2R/dS^2) were obtained from R through the taking of first and second differences. The value of K for the sample series was obtained through interpolation between $K = 0$ for the calibration series and K for the empty vessel. In general, it was found that the K^2 term was negligible in the sample series.

C. APPARATUS

Because of the limited heat flux available at the image plane, the vessel size determined the maximum temperature attainable and also the rate of cooling. For example, a 1.6-cm-diameter spherical body would have a maximum temperature of 2250°K and a maximum cooling rate of roughly 135°K/sec. Such cooling rates were too high to be measured accurately, and shielding was necessary in order to reduce them to more reasonable proportions. In order to keep the heat loss rate dependent only on the temperature, it was necessary to have a highly reproducible surface on the shield. This was achieved by casting of a heavy-walled gold shield, which was highly polished on the inside. The shield was given a graphite outer cover so that its total heat capacity would be increased and its temperature

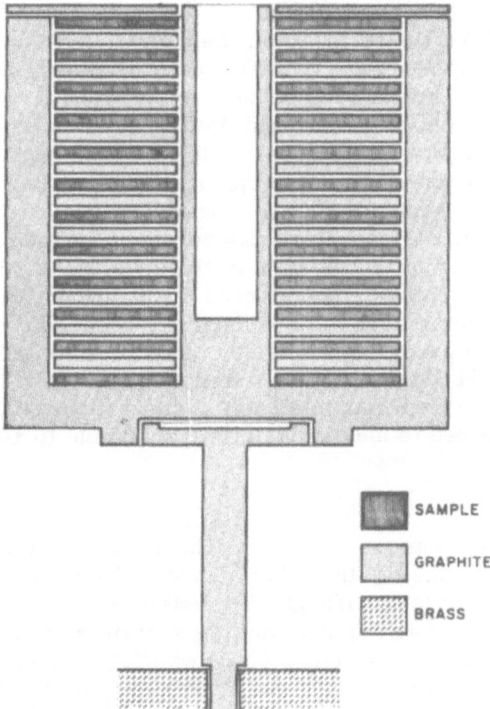

Fig. 21-2. Filled container and support—interleaved
discs of graphite and sample.

would be maintained constant. This was extremely effective and reduced cooling rates to 20-30% of their unshielded values.

The $\frac{1}{2}$ by $\frac{1}{2}$ in. standard container was made from ultrahigh-purity, high-density, SPK grade National Carbon graphite. Graphite was chosen because of its high emissivity (which was relatively insensitive to surface condition), excellent thermal shock resistance, good thermal conductivity, and ease of working. So that blackbody radiation could be obtained for accurate temperature measurement, a reentrant well was provided in the central post. The dimensions of the well were chosen according to the formula of de Vos ([4]) to have a length-to-radius ratio of 10. The container was seated on a graphite support, which also served to position the sample container exactly in space. The container and support are shown schematically in Fig. 21-2. The small cross-sectional area of the support served to limit conduction, and the massive brass sample holder acted as a constant-temperature heat sink.

The temperature measuring system consisted of a Leeds & Northrup total radiation pyrometer, whose output was amplified and recorded on a millivolt recorder. The pyrometer was able to sight on an area 0.01 cm^2 at 12 in. and had a response time of 0.13 sec. for 63% of a change. The output of the instrument was calibrated in position, with the freezing point of gold ([5]) at 1336°K, and the platinum carbon eutectic point ([6]) at 2003°K being used as the primary calibration points. The standard container held roughly 12 g of gold or platinum and gave well-defined plateaus, from which the purities were calculated to be 99.8 to 99.95 mol.%. The outputs for pure systems were calculated and used for the calibration; the overall reproducibility of the calibration of the pyrometer was ±1°K. The freezing times, with the gold shielding system, were 17 sec at 1336°K and 7 sec at 2003°K.

The system is shown assembled in Fig. 21-3. The glass envelope was held rigidly in position at its upper end; the brass support fitted exactly in the open base of the envelope. The upper shield was located by meshing the brass cone with the graphite cover. The glass envelope was open at the top so as not to affect the temperature measurement. Argon gas flowed through the system, entering near the base of the envelope and leaving through the open top. The pyrometer was locked in position above the envelope and adjusted to maximum output while sighted on the freezing point of gold. Once the system was aligned, the container could be removed and replaced with ease and its blackbody cavity was automatically in correct alignment. The system gave highly reproducible cooling curves, the cooling time from one specified temperature to another being reproducible to ±0.2%.

D. PROCEDURE

The procedure was as follows. The container was weighed and then placed in position inside the glass cover. The air was flushed out with argon, which was passed through the system for at least 10 min. The gold shield was raised, the arc struck, and the container irradiated by opening of a shutter. The rising temperature trace was followed on the recorder, until the trace leveled off; the shutter was then closed, as the shield was dropped. The cooling curve was followed, with periodic range changes, until the container cooled below 1300°K. The whole operation, from raising of the shield to ending of the cooling curve, required only 2 min. The system was allowed

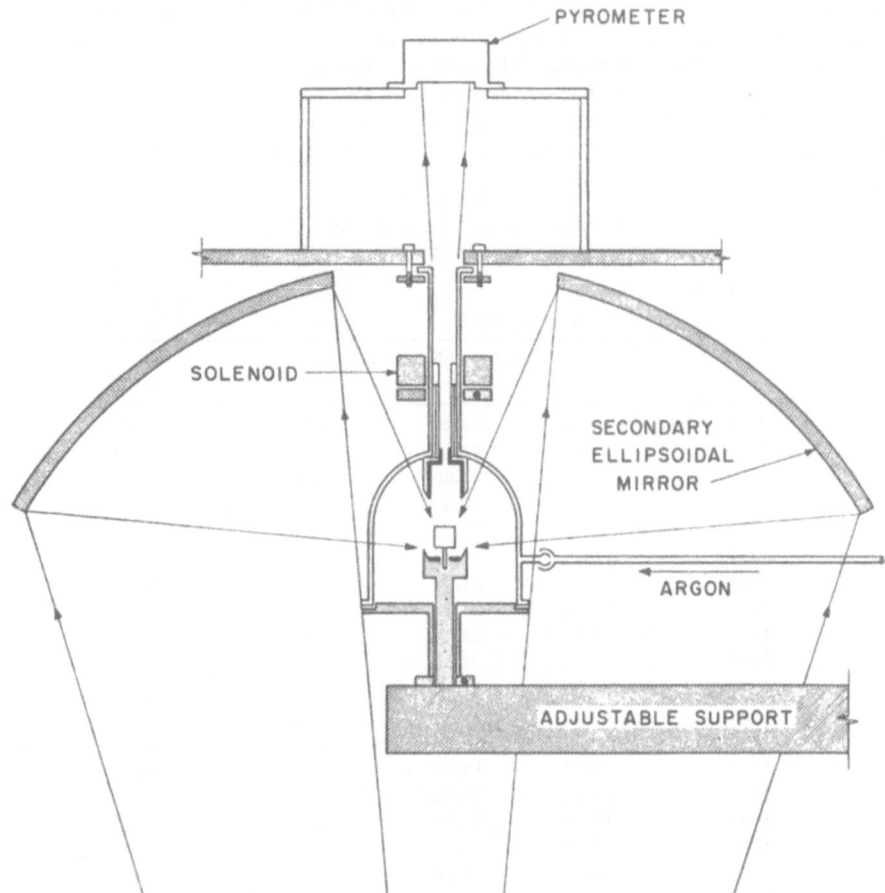

Fig. 21-3. Assembled system—upper gold shield raised.

to cool to room temperature, i.e., for roughly 30 min, before the next heat input. The procedure was repeated until two identical cooling curves were obtained. This usually took three determinations, due to the fact that the first cooling curve after assembly always showed much faster decline than subsequent curves. The reason for this phenomenon was not entirely clear, but it was believed to be due to water condensation on the shields.

The rates of cooling were determined from the cooling curve trace as follows. First, the outputs were read off at even intervals of roughly 1.3 sec and were converted to temperature readings. These temperatures were then fitted by computer to an equation of the form

$$\text{Time} = a + b(T - T_0) + c(T - T_0)^2 + d(T - T_0)^3 + e/(T - T_0)^2$$

where T_0 is an arbitrary temperature of the order of 1000°K, used to obtain improved fits. The cooling rate, or slope of the line, was calculated at specified intervals. Unfortunately, the power form of the equation allowed the slope to flex slightly through the points, and this caused the derivative to flex to an even greater extent. Thus, the values given by the computer were not accurate enough, and so they were plotted and a perfectly smooth line was drawn through them. This part of the procedure was most unsatisfactory, but so far no better way of determining the cooling rates has been found. If

TABLE 21-I
Cooling Rates (°K/unit time)

Temperature, °K	Vessel empty		Vessel + graphite	Vessel + graphite + compound
A. Boron Nitride				
	1	2	3	4
2200	51.68	55.25	30.63	33.75
2150	46.73	49.84	28.16	31.02
2100	42.46	45.35	25.91	28.51
2050	38.62	41.27	23.84	26.20
2000	35.27	37.67	21.98	24.07
1950	32.25	34.40	20.23	22.11
1900	29.58	31.48	18.61	20.29
1850	27.16	28.78	17.08	18.61
1800	24.90	26.29	15.65	17.05
1750	22.84	24.03	14.34	15.64
1700	20.95	21.95	13.13	14.33
1650	19.19	20.07	12.02	13.11
1600	17.57	18.34	11.00	11.97
1550	16.05	16.73	10.04	10.90
1500	14.62	15.22	9.14	9.90
1450	13.24	13.76	8.30	8.97
1400	11.91	12.35	7.44	8.06
1350	10.59	10.98	6.61	7.15
1300	9.30	9.61	5.77	6.25
B. Aluminum Oxide				
		5	6	7
2000		36.07	23.01	25.06
1950		33.09	21.11	22.97
1900		30.33	19.36	21.05
1850		27.77	17.74	19.27
1800		25.40	16.24	17.63
1750		23.21	14.85	16.11
1700		21.19	13.56	14.70
1650		19.32	12.36	13.39
1600		17.59	11.25	12.18
1550		15.98	10.22	11.06
1500		14.47	9.25	10.00
1450		13.05	8.34	9.01
1400		11.70	7.47	8.07
1350		10.25	6.60	7.11
1300		8.81	5.65	6.08

the temperatures could be read out with greater precision, then a least-squares fit of successive differences at the mean temperature would probably be more accurate.

E. RESULTS

The heat capacities of hot pressed boron nitride and sintered aluminum oxide have been measured. The samples were in the form of thin annular discs 0.010 in. thick and interleaved with graphite in order to maintain temperature equilibrium in the system. As a check, the apparent heat capacity of a single solid piece of BN was also measured; the result was found to be 8 to 10% lower than the value from the equilibrated discs. The boron nitride was 98% pure and had a chief impurity (1.7%) of boron oxide, which was removed by preheating the discs to 2300°K in an argon atmosphere. The aluminum oxide was 98.7% pure and had a chief impurity of 1.0% SiO_2; no correction was made for this impurity as the thermal effect would be negligible, 1 g of Al_2O_3 having almost the identical heat capacity of 1 g of SiO_2 at high temperatures. The uncorrected cooling rates, or (dS/dt) values, obtained for the two systems are given in Table 21-I, unit time being roughly 1.3 sec. Table 21-II shows the weights of material in the systems during the determination and calibration. In Table 21-III, the heat capacities are given as an average of the two sets obtained from initial and final calibrations. The data are plotted with drop calorimetry values in Figs. 21-4 and 21-5.

The whole method depends on the comparison with the calibration material, in this case National Carbon SPK graphite. The heat content and heat capacity of this material was determined separately in a drop calorimeter by R. A. McDonald of this laboratory. The heat capacity was measured from 300° to 1720°K and was extrapolated according to the Shomate function ([7]) to 2200°K; the values are given in Table 21-IV. The estimated error of the cooling rate method, including the measurement and extrapolation of the calibration graphite, would be ±2% below 1800°K and ±3% above 1800°K. It is thought that the advent of much more rapid and precise temperature measuring systems could reduce the errors to the order of ±0.5%.

The fact that the method worked as well as it did was due mainly to the calibrating out of many inherent errors. When a temperature gradient existed within the container, it changed the output of the blackbody cavity slightly.

TABLE 21-II
Weights of Material Used (grams)

Series	Vessel	Graphite	Boron nitride
1	1.79310	—	—
2	1.74715	—	—
3	1.78409	1.03079	—
4	1.80022	0.34520	0.47347
			Aluminum oxide
5	1.82038	0.01987	—
6	1.81336	1.02884	—
7	1.83308	0.41331	0.61110

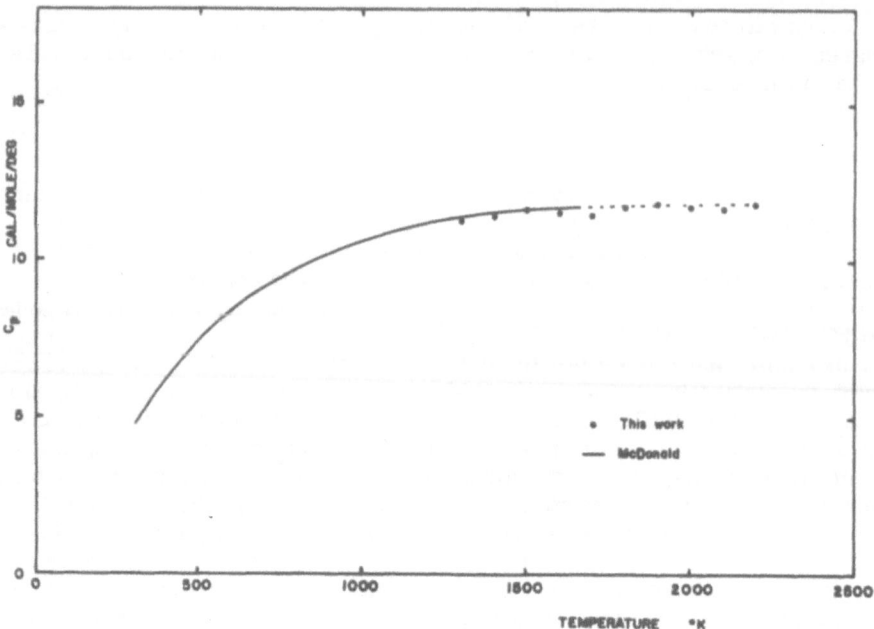

Fig. 21-4. Heat capacity of boron nitride. The comparison data points are from reference (8).

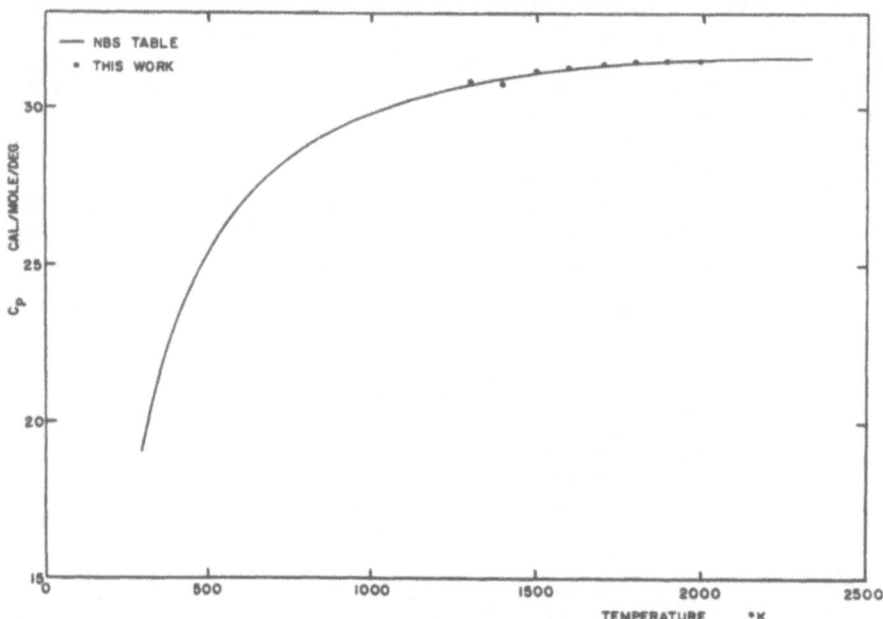

Fig. 21-5. Heat capacity of aluminum oxide. The comparison points are from reference (9).

TABLE 21-III
Calculated Molar Heat Capacities

Temperature, °K	C_p, cal/mole-deg
A. Boron Nitride	
1300	11.25
1350	11.17
1400	11.41
1450	11.66
1500	11.63
1550	11.60
1600	11.54
1650	11.46
1700	11.47
1750	11.58
1800	11.72
1850	11.79
1900	11.83
1950	11.74
2000	11.73
2050	11.60
2100	11.66
2150	11.69
2200	11.81
B. Aluminum Oxide	
1300	30.81
1350	30.85
1400	30.54
1450	31.02
1500	31.10
1550	31.14
1600	31.16
1650	31.21
1700	31.37
1750	31.36
1800	31.34
1850	31.45
1900	31.45
1950	31.35
2000	31.37

However, the effect was made nearly identical in calibration and determination by using similar cooling rates and thermal diffusivity. Also, although the temperature difference between the inside and outside of the container rendered the equations for heat loss not strictly applicable, any error was calibrated out. Lastly, the use of very thin discs increased the thermal diffusivity by both conduction and radiation between the large surface areas. In a system of 20 sample and 20 graphite discs, the total radiated energy

TABLE 21-IV

The Heat Content and Heat Capacity
of SPK Graphite

Temperature, °K	H_1-$H_{298.15}$, cal/mole	C_p, cal/mole-deg
300	3.78	2.05
400	248.10	2.82
500	564.56	3.49
600	941.96	4.03
700	1365.7	4.44
800	1825.4	4.74
900	2310.8	4.97
1000	2817.5	5.15
1100	3339.8	5.30
1200	3876.0	5.43
1300	4425.2	5.56
1400	4988.5	5.67
1500	5560.7	5.77
1600	6143.4	5.86
1700	6730.7	5.93
1800	(7327.5)	(6.00)
1900	—	(6.06)
2000	—	(6.11)
2100	—	(6.16)
2200	—	(6.20)

Source: Private communication, R. A. McDonald,
Thermal Lab., Dow Chemical Co.

transferred per second would be 5 to 10 times as great as the heat loss of the system.

The most serious limitation of the method is reaction between the sample and graphite; this reaction puts an effective limit of 2300°K on the system. However, the very short hot contact times may allow many reactive systems to be measured. Generally, less than 20 sec per heat would be spent above 2000°K and only 100 sec would be spent above 1000°K.

REFERENCES

1. Null, M. R., and Lozier, W. W., Rev. Sci. Inst. 29:163 (1957).
2. Davis, T. P., "The Carbon Arc-Image Furnace," High Temperature—a Tool for the Future, Stanford Research Institute, Menlo Park, California (1956), p. 10.
3. Glaser, P. E., J. Electrochem. Soc. 107:226 (1960).
4. de Vos, J. C., Physica (The Hague) 20:669 (1954).
5. Stimson, H. F., J. Research, Natl. Bur. Standards 42:209 (1949).
6. Collier, L. J., Harrison, T. H., and Taylor, W. G. A., Trans. Farad. Soc. 30:581 (1934).
7. Shomate, C. H., J. Phys. Chem. 58:368 (1954).
8. McDonald, R. A., and Stull, D. R., J. Phys. Chem. 65:1918 (1961).
9. Preliminary Report on the Thermodynamic Properties of Selected Light Element Compounds National Bureau of Standards Report 6928, July 1, 1960.

Chapter 22

A Radiation Technique for Measuring the Freezing
Points of Refractory Oxides

D. F. Comstock, Jr., and A. L. Camus
Arthur D. Little, Inc.
Cambridge, Massachusetts

A. INTRODUCTION

In the course of developing high-temperature equipment at Arthur D. Little, Inc., using an arc imaging furnace, we have observed a phenomenon that appears to be potentially useful as a technique for determining the freezing points of unknown refractory materials. While studying alumina samples, we noted that after the heating radiation had been removed, the temperature decline of the sample was temporarily arrested at a point roughly corresponding to the melting point of the sample. After the sample temperature stayed at this value for a short time, it continued to decline, and eventually reached ambient.

We believe that this phenomenon can be made the basis for a convenient technique for measuring the freezing points, and thereby inferring the melting points, of high-melting refractory oxides, and possibly other materials. The advantage, common to all image heating methods, is that the technique does not contaminate the sample, and is applicable to materials having very high melting points.

Measuring properties of high-temperature samples by nonequilibrium methods, particularly measuring melting and freezing points by observing heating and cooling curves, has been discussed by Kingery [1].

In a typical experiment, a sample at one temperature is placed in an oven at a different temperature and the temperature of the sample, usually measured by imbedded thermocouples, is observed as a function of time. If the sample is liquid at any time in the experiment, it is contained in a vessel of higher-melting-point material.

The temperature–time curve rises (or drops) smoothly, except for a central portion in which a flat plateau occurs. This plateau corresponds to a temperature at which a phase change in the sample is supplying heat to (or withdrawing it from) the sample, thereby temporarily arresting its temperature rise (or drop). Temperature gradients within the sample are assumed to be negligible, the major temperature difference existing between the sample and oven. The high-temperature limits of this method are determined by the high-temperature limits of the oven and the chemistry of the sample within its container.

In our method, the sample contained within its own material is heated at one spot on the face of an otherwise cool sample brick.

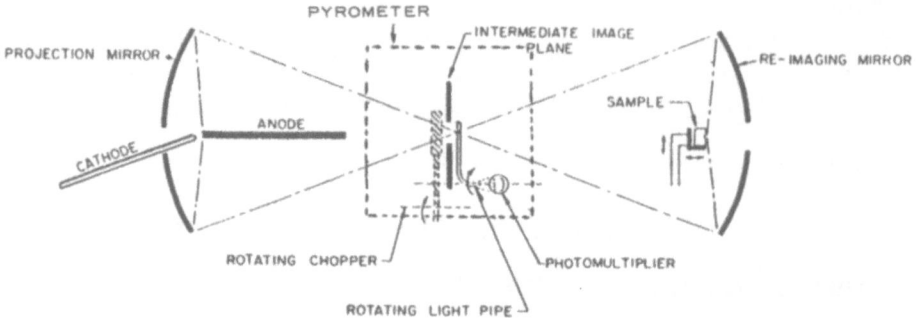

Fig. 22-1. Basic configuration of the imaging furnace and measuring apparatus.

B. APPARATUS

The equipment consists of two major components: an imaging system and an arc source for heating the sample, and an optical measuring and recording system for measuring its temperature.*

The sample is heated in an ADL strong arc imaging furnace. In this furnace, an arc is imaged by two identical elliptical mirrors onto a sample 7 ft away. Radiation from the arc is collected by a 21-in. ellipsoidal reflecting dish, and cast onto a similar ellipsoid that refocuses radiant energy onto a 1-cm-diameter spot on the face of the sample, the image being the same size as the arc. The thermal power arriving at the sample is about 150 cal/cm^2-sec. Well over 50% of the heating radiation from the arc is in the visible spectrum, with its peak in the blue. A schematic drawing of the heating and measuring system is shown in Fig. 22-1. This apparatus has been described elsewhere [2,3], as has its application to this type of experiment [4-6], and need not be described in further detail here.

The measuring apparatus, also illustrated in Fig. 22-1, consists of a rapid electronic pyrometer capable of alternately reading the self-radiation from the hottest point on the sample, the reflected radiation from the same point, and the radiation illuminating the sample at the same point. These three quantities are recorded permanently on "Visicorder" paper every 66 msec.

To measure emission and reflectance of the sample, the apparatus first measures emitted plus reflected light. Shortly thereafter, the arc light illuminating the sample is briefly interrupted so that only the emitted light from the sample is measured. The difference between the emitted-plus-reflected light and the emitted light is the reflected light, the quantity of interest.

Another necessary quantity, the incident radiation, is determined as follows: In a separate experiment, an overall combined constant of the apparatus is determined and is recorded by the pyrometer using a completely "white" sample. This overall combined constant is a measure of the quantity of arc radition that eventually falls on the sample. In effect, all sample reflectance readings are referred to the reflectance of a "white" sample. To obtain a "white" sample, 97% reflecting magnesia is smoked onto a water-cooled copper block, and a slight correction applied to compensate for the

*C. P. Butler mentions a related technique—heating a sample by imaging methods—for measuring the thermal diffusivity of unknown materials [3].

fact that the block is not absolutely "white." The reasoning and procedure are described in Reference (4).

A means is provided for avoidance of errors due to drifts in the electronic gains and sensitivities of the apparatus. A stable reference lamp (not the standard lamp) operating from a well-regulated supply is operated continuously during readings and calibration. The measuring optical collector (a light pipe) is arranged to be exposed to radiation from this reference lamp every 33 msec, at a point in the cycle when a reading is not being made on the sample. The reference lamp provides a steady signal arriving through the same optical path as the real signal and through the same electronic channel. This reference signal is one against which all others may be normalized. Thus signals taken at different times, even hours apart, may be compared, despite changes in electronic gains and power supply voltages.

Briefly, the theory of the equipment is as follows: The hemispherical spectral emittance of the sample at its hottest point is determined from the equation

$$\epsilon_\lambda = 1 - \frac{r_\lambda}{i_\lambda} = 1 - K \frac{(R_\lambda + E_\lambda) - E_\lambda}{A_\lambda} \tag{1}$$

where r_λ is the reflected flux from an element on the sample surface, i_λ is the incident flux on the same element on the sample surface, K is a combined constant of the arc imaging furnace and measuring apparatus, R_λ is the oscillogram indication of reflected light from the sample, in arbitrary units (for example, inches), E_λ is the oscillogram indication of emitted light from the sample, in the same arbitrary units, and A_λ is the oscillogram indication of arc light, in the same arbitrary units.

The temperature measurement is then deduced from a comparison of the radiation emitted from the sample to the radiation from a standard lamp of known temperature, as shown in equation (2):

$$T = \frac{T_L}{1 - (\lambda/C_2)T_L \ln(E_\lambda/E_L \epsilon_\lambda)} \tag{2}$$

where T is the absolute temperature of the hottest point on the sample, T_L is the blackbody temperature of the standard lamp (in these experiments, 2600°K), λ is the effective operating wavelength of the pyrometer, in microns (in these experiments, 0.7μ), C_2 is the second constant of Planck's equation, $1.4385 \times 10^4 \mu$-°K, E_λ is the oscillogram indication of emitted light from the sample, in arbitrary units, as defined for equation (1), E_L is the oscillogram indication of emitted light from the standard lamp, in the same arbitrary units, and ϵ_λ is the hemispherical spectral emittance of the sample, as determined from equation (1). The methods and procedures discussed in this section are dealt with in further detail in Reference (6).

C. EXPERIMENTAL PROCEDURE

The sample brick is heated until a well-established molten puddle is produced in the center of the sample face. In these experiments, the sample was heated for 2 to 15 sec, depending on whether a growing or an equilibrium puddle is desired. At the end of the heating cycle, the heating radiation is removed by a mechanical shutter within a period of 50 msec, and the sample is allowed to cool. The radiation from the sample in a wavelength band centered at $0.7\,\mu$ is recorded by the automatic pyrometer. Since the value of the emission is thus measured every 66 msec, a continuous history may be kept of the spectral emission of the sample as it goes through a heating and cooling cycle. The pyrometer also measures the sample's reflectance for that portion of the experiment during which arc radiation is falling on the sample.

D. RESULTS

Alumina was chosen because it is convenient and relatively stable in air [7]. We obtained dense alumina samples from the Norton Company (RA5190; approximately 99.5% Al_2O_3). These samples were 2 by 2 by $^3/_4$ in. smooth,

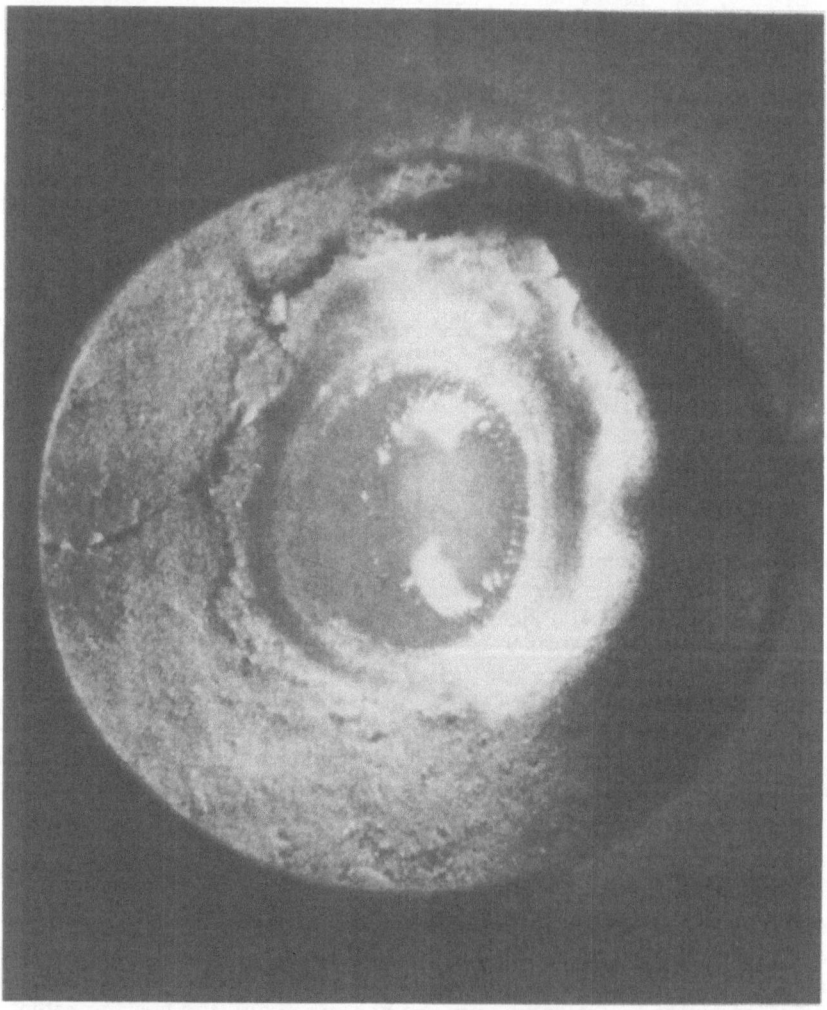

Fig. 22-2. Photograph of a typical molten puddle.

white, unpolished bricks. When the brick was heated in the arc imaging furnace, a puddle about 7 mm in diameter formed at the midpoint of the face of
the brick. A photograph of a typical molten puddle is shown in Fig. 22-2.
We found that this puddle could be remelted a number of times apparently
without changing the results and without cracking the brick.

When the heat was removed from the sample, we obtained a cooling curve
appearing typically as shown in Fig. 22-3. The plateau shown represents a
temperature of 2515°K.

Five separate determinations with the same method and apparatus and
with varying times of initial heating gave the following plateau temperatures:

Heating time, sec	Cooling plateau temperature, °C
2	2237
4	2242
12	2272
15	2287
15	2327

The average of these five temperatures is 2273°C, and the standard deviation is 33°C. There seems to be a marked relation between plateau temperature and heating time.

Another result necessary in calculating the temperature of the samples
is the spectral emittance of the samples at the temperatures of interest. In
these experiments, we did not illuminate the samples during the cooling
period. Therefore, the reflectance, and consequently the emittance, could
not be determined directly. To obtain emittance data applicable to the cooling
portion of the curve, we measured the emittance during the heating (melting)
portion of the curve when the sample could be expected to be at the same
temperature. These emittance values were used for calculating cooling temperatures.

Spectral reflectance and emittance of the alumina samples (dense 99.5%
alumina brick) measured as functions of temperature are shown below:

Temperature, °C	Spectral reflectance	Spectral emittance
20	0.74	0.26
2242	0.31	0.69
2459	0.20	0.80
2707	0.23	0.77

It is clear from these results that in the region of melting temperature and
higher, emittance becomes a weak function of temperature, and has values
in the region of 0.75. Further, it can be shown from equation (2) that the
determined temperature is a weak function of emittance. Therefore, we are
safe in assuming that the value of spectral emittance measured during the
heating stage of the experiment in the region of 2250°C is applicable with
small error to the calculation of the cooling plateau temperatures.

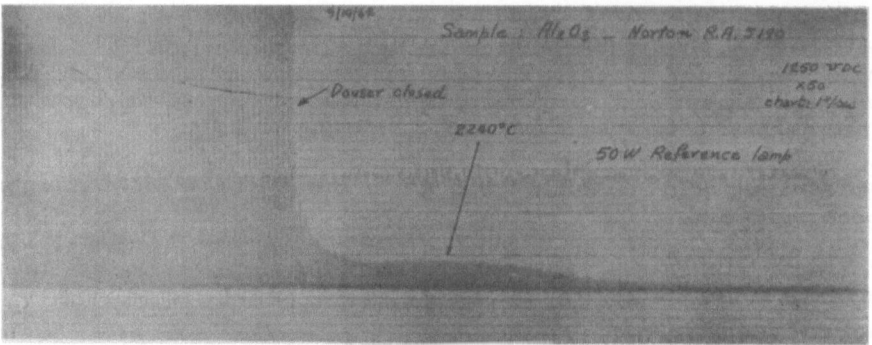

Fig. 22-3. Cooling curve of alumina.

E. DISCUSSION

The results of the experiments suggest two alternate interpretations, and further work will be required to determine what is taking place in the sample. First, the measured plateau temperature could be interpreted as corresponding to the freezing point of alumina. On the other hand, the dependence of measured plateau temperature on heating time suggests that the temperature depends on the depth of the puddle. A sample that has been heated a longer time has a deeper (and wider) puddle formed in its face, and we are looking at a thicker layer of molten alumina.

If this second interpretation applies, then either (1) the puddle is somewhat transparent, and a thicker puddle emits more radiation, or (2) the liquid of the puddle is just sufficiently transparent that the pyrometer, in measuring the puddle temperature, views a certain depth into the puddle, where higher internal temperatures prevail, and thus the effective temperature read is an average. There is some support for this second argument, since it is reasonable to suppose that with the front face brought to about 2700°C before the removal of heat, the inner regions of the puddle will reach an intermediate temperature (in the range of 2400°C) before cooling commences. The outer surface of the puddle is presumably at the freezing point, an assumption that is supported by visual observations (slow motion movies). A crust of solid alumina is seen to form on the outer surface of the puddle. The correctness of this interpretation depends on the transparency of the molten alumina, a factor presently unknown.

F. ERRORS

The errors associated with the pyrometer have not been established, because of the lack of high-temperature reference points. Nevertheless, an accuracy of ±15°C is attained when a standard lamp is used in place of the sample, and the reproducibility for emittance determinations of alumina samples is within 7%. An emittance variation of this magnitude gives rise, by the application of Wien's (or Planck's) radiation law, to an error of the order of 1% of the absolute temperature, i.e., 25°C at the temperatures (≈2500°K) we are considering.

ACKNOWLEDGMENTS

The authors wish to acknowledge the support and suggestions of P. E. Glaser, as well as the cooperation of other members of the Arthur D. Little, Inc., staff.

REFERENCES

1. Kingery, W. D., Property Measurements at High Temperatures (John Wiley & Sons, New York, 1959), Chapters 10.2 and 11.2–11.4.
2. Glaser, P. E., J. Electrochem. Soc. 107(No. 3) (March 1960).
3. Butler, C. P., "Image Furnace Research," International Symposium on High Temperature Technology, (McGraw-Hill, New York, 1960).
4. Comstock, D. F., Jr., "Method for Temperature and Reflectance Determination in an Arc Imaging Furnace," Temperature—Its Measurement and Control in Science and Industry, Vol. 3, Part 2, (Reinhold, New York, 1962), pp. 1063–1071.
5. For a description of a technique similar to that described in this paper, see J. C. Cook, "A Scanning Radiation Sampler for Imaging Furnaces," 4th Symposium on Temperature, Ibid., pp. 1051–1061.
6. Comstock, D. F., Jr., "A Radiation Technique for Determining the Emittance of Refractory Oxides," Preprint, Symposium on Measurement of Thermal Properties of Solids, Sept. 5, 6, 7, 1962, Dayton, Ohio.
7. Kingery, W. D., "Oxides for High Temperature Applications," International Symposium on High Temperature Technology, (McGraw-Hill, New York, 1960).

INDEX